Regulated Power Supplies
4th Edition

Regulated Power Supplies
4th Edition

Irving M. Gottlieb

TAB Books
Division of McGraw-Hill

New York San Francisco Washington, D.C. Auckland Bogotá
Caracas Lisbon London Madrid Mexico City Milan
Montreal New Delhi San Juan Singapore
Sydney Tokyo Toronto

© 1992 by **TAB Books**.
TAB Books is a division of McGraw-Hill, Inc.

Printed in the United States of America. All rights reserved. The publisher takes no responsibility for the use of any of the materials or methods described in this book, nor for the products thereof.

pbk   5 6 7 8 9 10 11 12 13 FGR/FGR 9 9 8 7
hc   1 2 3 4 5 6 7 8 9 10 FGR/FGR 9 9 8 7 6 5 4 3 2

Library of Congress Cataloging-in-Publication Data

Gottlieb, Irving M.
    Regulated power supplies / Irving M. Gottlieb. — 4th Ed.
      p.   cm.
    Includes index.
    ISBN 0-8306-2540-2 (hard)   ISBN 0-8306-2539-9 (paper)
    1. Electronic apparatus and appliances—Power supply. I. Title.
TK7868.P6G67 1992
621.381′044—dc20   91-34963
                                                                          CIP

Acquisitions Editor: Roland S. Phelps
Book Editor: Andrew Yoder
Director of Production: Katherine G. Brown
Book Design: Jaclyn J. Boone   EL1
Cover Design: Graphics Plus, Hanover, Pa.   3991

Contents

Introduction *xi*

1 Why use regulated power supplies? *1*

Pro and con *1*
Characteristics of the giant storage battery *2*
Improvements in hi-fidelity amplifiers *4*
The voltage-regulated power supply as a simulated capacitor *8*
The ideal voltage-regulated power supply *9*
Summarizing the voltage-regulated power supply *10*
The current-regulated power supply *10*
Other applications for the current-regulated power supply *11*
Electrochemical reactions *11*
Linear voltage ramps *14*
Simulation of high resistance *14*
Constant current in voltage-regulated supplies *15*
Zener diode voltage reference *15*
Stabilized emission in vacuum tubes *15*
Current regulation for testing purposes *15*
Unique application of the shunt-current regulator *17*
Stabilization of light output from LEDs *18*
Constant-current sources for photon-emitting diodes *19*
Dynamic load *20*
Voltage and current regulation combined *21*
Accurate power regulation *23*
Other applications of regulation techniques *26*
Power supplies for microprocessors *29*
High-voltage regulated power supplies *30*

The uninterruptible power supply or system (UPS) *32*
Other uses for regulated power supplies *33*
Using the regulated power supply to stabilize light intensity *36*
Using a regulated dc supply to stabilize RF output level of a TWT *37*
Prevention of remote-sensing problems *39*
Using a regulated power supply to extend transmitting-tube life *40*
Other applications using regulation circuitry *41*
A problem on the horizon *43*
Overall view of regulated supply features *45*

2 Static characteristics of regulated power supplies *47*

A problem in semantics *47*
A few examples of the tyranny of words *48*
The basic concept of regulation *50*
Beware—a trap *50*
After the trap, a pitfall *51*
An ambiguity—but whose? *52*
Load-current regulation of a current-regulated power supply *53*
Line-voltage regulation for the voltage-regulated power supply *55*
Line-voltage regulation for the current-regulated power supply *57*
Temperature coefficient *58*
Temperature rise and heat removal in a working environment *58*
Line and load regulation combined *59*
Total combined regulation *61*
Stability *61*
Additional regulation criteria *62*
Protection techniques *63*
Special considerations for transient protection *68*
Special IC for overvoltage protection *73*
Using the crowbar technique when supply and load are far apart *75*
Comments on current-limiting modes *77*
Protection techniques for the three-terminal IC regulator *81*
Evaluation and selection of power transistors *82*
Interconnecting regulated power supplies *84*
    Parallel operation by means of separate pass-element circuits *85*
    Parallel operation by parallel programming *85*
    Automatic parallel operation of two voltage-regulated power supplies *87*
    Automatic tracking operation of two voltage-regulated power supplies *88*
    Series-connected voltage regulators *88*
    Automatic series operation of two voltage-regulated power supplies *89*

        Dual-output tracking operation of two voltage-regulated power
           supplies   *91*
        Parallel operation of current-regulated power supplies   *91*
        Other situations   *92*
Basic aspects of heat removal   *92*
        A practical example of the use of the thermal circuit   *96*
The effect of forced air cooling   *99*
The effect of heat radiation   *99*
The all-important matter of thermal conductivity   *101*

3 Dynamic characteristics of regulated power supplies   *107*

A look at dynamic output impedance   *107*
Operation of the voltage regulator   *108*
Dynamic behavior from static characteristics   *110*
Regulator as an ac feedback amplifier   *110*
Some comments on the output capacitor   *114*
Another feedback arrangement for the voltage regulator   *114*
Transconductance and the regulator circuit   *115*
The transconductance viewpoint applied to tube and transistor
   circuits   *115*
        Voltage gain for triodes   *116*
        Gain for pentodes   *117*
        Transconductance of transistors   *117*
A look in retrospect at the voltage regulator   *119*
A closer look at the error signal   *120*
A still closer look at the error signal   *120*
Ripple   *121*
A new look at ripple   *121*
        Pard   *122*
        A practical consideration where clean dc is of importance   *122*
Transient responses   *123*
Dynamics of the constant-current supply   *125*
Two ways of specifying programming speed (slew rate)   *127*
Specifications of regulated power supplies   *129*
EMI and RFI   *134*
The output capacitor as a contributor to EMI   *141*
Noise benefits from synchronization of regulator switching rate   *143*
Some notes on shielding   *144*
Attenuation of noise spikes with ferrite beads   *145*
Feedthrough and connector-pin filters   *148*
Electrical noise in high-voltage supplies   *149*
Synchronous rectifiers   *152*
A subtle pitfall with regard to regulation specifications   *161*

4 Implementation of regulation techniques 165

Dissipation control *165*
    Dissipation limiting techniques *166*
    Double regulation *170*
    Dissipation control with a preregulating transistor *171*
The ferroresonant constant-voltage transformer *172*
The bipolar voltage-regulated power supply *174*
    Dual output tracking *175*
The shunt voltage regulator *177*
Junction field-effect transistors in dissipative regulators *179*
    FET characteristics important in regulated supplies *179*
    Using FETs in the regulator circuit *180*
Switching-type regulators *185*
    General characteristics and use of switching-type regulators *186*
    Combined use of dissipative- and switching-type regulators *186*
Phase-controlled switching regulators *186*
The dc switching regulator *189*
Some comments on inverter circuits and power switches *190*
The driven inverter *194*
Bridge-circuit inverters *196*
Symmetry correction circuit *198*
The three basic power switches *200*
Regulators based on parametric power conversion *203*
Unusual characteristics of the Paraformer transformer *206*
The RF power supply *210*
The solid-state RF high-voltage supply *212*
Developing a voltage reference via the energy-gap principle *214*
The off-line switching regulator *216*
Pulse-width modulation *218*
Line-operated power supplies using both linear and switching techniques *221*
Digitally controlled power supplies *222*
The systems-oriented digital control of dc power *228*
Multiple output supplies *229*
Hold-up time—uninterruptible operating power for awhile *230*
Regulation of ac *231*
An ac voltage regulator with electronic sensing *234*
Regulating a permanent-magnet alternator's charging current *236*
Improving long-accepted inverter design *237*
Ballparking the series inductor in the current-driven inverter *244*

5 Devices and components 247

Zener diodes *248*
    The LVA zener diode *251*
    The temperature-compensated voltage reference *252*
    The zener diode as a transient suppressor *255*
The Schottky diode *257*
The LM105 positive-voltage regulator *262*
    The regulator circuit *263*
    Voltage comparator *264*
The LM113 energy-gap reference diode *266*
The three-terminal IC regulator *267*
    Special ICs for inverter-type switching regulators *270*
    The Silicon General SG1524/SG2524 regulating pulse-width modulator *272*
Giant Darlington transistors *276*
The MOSPOWER FET or power MOSFET *278*
High-rating MOSFETs *282*
The gate turn-off silicon-controlled rectifier *287*
Transcalent power devices *291*
General considerations for choke and inductor selection *292*
What the hysteresis loop tells us *296*
Magnetically biased chokes and transformers *298*
The ferroresonant constant-voltage transformer *301*
Filter capacitors *303*
    Tantalum electrolytic capacitors *308*
    Pitfalls in the use of electrolytic capacitors *310*
    A technological breakthrough in filter capacitors *312*
Thermoelectric heat pump *315*
Keep your eye on the IGBT, a new solid-state power device *318*

6 Linear regulating supplies that use integrated circuits 327

Regulators that use IC operational amplifiers *327*
Simple op-amp regulators *330*
Voltage regulators that use the current-mirror op amp *331*
Op-amp voltage regulators with current-booster stages *333*
Op-amp regulator with added features *335*
High-performance voltage-regulated supply that uses ICs and discrete elements *336*
Regulators that use power op amps *337*
A 5-A voltage/current regulator *339*
Linear regulators that use the LM105 IC *340*
The type 723 voltage regulator *340*
Three useful regulators that are based on the MC1560/1561 IC *344*
A simple ±15-volt dual regulated supply *345*
Some typical applications of fixed-voltage three-terminal IC regulators *347*

Specially designed three-terminal regulators for adjustable voltage applications *349*
Versatility of three-terminal IC regulators *353*
Power MOSFET battery charger *354*
Voltage and current regulation for low-voltage and/or low-current circuits *335*
A 1.2-V, 200-µA voltage-regulated supply *358*
Low voltage from an LED *359*
Low voltages from Zener regulators *360*
Low voltages from reverse-biased base-emitter diodes in transistors *361*
Miscellaneous applications of linear IC regulators *361*
Dual-polarity outputs from single-ended power supplies *364*
Dual-polarity regulator with independently adjustable outputs *366*
Automobile voltage regulator with dedicated IC *367*
Paralleling regulator ICs *370*
Distributed power systems *372*
A unique backup source of dc power *377*

7 Switching-type regulators that use integrated circuits *379*

ICs for switching regulators *380*
The IC switching-type regulator *381*
    Circuit details and operation *381*
        Determination of the output-filter inductor and capacitor *383*
Switchers that use other popular ICs *387*
Negative switching regulators *390*
The constant-current switching regulator *391*
Stacked switching regulator *392*
A 250-W switching-mode power supply *395*
A 100-W switching regulator that uses the flyback principle *401*
An off-line 5-V 200-A switching supply *406*
    The 20-kHz clock *408*
    The pulse-width modulator *409*
    The error amplifier *410*
    Driver logic *410*
    The soft-start and overload circuit *412*
A 200-kHz 50-W switching regulator that uses a power MOSFET *416*
A 500-kHz flyback-circuit switching power supply *418*
Experimenter's prototype switching supply *422*
The current-mode IC controller *425*
The resonant-mode IC controller *430*
An example of a resonant-mode regulated power supply *437*
Sample list of dedicated ICs for switching supplies *444*

Index *455*

Introduction

The dominant theme of the previous three editions of *Regulated Power Supplies* had been response to the notion that the power supply was little more than an unfortunate, but nonetheless necessary appendage to electronic circuits and systems. In demonstrating the truly significant role that is played by power supplies, it became all too obvious that the power supply had to be recognized as much more than simply the source of dc power to the load; indeed, the long-enduring concept that the power supply and its empowered load were separate and independent entities had to be put to its overdue rest. The intimate linkage between circuit performance and the nature of its dc source paved the way for the development and use of the regulated power supply.

This fourth edition of this title fully accepts this evolutionary process of design, taking for granted the now-universal use of regulated power supplies. Now it is important to be prepared to further refine the proven techniques of regulation. This is composed of new ways to achieve regulation, the extension of identifiable trends, such as the use of ever-higher switching rates, the deployment of newer control ICs and power switches, improved system philosophies, and last but not least, more sophisticated "bells and whistles."

Interestingly, the newer state-of-the-art regulated power supplies do not necessarily make the previously used regulators obsolete; instead, we now have a much wider range of options from which to choose. This is because many applications do not require utmost consideration be bestowed on size, weight, operation frequency, efficiency, or even on inordinately tight regulation. Accordingly, it is fitting that this fourth edition of *Regulated Power Supplies* should expand the art as depicted in previous editions to include more recently developed practices.

This book on the intriguing subject of regulated power supplies has been written to reflect the manner in which a design engineer might hold an informal discussion with another technically versed person. Thus, the reader will be conveniently able to extract useful information without being "snowed" by forbidding mathematics; at the same time, resort to the simplistic has not been for the sole purpose of making the text easily digestible. To avoid these extremist editorial formats, the emphasis has been on the readable presentation of practically useful material.

Accordingly, it is of the reader's best interest to use this work as practical guide for designing, operating, building, and repairing regulated power supplies. It is, indeed, for this purpose that much focus has been directed to basic theory and to principles of operation. The intent is to provide essential insights and to sharpen the intuition. Thus, both the time-proven and the very recently developed devices and circuit techniques share space in the pages ahead.

Although educational purposes can best be served by sequentially reading the chapters, please relax and follow your individual inclinations and needs; no regimented procedure is mandatory—just consult the index and refer to the chapter or section that deals with the topic of interest at the moment. With just a bit of luck, you might find the specific answer to your unique problem. If such luck, however, is not the order of the day, good fortune will not have eluded you—you should prosper nonetheless via enhanced intuition and stimulated creativity.

Irving M. Gottlieb
Redwood City, California

1
Why use regulated power supplies?

DURING THE PAST SEVERAL DECADES, A MULTITUDE OF ELECTRONIC devices and systems have obtained dc operating power from simple, unregulated power supplies. Probably more often than not, these supplies have been hastily provided under design schedules that demand the very least expenditure of time and money for such "auxiliary circuitry." A transformer, a couple of rectifying elements, perhaps a filter choke, and one or more inexpensive electrolytic capacitors have constituted a kind of "brute-force" approach for a host of applications. It must be conceded that satisfactory performance has generally been attained. Moreover, even a superficial acquaintance with regulated power supplies reveals a considerably higher level of complexity than is encountered in the unregulated types. Surely, this must also be accompanied by additional cost. So, why not leave well-enough alone? Indeed, why use a regulated power supply?

Pro and con

The voltage-regulated power supply is the most common type of regulated power supply. In it, the dc voltage applied to the load is automatically maintained at a near-constant value, despite variations in ac line voltage or in the current demand of the load. It might be argued that such behavior is not needed because the powered circuitry can be designed to be noncritical with respect to impressed voltage. Such contention is immediately subject to the counter logic that freeing design from the effects of voltage variation is already a worthwhile accomplishment. Although both philosophies are valid, the most important attribute of the voltage-regulated supply is of a more subtle nature. So desirable is this not-so-obvious feature, that the use of voltage-regulated power supplies would often be justified by

2 Why use regulated power supplies?

this single attribute—despite the fact that upgrading of circuit performance is generally obtained by stabilizing the dc voltage level alone. Don't be surprised even to encounter cases where "regulation" is so implemented to provide this feature alone, and the dc voltage level is allowed to be stabilized poorly (sometimes it is altogether unregulated).

To dispel suspense, this mystical feature of the voltage-regulated power supply is its *low dynamic impedance*, that is, its tendency to approach short circuit for alternating currents. The attributes of mystery and subtlety exist only because inordinate attention is often focused upon dc behavior, which is seemingly natural for a dc power supply. Paradoxically, it is usually the ac characteristics that merit concentrated attention; static dc stabilization of the voltage level is satisfactory for the operation of most circuits in the most rudimentary voltage-regulated supply. Even in those cases where exceedingly close static regulation is important, the also important effect of dynamic regulation is often overlooked. Accordingly, it is best to devote separate chapters to static and dynamic regulation. First, however, it is appropriate to consider some aspects of the voltage-regulated power supply. For this purpose, consider a device that, without regard to practicality, might be employed if electronically regulated supplies were not available. The device is a storage battery—a giant one. Such a monstrosity is depicted in Fig. 1-1.

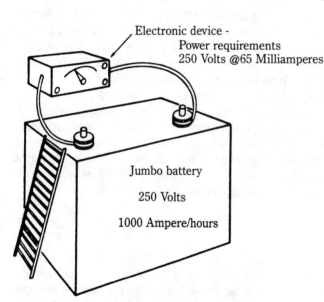

1-1 The voltage-regulated power supply is comparable to giant storage battery.

Characteristics of the giant storage battery

By a giant storage battery, we imply one with an ampere-hour that rates hundreds (or thousands) of times greater than is required to supply current for reasonable intervals to electronic equipment. Among other things, the terminal voltage of such a battery would suffer a negligible drop when the equipment was turned on. This follows from the fact that such a relatively large battery would exhibit a correspondingly low internal resistance. In a more general sense, such a battery has a low internal impedance. That is, for a considerable range of ac frequencies, the impedance that is effectively in series with the terminals is very low. So, it is proper to think of the giant battery as simulating a voltage-regulated source. It is true that its characteristics do not arise from any automatic corrective action. However, from the viewpoint of the load, either the giant battery or the voltage-regulated power supply can provide the requisite stabilization of dc voltage level and the low impedance for ac. The beauty of the voltage-regulated supply is that it can thus simulate the static and dynamic characteristics of the giant battery without having to simulate its tremendous current capability.

A quick look at Fig. 1-2 should pave the way for ways in which regulation can upgrade performance of circuit and system performance. As shown in Fig. 1-2A, the internal resistance, R_O, of the regulated supply is very small compared to an unregulated supply. R_O is not likely to cause instability by behaving as a mutual impedance to supply and amplifier stages. In Fig. 1-2B, the "electronic filtering" of the regulated supply is equivalent to a large value of LC filtering. Fig. 1-2C illustrates that the use of a regulated power supply enhances the accuracy of calibrated equipment.

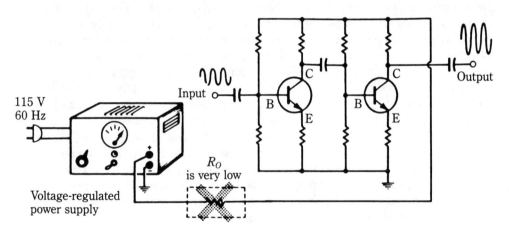

(A) Low internal resistance.

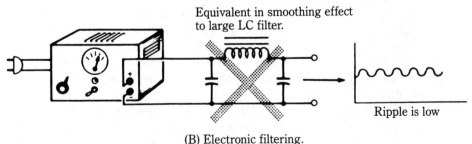

(B) Electronic filtering.

1-2 Advantages of a voltage-regulated power supply.

4 Why use regulated power supplies?

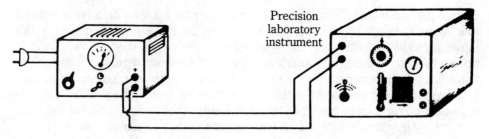

(C) Improves accuracy of test equipment.

1-2 Continued

Improvements in hi-fidelity amplifiers

The hi-fidelity amplifier is an excellent example of the advantages of a regulated power supply because distortionless amplification is a requisite achievement in many diverse instruments and pieces of equipment. The three graphs of Fig. 1-3 are typical of the improvements in performance that can be obtained from an audio amplification system by substituting a voltage-regulated power supply for a simpler nonregulated type. The production of such audio equipment is preceded by extensive development, in which much effort is expended upon the design of feedback networks, special output transformers, and specifications that must meet real and alleged claims of the competitor's product. Unfortunately, preconceived notions and a false sense of economy might preclude consideration of one of the most rewarding techniques of all, the use of a voltage-regulated supply.

The curves of Fig. 1-3 show a worthwhile extension of range of the performance parameters that are considered important in a hi-fidelity audio amplifier. Much of this enhancement accrues from the low ac impedance that is provided by the voltage-regulated power supply. In any event, it is instructive to investigate these improvements somewhat in detail. Just how significantly is the source of dc power involved in the operation of this audio circuitry?:

- Although most hi-fidelity amplifiers use a push-pull output stage, perfect balance throughout the dynamic and spectral range is not attainable in practice. This means that audio-frequency currents will circulate in the power supply. The ac impedance of the supply causes distortion, degrades frequency response, and limits available power. The power supply presents both resistive and reactive components to the audio "return" current. This would already produce frequency discrimination and limit peak power. Even worse, such impedance is nonlinear throughout the audio-frequency cycle. This must further increase distortion.

    Rather than attempt to make power-supply impedance purely resistive or extremely linear, a more practical approach is to make it very small. Then, the effects of reactance and nonlinearity become correspondingly small. Notice, too, that with the resistive component at a relatively low

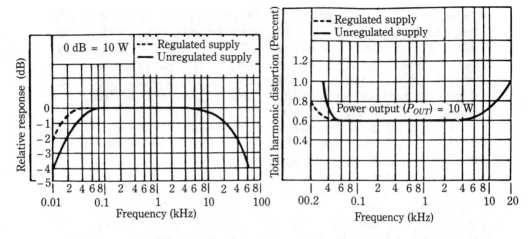

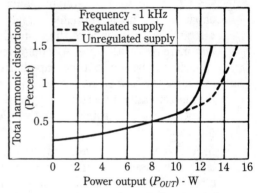

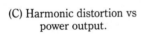

(A) Relative response vs frequency.

(B) Harmonic distortion vs frequency.

(C) Harmonic distortion vs power output.

1-3 Performance curves for an audio amplifier with regulated and unregulated supplies.

value, the true output-power capability of the amplifier can be expected to be approached closely.

- The driver and preamplifier stages are often single-ended, rather than push-pull circuits. In a single-ended stage, the power supply is most certainly part of the audio circuit: the entire audio current that circulates in the input and output portions of a single-ended stage passes through the power supply as well. Here, it is even easier for the power supply to contaminate amplification with frequency discrimination, rectification ripple, and distortion because of nonlinearity.
- The aforementioned effects of power-supply impedance lead to consideration of the even worse situation that prevails when audio currents of two or more amplifier stages mutually circulate through the supply. The simultaneous presence of such alternating currents in a common impedance constitutes *feedback*. Such unintended feedback can be either negative or positive.

Inasmuch as it generally varies in degree, and often in sign, throughout the frequency range, distortion and frequency discrimination tend to worsen. The positive feedback thereby produced can manifest itself as a low-frequency "motorboating," as an ear-splitting howl, or as severe distortion because of bias changes from rectification of ultrasonic oscillations. Again, the severity and probability of such malperformance is reduced if the power supply exhibits low impedance across its output terminals.

Figure 1-4 is an interesting study that shows that a two-stage amplifier can effectively become a *multivibrator* as a consequence of power-supply impedance. It is not uncommon for an unregulated supply to provide the dc power for high-gain amplifier stages. The standard Eccles-Jordan multivibrator circuit is shown in Fig. 1-4A. Figure 1-4B is a slight modification of this circuit. The conventional RC-coupled amplifier in Fig. 1-4C is modified in Fig. 1-4D by adding the impedance of the power supply. Notice the resemblance to Fig. 1-4B. C_O represents the effective output capacitance of the power supply and R_O represents the output resistance.

Often the sole protection against inadvertent oscillation is the output capacitor of the unregulated supply. Because of temperature and aging effects, the impedance of the commonly used electrolytic capacitor generally increases with time. Although regulated supplies usually have output capacitors, the low output impedance is predominantly derived from the *electronic feedback* of the circuit. This holds for most of the frequency range that is generally important to operation of audio amplifiers. At higher frequencies, low output impedance via the output capacitor is a relatively simple achievement.

- A sufficiently strong case has already been presented for the voltage-regulated power supply, but we can easily cite additional advantages to those already discussed. Consider, for example, that most push-pull output stages operate in class AB. In such operation, the average value of dc current consumed from the power supply varies with the amplitude of the audio signal. If the dc voltage delivered by the supply drops with increased current demand, fidelity of reproduction is impaired. The result is flattened signal peaks. Thus, an amplifier that is powered by an unregulated supply will, on this premise alone, exceed acceptable distortion at lower output than would be the case with operation from a voltage-regulated supply.

- Hum injection from imperfect dc is quite common when unregulated supplies are used. Usually, the electrolytic filter capacitors are chosen as a compromise from among "acceptable" output ripple, size, and cost. Because of internal chemical changes from time and temperature effects, such capacitors often lose filtering action before any great length of service. This is generally the main cause of an annoying increase in background hum heard from a hi-fi amplifier.

Conversely, the voltage-regulated power supply does, for most practical purposes, simulate the giant battery in production of clean dc. This tremendous suppression of the ac power-line frequency is not dependent upon the

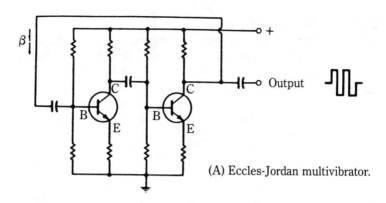

(A) Eccles-Jordan multivibrator.

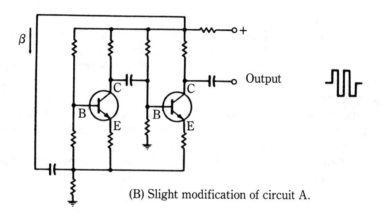

(B) Slight modification of circuit A.

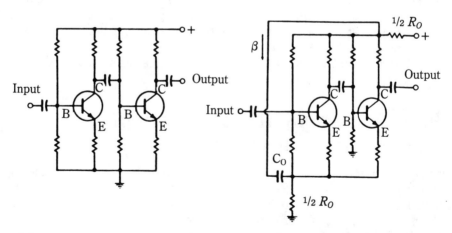

(C) Conventional two-stage RC-coupled amplifier.

(D) Circuit C with power-supply impedance added.

1-4 Circuits that illustrate how oscillation can be produced by power-supply impedance.

output capacitor. Rather, it occurs as a result of the following action: a sampled portion of the output voltage is compared with a stable reference voltage. If these two voltage levels differ, an error signal is produced. The polarity and magnitude of the error signal is, following amplification, used to alter the resistance of a series control element in the direction that will correct the output voltage deviation that caused the original error signal.

The error signal is, by such action, always extinguishing itself; the output voltage is caused to correct any deviation from its preset level. Thus, it is a sort of electronic "brain" that is programmed to be ever vigilant to any tendency for change of output voltage. Fortunately, such a circuit is "stupid" in the sense that it will act upon *any* deviation in output voltage, provided that the internal reference voltage remains constant. Thus, it makes no difference what causes attempted change in output voltage. Whether the cause is a fluctuation in ac line voltage, a variation in load current, or residual ac from the rectifier-filter circuit, the corrective action is the same.

Inasmuch as the function of the regulating circuit is to maintain constant dc output voltage, this precludes the presence of ac superimposed upon the dc output level. Thus, another viewpoint of the voltage-regulated power supply is that it behaves as an *electronic filter*. Significantly, the filtering effect greatly exceeds that which can be obtained from unregulated power supplies with practical filter choke and capacitors. For this reason, it is often permissible to dispense entirely with the use of the heavy and costly filter choke in voltage-regulated supplies. Surprisingly, this can, in some instances, lead to consideration of the regulated supply as an economy measure!

- Finally, because the voltage-regulated power supply isolates the powered equipment from changes in line voltage, a number of design and operational advantages, in addition to ripple suppression, are gained. These include:
  (a) Circuit adjustments, such as gain, balance, and biasing values, tend to remain at desired or optimum settings when a regulated power supply is used.
  (b) High immunity is attained against electrical noise that rides on the power line.

The voltage-regulated power supply as a simulated capacitor

Considerable emphasis has been placed on the low impedance that is provided by a voltage regulator. In the past, such low impedance depended on a large output capacitor in an unregulated supply. This system was often satisfactory for low and medium audio frequencies. For very low frequencies, however, it was always a losing battle. The nature of the impedance function of capacitors precludes any hope of attaining low impedance at these very low frequencies. This is why the term *motorboating* was so much in vogue. Motorboating refers to the low-frequency oscillations that plagued high-gain amplifiers because of the lack of bypass action at low frequencies.

The plots in Fig. 1-5 show the impedance characteristics of power supplies at low frequencies. Notice that the voltage-regulated power supply (so often likened to an electronic filter or capacitor) outperforms, rather than simulates, any capacitor. Not only can the impedance of such a regulator be made very low at the higher frequencies, but the low impedance prevails right down to zero frequency; that is, dc!

Although it is not shown in Fig. 1-5, practical regulators, as well as practical capacitors, exhibit a rising impedance at high frequencies. This is because of lead inductance in the capacitor. In the regulator, it is caused by lead inductance in the output capacitor and by diminishing amplifier gain. Such effects usually begin to set in between 10 and 50 kHz. However, remedial measures are available to extend the frequencies at which this impedance rise occurs so that no undesirable effects are produced (often, stray circuit capacitance provides sufficient bypass action). Here, the low-frequency behavior of voltage regulators is of concern. It is a gross understatement to merely say that voltage regulation *simulates* the bypass action of a capacitor.

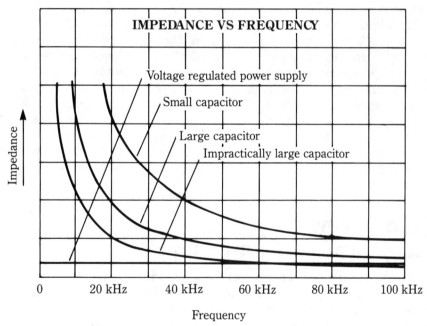

1-5 The impedance of a voltage-regulated supply, compared to that of the output capacitors in unregulated supplies.

The ideal voltage-regulated power supply

Having considered some of the important benefits from the use of a voltage-regulated power supply, it is only natural to postulate the ideal supply of this type for operating electronic equipment. As might be expected, the ideal power supply is approached as the ultimate extension of the attributes that have been discussed.

Thus, the ideal voltage-regulated power supply would have the following features:

1. Zero output impedance at dc and at all frequencies.
2. Zero regulation (constant dc output voltage) for a wide range of ac line voltages and over the entire range of load currents that are required by the powered circuitry.
3. Zero power dissipation.
4. Instantaneous recovery from changes in line voltage and load current.
5. Overload protection so that normal operation is automatically restored when the excessive current demand is removed.

Some of the above features are related; others are incompatible. Features 1, 3, and 4 cannot be achieved in actual practice. Zero power dissipation (3), would not allow the power supply to absorb additional power delivered by high line voltage. As a matter of fact, the most popular type of regulator, the series-losser type, *depends* upon power dissipation for its operation. Zero regulation can actually be accomplished. In most practical cases, however, the output voltage drops negligibly as load current is increased from zero to the rated value. Low output impedance (1), is readily approached at very low frequencies, where a voltage-regulated power supply "looks like" a tremendous capacitor. At higher frequencies, adversities set in from reduced gain and from the inductance of connecting wires. At such frequencies, electronic feedback no longer keeps the output impedance low. This problem is partially remedied by the use of an appropriate output capacitor. This capacitor is small compared to the output capacitor used in unregulated power supplies. As the frequency is further increased, various lead and parasitic inductances become significant, and output impedance rises rapidly from a low value. Fortunately, however, practical voltage-regulated supplies can be made to exhibit very low output impedances throughout (and well beyond) the normal range of frequencies involved in operation of the powered load.

Summarizing the voltage-regulated power supply

Considerable evidence has been presented to show that the use of a voltage-regulated power supply is a worthwhile consideration. True, the discussion did not exhaust the manifold aspects open to investigation and argument concerning the subject. However, the power supply is clearly part of the powered circuitry. This being the case, it is only appropriate that the voltage-regulated power supply should be considered early in the design and installation phases of circuits and systems. Also, we are led to investigate those applications where other types of regulated supplies better comply with the needs of the load.

The current-regulated power supply

Electromagnetic devices often perform more precisely when the *current* through them, rather than the voltage across them, is held constant. This follows from the fact that the strength of the magnetic field produced by a solenoid is proportional to three factors: the number of turns, the current through these turns, and the magnetic permeability of the medium through which the magnetic flux lines complete their closed paths. Usually, the number of turns and the medium are fixed. Therefore, the field strength will be a function of current. In an attempt to maintain the current at a constant value, best results are not obtained through the use of a voltage-regulated supply. Constant applied voltage will not ensure constant current because of the temperature-dependent resistance of the winding. For this reason, where a stabilized magnetic field is required from a solenoid, the current-regulated power supply is a better performer. An example of such a situation is the focusing solenoid for certain traveling-wave tubes (Fig. 1-6A).

In the process of supplying constant current, the current-regulated power supply becomes a dc power source with extremely poor voltage regulation. However, this does not infer that unregulated power supplies can be used to deliver constant current through a load. The voltage variation that appears at the output of a constant-current supply, though severe, changes in a certain discrete way. This behavior is brought about by a negative feedback loop and an internal voltage reference, much as in the voltage-regulated power supply. However, the comparator senses the voltage drop across a small resistance that is connected in *series* with the load. This voltage drop is proportional to, and therefore representative of, the load current. This contrasts with the sensing of the voltage *across* the output terminals (or load) in the case of the voltage-regulated supply. The stabilization of the desired output parameter is accomplished in both types of regulated supplies by actually controlling the *load current*. How the load current is controlled determines whether the supply behaves as the constant-voltage or as the constant-current type. This, in turn, is determined by the sensing techniques (just described).

Other applications for the current-regulated power supply

The control of speed and torque in a dc motor can often be accomplished advantageously with the aid of a current-regulated supply. In particular, when the supply is programmed, more precise control of the motor is generally possible than with field or armature rheostats. This follows from the fact that basic motor action is more directly a function of current than of impressed voltage. The impressed voltage required to duplicate a certain speed under given load conditions might vary as a result of the rather erratic nature of brush resistance and sparking from commutation (Fig. 1-6B).

Electrochemical reactions

Provided that the condition of constant temperature is met, the rate of electrochemical reaction is generally current dependent. Thus, the state of an electrochemical reaction can be accurately timed by providing a fixed current. Battery charging and electroplating represent two such cases (Figs. 1-6C and D). In both

12   Why use regulated power supplies?

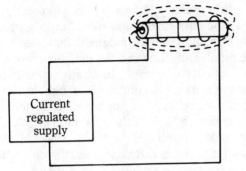

(A) Maintenance of constant magnetic field strength in a TWT solenoid.

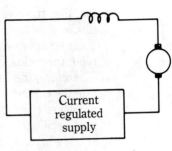

(B) Motor speed control.

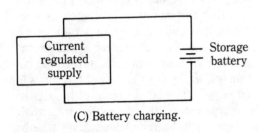

(C) Battery charging.

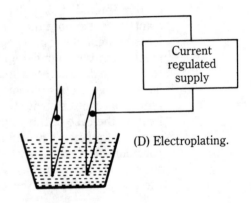

(D) Electroplating.

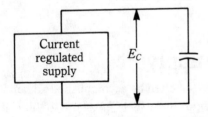

1-6 Typical applications of current-regulated power sources.

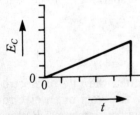

(E) Linear-time-base ramp.

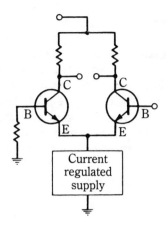

(F) Emitter resistance for a differential amplifier.

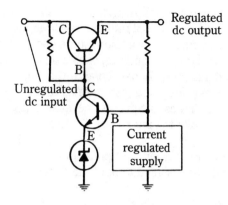

(G) Substitute for potentiometer dividing network in voltage-regulated power supply.

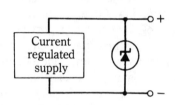

(H) Stable voltage reference with zener diode.

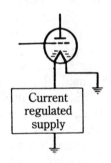

(I) Stabilization of thermionic emission in a vacuum tube.

1-6 Continued

instances, the voltage impressed across the cells is not a reliable parameter for control of the internal chemical reactions. Once sufficient voltage is available for forcing current through such cells, knowledge of the magnitude and time of the current defines the chemical state of the cells. This stems from one of Faraday's laws: *The mass of material liberated in an electrochemical cell is directly proportional to the quantity of electricity circulated through the cell.*

For practical purposes, the quantity of electrode material liberated (or collected) is directly proportional to the product of time and current. If, for example, a given cell deposits a 0.001-inch thickness of copper on its anode with one ampere of current over a period of 100 s, a plating of 0.003-inch thickness can be accomplished with the same current for 300 s, or one-half ampere for 600 s, etc. Here again, a programmable current-regulated supply can be used for precise control.

Charging storage batteries is also best achieved with constant-current supplies—even though the maintenance of a fixed charging rate throughout the charge cycle is not always necessary. Although a certain terminal voltage accompanies the fully charged condition of a storage battery, the mere attainment of such a voltage level is not a reliable indication of completed charge. Also, the application of such a voltage (or a slightly higher one) to the battery does not necessarily lead to any reasonable estimate of the time required for charging. Conversely, the application of a known current for a known time does provide some knowledge of the charge state of the battery (assuming, of course, that it is known that the battery will accept and hold a charge). In particular, if the ampere hours involved in discharge are known, the basic idea is to restore this product during charge. The current-regulated supply facilitates this process because charging current will remain a known parameter, despite the internal resistance and emf of the battery during the charge.

Although actual chargers can involve various sophisticated features, such as automatic turnoff and protection against reverse polarity, the better chargers are basically designed around the constant-current regulating mode.

Linear voltage ramps

The circuit shown in Fig. 1-6E can be used in the time base for oscilloscopes, digital voltmeters, and in other applications that require a linear voltage ramp. If the current involved in the charging of a capacitor is allowed to follow its own course, the resulting voltage developed across the capacitor will increase exponentially with respect to time. This, of course, is a nonlinear function. An approximation to linearity can be made by using only a small portion of such a charging cycle. A much better technique is to provide a constant current for the capacitor, in which case the voltage across the capacitor will increase linearly with time. Sometimes, this is brought about by charging through a very high resistance. Besides being an approximation to constant current, this approach constitutes an inefficient use of the voltage source. Also, it suffers from the disadvantage that independent control of time constant is lost. On the other hand, the use of a current-regulated supply for this purpose incorporates the feature that the rate of voltage change across the capacitor can be controlled by setting (programming) the value of the constant current supplied to the capacitor. This practice enables precise timing periods to be selected and it avoids conflicts caused by impractical capacitor sizes or awkward voltage sources.

Simulation of high resistance

A *differential amplifier* exhibits improved common-mode rejection as its emitter (or cathode or drain) resistance is made higher. This technique is limited by the maximum dc voltage that can be conveniently applied. If, however, a current-regulated source is substituted for the resistance, extremely high common-mode rejection can be had with a nominal value of dc voltage applied to the circuit. Such a constant-current source does not involve more than one or two additional transistors, a zener diode, and several resistances in a circuit (similar to the one in Fig. 1-6F). Often, the inherent constant-current characteristics of a single bipolar, or field-effect transistor, suffice for this application.

Constant current in voltage-regulated supplies

It so happens that the performance of voltage-regulated supplies can often be improved by incorporating constant-current sources within them. As a case in point, a constant-current source can substitute for one arm of the sampling divider in a voltage-regulated power supply (Fig. 1-6G). This practice largely avoids the attenuation of the loop gain that would otherwise occur. In many cases, this scheme can improve the line- and load-voltage regulation and can provide a lower output impedance.

Zener diode voltage reference

In Fig. 1-6H, a zener diode is operated from a constant-current source rather than from a resistance fed by an unregulated voltage source (the "conventional" method). The voltage developed across the zener diode is considerably more stable than that in the resistance-fed circuit. This is particularly true when the current supplied to a load is small compared to the current through the zener diode. Indeed, if the load is high impedance and demands negligible current, the circuit shown in Fig. 1-6H is an excellent voltage-reference source. An added refinement is the use of one or more forward-conducting diodes in series with the zener diode. This provides temperature stabilization because the temperature coefficient of the reverse-breakdown mode (over a certain voltage range) bears the opposite sign to the temperature coefficient of a forward-conducting diode. This technique is best left to the manufacturer—simply ask for a *voltage-reference source* and specify the desired temperature coefficient and voltage (while bringing the temperature coefficient close to zero, the choice of voltages is considerably diminished).

In sophisticated power supplies and where tight specifications must be met, it is not uncommon to find one or more of the auxiliary constant-current sources (shown in Figs. 1-6F, G, and H).

Stabilized emission in vacuum tubes

Current-regulated supplies are sometimes used for operating the filament or heater of vacuum tubes when it is necessary to control or stabilize electron emission. Although a voltage-regulated supply might be superior to an unregulated source in this respect, best results are generally obtained with current regulation because electron emission is more closely associated with the thermal effect of current flow (Fig. 1-6I).

Current regulation for testing purposes

Current-regulated power supplies are very useful in testing and measuring techniques. It is often safer, easier, and more efficient to *supply* current, and *monitor* voltage. This stems from the fact that the inherent high ac impedance of constant-current sources enables you to superimpose dc on a device undergoing evaluation for its dynamic behavior without interfering with the ac operation of the device. The calibration of meters is a natural application for the constant-current sources. Such a source likewise lends itself admirably to the evaluation of pull-in and drop-out of relays. Many semiconductor tests are best accomplished via the self-protection conferred by constant current. Included are the measurement of transistor parameters, the evaluation of zener-diode dynamic impedance, and the testing of ordinary diodes. Electrolytic capacitors are conveniently checked for leakage by charging them from a constant-current source. An excellent way to determine the capacitance of a large electrolytic capacitor is to measure the time that is required to develop voltage E across the capacitor when charging it from a constant-current source that supplies I amperes. The capacitance is obtained from the relationship:

$$C = \frac{I \times t}{E}$$

where:

C is the capacitance in farads,
E is the voltage across the capacitor in volts,
I is the current supplied by the constant-current source in amperes,
t is the time in seconds required to charge the totally discharged capacitor.

An interesting application of a constant-current source is illustrated in Fig. 1-7. The device symbolized is an LM134 programmable current source, made as a three-terminal monolithic IC. This current source supplies extremely well-regulated current over an adjustable range of 10,000 to 1. The particular implementation shown, as a sensor for remote temperature readout, stems from another operating feature, however. The temperature coefficient of current, with respect to absolute temperature, is both predictable and linear, and is selectable by the choice

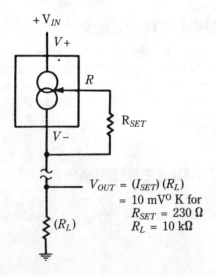

1-7 A current source that is used to provide remote-sensor thermometer function.
National Semiconductor Corp.

of two resistances. One of these resistances, R_{SET}, is situated with the current source, itself. The other resistance, R_L, can be located dozens, hundreds, or more feet distance from it. The voltage analog of temperature is then developed across R_L so that we have an *electronic thermometer*. Either analog or digital temperature readout can be arranged and the indicated temperature can be scaled in terms of Fahrenheit or Celsius, as desired.

With the resistance values shown for R_{SET} and R_L, 10 mV per degree is developed across R_L. In the interest of precision, these resistances should be stable, with low temperature coefficients. Changing the value of these resistances yields other scaling factors, but does not alter the linearity of the response.

Instead of deploying the predictable and linear temperature coefficient for a thermometer, a voltage-reference source for a regulated power supply can be made from this current source. This voltage reference can be made to exhibit a zero temperature coefficient. Still better, in some cases, a voltage reference can be designed with a temperature coefficient so that the *overall* temperature coefficient of the entire power supply is negligible over a wide temperature range.

Unique application of the shunt-current regulator

Voltage regulation is attainable in circuits that use either series-pass elements or in circuitry where the dissipative element is in shunt with the load. Moreover, a slight modification of the series-pass voltage regulator readily converts it to a current regulator (it is only necessary to have the sensing lead sample a voltage that represents the load current, rather than a fraction of the load voltage). Accordingly, it would appear reasonable to think of it as a shunt current regulator. On second thought, however, such a scheme would be inordinately inefficient because the shunting element would absorb excess current from the load. But, unlike the somewhat similar action in the shunt voltage regulator, such a shunt current regulator would dissipate greater power as load resistance is decreased. In ordinary applications, this would be undesirable and the series-type current regulator would be more appropriate.

A unique application where current stabilization via dissipation in a shunt element yields desirable performance is shown in Fig. 1-8. The scheme is used in an opto-coupler used as a line receiver for digital data. In this function, faithful reproduction of the input data stream is hampered by excess current through the LED. Such excess current can arise from transients, reflections, and system fluctuations in power supply voltage at the transmitter (another argument for regulated power supplies). In any event, the less variation in the light intensity produced by the LED, the better. In Fig. 1-8, the sensing (base) lead of shunt transistor Q1 samples a forward-bias voltage that is proportional to load current. This voltage is developed as a potential drop across series resistance R_1. Notice that the LED is, at once, both load and "voltage reference" (the other diode in the circuit, D1, is not directly involved in the current regulation—it merely serves to absorb negative transients from the data line).

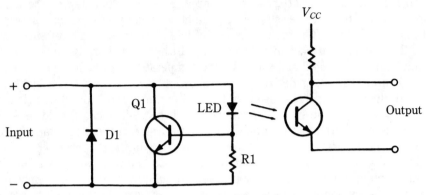

1-8 Shunt current regulation scheme in a data system optocoupler.

Inasmuch as the forward characteristics of an LED are not comparable in voltage constancy to the reverse voltage that is developed across a zener diode, it might be supposed that only a sloppy approach to current regulation is obtained from this arrangement. The performance improvement of the opto-coupler line receiver is quite dramatic, however. Inasmuch as the clamping action of the transistor is roughly proportional to "load" current, the LED current and its reference voltage are forced to be fairly constant, despite fluctuations in the "high" state of incoming voltage pulses. In actual practice, this technique can provide incoming noise immunity to the extent that twisted-pair transmission line can sometimes be used in place of costlier coaxial cable.

Stabilization of light output from LEDs

Solid-state lamps, otherwise known as *LEDs (light-emitting diodes)*, are commonly operated from a voltage source and a series current-limiting resistor. The voltage source can be "sloppy" or regulated. In either case, the light emitted from the LED is subject to variation from at least such causes as temperature and aging. Tolerances in the LED, itself, in the limiting resistance, and in the voltage source render it difficult to obtain uniform optical performance in production runs. Of course, all these variables become more pronounced if an unregulated voltage supply is employed. Although it is true that in many applications, such as the common indicator light, constant light intensity is of little consequence, it is nonetheless better to have a regulated current source for device longevity. Of course, in certain optical applications, constant light intensity enhances system performance. Optocouplers frequently fall in this category.

Figure 1-9 shows two schemes where the LEDs are operated from constant-current sources. The circuit using the FET is noteworthy because of its inherent simplicity—no zener diode or other voltage reference is needed. These "two-terminal" constant-current sources are available at specified currents. Some manufacturers of LEDs actually fabricate the LED and its constant-current source as a single package. Such an LED does not need a series resistor and it can be safely operated from a wide variety of voltage sources.

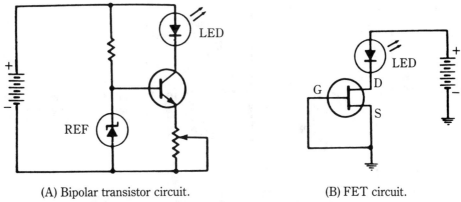

(A) Bipolar transistor circuit.   (B) FET circuit.
1-9 Constant-current sources for stabilizing LED output.

Constant-current sources for photon-emitting diodes

Semiconductor photon-emitting devices often rely on current-regulated power sources for proper operation. Although a rigorous interpretation of this terminology would necessarily include ordinary LEDs, general usage of the term is confined to infrared diodes and injection laser diodes. These diodes can be made to produce their radiant energy by means of two different operating modes—via high energy, short duration pulses, or under constant-current continuous-duty conditions. Because the requisite dc source falls within the scope of regulated power supplies, the latter operating mode is of most concern here (generally, special design is accorded to laser-injection diodes according to whether they will be pulsed or operate at continuous duty. Many infrared diodes can successfully be operated in either mode).

Figure 1-10 shows a constant-current source for a continuous-duty infrared diode, such as the RCA SG1002. This GaAs device develops its peak radiant intensity at 940 nm wavelength. The constant-current regulator configured about the RCA CA3085A IC is straightforward, despite appearance to the contrary. The electrical drafting techniques used in the layout of the circuit diagram deviate somewhat from common practice. Although a triangle is depicted, do not think of the CA3085A as an operational amplifier. Rather, it is an integrated subsystem: a low-power, but complete series-pass regulator. Comparator, voltage reference, and control and protection circuitry are self-contained. Accordingly, the enumeration and placement of the terminals do not comply with those that are ordinarily associated with triangle-symbolized op-amps.

In the circuit of Fig. 1-10, R_L is the current-sensing resistance and terminal 6 is the sensing lead of the CA3085A. Resistance R_{SS} governs the maximum current that can be delivered under short-circuit conditions. Under ordinary operation, short-circuit current is limited by the voltage sensed across resistance R_L. However, R_{SS} provides a back-up current limit within the IC itself in the event that

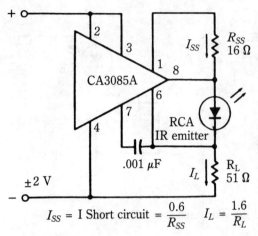

$I_{SS} = I\text{ Short circuit} = \dfrac{0.6}{R_{SS}} \qquad I_L = \dfrac{1.6}{R_L}$

1-10 Regulated-current sources for an infrared-emitting diode.
RCA Electro-Optics and Devices

something goes wrong in the external circuitry. This feature is worthwhile because infrared diodes are easily destroyed. The regulated current is $\dfrac{1.6}{51}$ (31.4 mA) and the back-up current limit is $\dfrac{0.6}{16}$ (37.5 mA).

Dynamic load

Somewhat related to the application of Fig. 1-6B is the high-impedance dynamic load shown in Fig. 1-11. Here, a constant-current source, in the form of an FET, constitutes the load impedance for another FET, which is the amplifier proper.

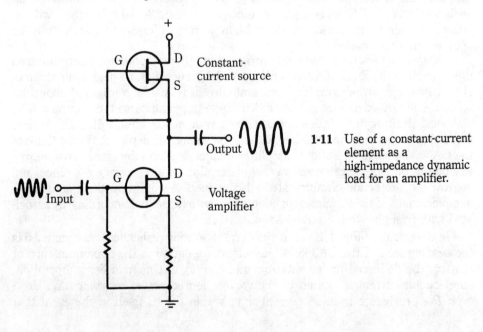

1-11 Use of a constant-current element as a high-impedance dynamic load for an amplifier.

This scheme enables a high load-impedance to be realized without the relatively high dc operating voltage that would be required for the same allowable voltage swing if the conventional load-resistance was used. Accordingly, high-voltage amplification and high output levels are attained. Similar arrangements can be made with bipolar transistors or with mixtures of FETs and bipolars. The FET, however, is uniquely endowed with excellent constant-current characteristics. Two terminal constant-current "diodes" are often FETs that are internally connected (Fig. 1-11). Also, more sophisticated IC constant-current devices can be readily adapted for use as a dynamic load.

In direct contrast to the voltage-regulated source, the *ideal* regulated-current, or constant-current source, behaves as an infinite resistance. In practice, when used as a dynamic load, it provides the attributes of an inductor load, but without the frequency discrimination of the inductor. Also, it can equal or exceed the broadband performance of a high-resistance load because of the vulnerability of such a load to stray capacitance.

Voltage and current regulation combined

You might think that it would be advantageous for certain purposes if voltage and current regulation were combined in the same supply. This is indeed the case, and four unusual combinations of voltage and current regulation are shown in Fig. 1-12. In these supplies, various degrees of regulation or control can exist simultaneously in order to provide certain useful results.

Figure 1-12A depicts the characteristics of a power supply that is voltage-regulated until a certain load current is demanded from it. Then, it abruptly changes to a constant-current supply. Such a dual-performance power supply automatically protects itself from overload. To remove an overload causes the supply to revert to its constant-voltage mode. Often, independent controls are provided for the regulated voltage level and for the current load, at which the supply becomes a constant-current source. In Fig. 1-12B, the available current from the supply is considerably restricted, but not held constant when the load demand exceeds a preset value.

The power-supply characteristic illustrated in Fig. 1-12C exhibits a more severe current limiting. The output current actually diminishes once a certain current demand is reached. At short circuit, such a supply will furnish very little current. This combination of voltage regulation and current limiting obviously constitutes excellent protection to the supply from the effect of an unintentionally low load resistance. Such a supply is said to have a *foldback characteristic*.

The foregoing power supplies also prevent damage to the load. For this reason, they are more suitable for use with breadboards and experimental work than with a supply that has no provision to prevent inadvertent and destructive load currents.

Some load requirements are best met in such a way that the power delivered to the load is maintained constant. Thus, an interplay of voltage and current regulation is required to stabilize the product of load voltage and current, despite the current demand (resistance) of the load. The characteristic of a constant-power supply

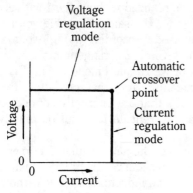
(A) Automatic crossover from voltage to current regulation.

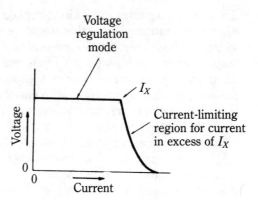
(B) Current limiting on high demand.

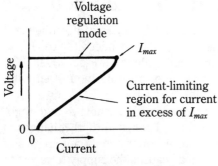
(C) "Foldback" characteristic on high demand.

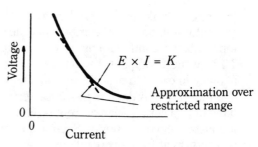
(D) Constant power to load.

1-12 Output characteristics of power supplies that have both voltage and current regulation.

is shown in Fig. 1-12D. Arc lamps, especially, can be beneficially operated by such a supply when constant light output is desired. Two methods for approximating constant power are shown in Fig. 1-13. In Fig. 1-13A, simultaneous voltage and current feedback are employed to cause the output voltage vs current characteristic to have a slope somewhere between that corresponding to constant voltage and constant current. This, in itself, does not produce constant power. However, if the load range is not too great, the approximation to constant power can be satisfactory for some applications. A closer approximation to the parabolic curve of Fig. 1-12D can be caused by inserting nonlinear elements in the feedback loops.

In Fig. 1-13B, the light output of an arc-lamp load is converted into feedback voltage by a photovoltaic cell. To the extent that the light emitted is proportional to the power consumed by the gaseous discharge, this method approaches accurate power regulation.

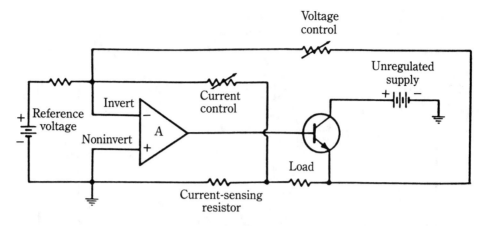

(A) Simultaneous voltage and current regulation.

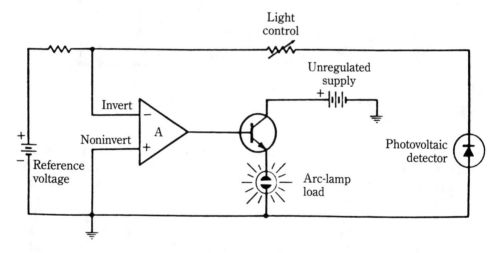

(B) Feedback control by power-dependent parameter (light).

1-13 Methods of approximating constant load power.

Accurate power regulation

The availability of inexpensive monolithic and hybrid circuit modules has made it relatively easy to construct supplies that accurately regulate load power. This contrasts with the two aforementioned schemes for wattage stabilization, which, at best, are approximate. As shown in Fig. 1-14, the heart of the accurate method is an analog multiplier module. As the name implies, these function-modules deliver an output voltage that is proportional to the product of two input voltages. Thus, it becomes feasible to generate an error signal that represents the product of load

24 *Why use regulated power supplies?*

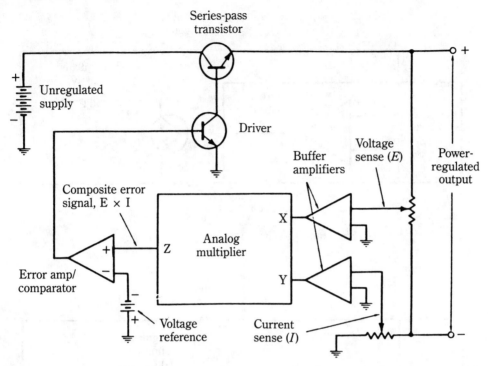

1-14 Simplified circuit of a power-regulated power supply.

voltage and load current: *load power*. Once this quantity (power) is obtained in the form of a representative voltage, the regulatory process locks on to it and stabilizes it in exactly the same manner that either voltage alone or current alone is stabilized in most regulated supplies.

The basic similarity of this circuit to the more common voltage- or current-regulated supply can be advantageously exploited by providing a panel-mounted three-way switch for selecting the operating mode of the supply, i.e., voltage regulation, current regulation, or power regulation. To accomplish voltage regulation, the switch would connect the sensing lead (+) of the comparator to X instead of to Z of the multiplier. For current regulation, the connection would be made to Y instead of Z. For power regulation, the arrangement would be just as depicted in Fig. 1-14.

Accurate power regulation is useful for certain lamps, for meter calibration, and for various scientific and industrial purposes. It merits consideration in dc motor control, and could conceivably find application with certain audio, servo, and RF amplifiers. Also, the very same concept can be implemented with switching-type, as well as with linear regulators (both types of regulators are "stupid" in the sense that they will stabilize any quantity that is represented by an appropriate error-signal). Several companies that have specialized in dc analog multipliers are Analog Devices, General Magnetics Inc., and Teledyne Philbrick.

The converse of the voltage-regulated supply with current limiting is the current-regulated supply with voltage limiting. This supply is often very desirable in order to protect voltage-sensitive loads. The current-regulating circuit, unless special precautions are incorporated in its design or application, cannot indefinitely behave as a constant-current source as the load resistance is increased. As the load resistance is raised, the voltage applied to the load (compliance voltage) increases and ultimately becomes very nearly equal to the unregulated supply voltage. In many applications, the current regulator does not need to provide constant current for lighter loads than would be encountered during normal system operation. For safety, however, it remains desirable to limit the voltage that is available from the supply for the light-load condition. This limiting is readily accomplished by inserting an appropriate zener diode in the feedback of the current-regulated supply (Fig. 1-15A). The resultant voltage limiting is shown in Fig. 1-15B.

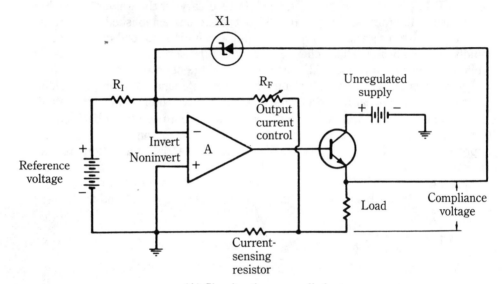

(A) Circuit using zener diode.

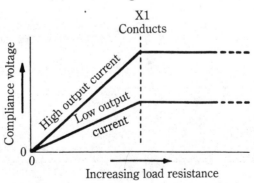

(B) Characteristics of current-regulated supply with voltage limiting.

1-15 Voltage limiting in a current-regulated power supply.

Other applications of regulation techniques

An intriguing aspect of regulated power supplies is the realization that innumerable useful results can be conferred not only to electronics in the ordinary sense of the term, but to many other applications. For example, the field of electric arc welding has long been a hazy mixture of art and science. Unfortunately, some manufacturers of these machines have contributed greatly to the black magic surrounding the alleged abilities of their products.

In both advertising and equipment specifications, the descriptions of the electrical characteristics often cater to the technically unversed. The actual fact of the matter is that the electric arc welder is a power supply. Because of the peculiar characteristics of the load, certain volt-ampere output characteristics provide optimum welding properties for specific types of work. For example, when the arc is being struck, a short circuit prevails, and it is necessary for the welder to be inherently current-limiting (see Fig. 1-16). After the arc has established itself, it displays a negative-resistance characteristic and it can alter its load characteristics rapidly and erratically. These changes are caused by several factors, such as the molten globules of work material, unsteadiness of the operator, and varying thermal properties of the incandescent metal. The maintenance of a proper welding arc involves what appears to be a tacit cooperation between machine and operator.

Although welding will continue to require skill, the use of feedback control techniques, such as those employed in regulated power supplies, can greatly improve welding. This technique leads to better repeatability and to more standardized practices. At this writing, such a trend is already evident with SCRs, giant transistors, and programmed control featured by the more progressive makers of welders. Also relevant is that some makers of regulated and controllable

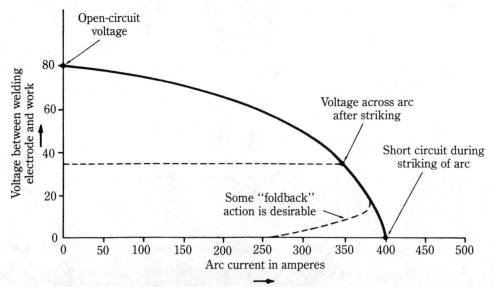

1-16 Generalized volt-ampere characteristic of an arc welder.

power supplies with 10- to 50-ampere ratings advocate their equipment for light welding.

The application of electronics to the automobile has had a longstanding advocacy. Until recently, traditional approaches have stubbornly held their ground; to what extent this has been caused by industrial inertia, vested interests, or valid objections to technological shortcomings is difficult to assess. However, various forces have finally converged to pave the way for many applications of electronic control functions to the automobile. Among these applications are a number that derive their operating principles from the very concepts that are used in regulated power supplies. For example, the voltage-regulator circuits (Figs. 1-17 and 1-18) replace the long-used regulators that employ electromagnetic relays. With these older devices, one accepted the inherent need for maintenance, the rather sloppy performance, and the possibility for required replacement after 40,000 miles or so of service. Conversely, solid-state voltage regulators are capable of much greater precision, are inherently very reliable, and can reasonably be expected to display considerably extended lifetimes relative to relay-type regulators.

The solid-state voltage regulator (Fig. 1-17) is one of a number of designs being used with automotive alternators. Potentiometer R1 is adjusted so that transistor stage Q1 is conductive for battery voltages below approximately 13.5 V. If the battery voltage rises above this value, the nominally 0.6-V base-emitter voltage

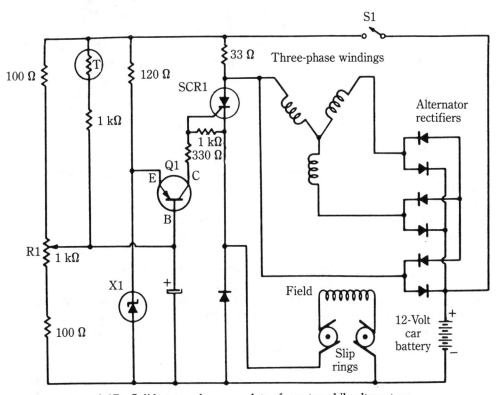

1-17 Solid-state voltage regulator for automobile alternators.

28 Why use regulated power supplies?

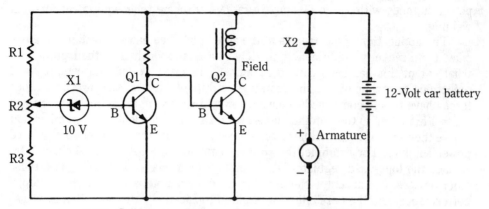

1-18 Solid-state voltage regulator for autos with dc generators.

of Q1 decreases. The voltage at the emitter of Q1 is held at 10 V by zener diode X1. As long as the battery voltage is above 13.5 V, transistor Q1 remains off and no gate current is delivered to SCR1. However, when the battery voltage falls below 13.5 V, transistor Q1 is turned on and SCR1 fires. The SCR then allows current to flow in the field winding of the alternator. The activated alternator charges the battery until its voltage rises sufficiently to again turn off the charging process. The action is similar to that of the electromechanical voltage regulator. The on/off charging cycle can occur at a rate of many times per second.

The conduction of the SCR is extinguished and initiated at the frequency developed in one phase winding of the alternator stator. This *commutation* permits the use of an SCR, rather than a power transistor. Commutation occurs at a much higher rate than the on/off cycle of the battery charging current. It is often more economical to obtain the required electrical ruggedness from an SCR, rather than a power transistor. Of course, access must be gained to the "hot" end of a phase winding to utilize this scheme. This is not necessary when a power transistor is used. The thermistor and the 1-kΩ resistor are selected to enable the threshold of regulation to occur at a slightly lower battery voltage when the ambient temperature is high.

A different design approach that accomplishes similar overall results is shown in Fig. 1-19. Because a Darlington transistor is used in this scheme to control field current, no commutation is required. Therefore, it is not necessary to gain access to the stator windings on the alternator. In this arrangement, input transistor Q1 conducts only when the battery voltage becomes high enough to break down the zener diode in its base circuit. When this happens, the Darlington output stage is turned off and the field winding of the alternator is deprived of current. Depending on the state of the battery and on the load, the field circuit will be energized and opened at widely varying duty cycles. The circuit can be easily used with semiconductor devices other than the types specified.

The voltage regulator shown in Fig. 1-18 is similar to the circuits shown in Fig. 1-17 in that it either provides charging current or it does not. This solid-state circuit is intended for older cars that use generators, rather than alternators in the

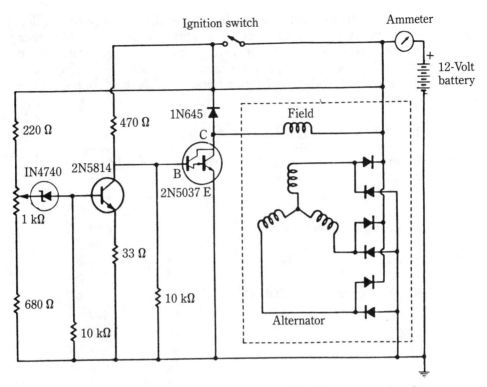

1-19 Alternator voltage regulator that uses a Darlington power transistor.

charging system. This circuit can also be used with alternators if diode X2 is replaced with a direct connection. Diode X2 substitutes for the generator cut-out relay used in the older generator charging systems. The function of the cut-out relay was to prevent discharge of the battery through the generator armature when the engine was not running. Such a provision is usually not necessary with alternators because isolation is automatically provided by the alternator rectifier diodes. An automobile voltage regulator with a dedicated integrated-circuit module is described in chapter 6.

Power supplies for microprocessors

Because of the rapid rate of technical evolution, the wide range in capabilities and complexities, and the different demands of competing semiconductor technologies, it is not feasible to state specific power supply requirements of microprocessors. In general, they require more than one voltage, although some companies feature single-voltage operation. A given microprocessor could conceivably require 5 V, ±12 V, and −5 V. The last designated voltage usually would be an auxiliary voltage (used to bias the semiconductor substrate) that would consume a small fraction of a milliampere. Auxiliary voltages are often used for other on-chip purposes (such as keeping alive static RAMs) where current consumption might be in the vicinity of a few tens of milliamperes. In general, the overall power requirement of a wide variety of microprocessor types could run the gamut of less than 100 to over 1000 mW. Other factors that would bear on both voltage and current needs are: operating speed, nature of associated peripherals, and I/O (input-output) compatibility matches. Table 1-1 shows voltage and current formats of some typical microprocessors.

Table 1-1. Voltage and current formats of typical microprocessors.

Manufacturer	Model	Voltage format	Current
Texas Instrument	9900	+5	125 mA
		−5	1 mA
		+12	30 mA
Rockwell International	6500/1	+5	100 mA
		+5	10 mA
Fairchild	F8	+5	80 mA
		+12	25 mA
NEC Microcomputers	μCOM-43	−10	30 mA
Motorola	6800	+5	100 mA
MOS Technology	650X	+5	50 mA
General Instrument	CP-1600	+12	70 mA
		+5	12 mA
		−3	0.2 mA
Intel	8086	+5	275 mA

Where single or several microprocessors are employed, the linear series-pass power supply is suitable. Where many microprocessors are used in a large system, or where microprocessors are used in conjunction with a number of other subsystems, the switching-type power supply merits consideration. Not only is the "switcher" more efficient, but it more readily provides multiple voltage formats and levels than does the linear supply. Whereas switching supplies used to be considered only at the several-hundred watt level, their deployment has become economically justified at 50 W and less. The poorer regulation, greater noise, and slower response of the switching regulator has not been detrimental in many microprocessor applications, although these less-than-ideal characteristics deserve careful scrutiny.

High-voltage regulated power supplies

The realm of high-voltage electronics can be arbitrarily designated as that which embraces the several-kilovolt to several-hundred-kilovolt range. Probably, the majority of such applications utilize voltages between 5 and 50 kV. In any event, these voltage levels are uniquely used in science and industry, as well as in consumer products. It has become increasingly appreciated that regulation is beneficial. Previous high-voltage techniques were predicated upon "brute-force." These approaches exacted a heavy toll in predictability, cost, reliability, and in various performance parameters. In contrast, regulation enables precise correlation between cause and effect and permits the manufacturer to "get a handle" on his product. Similarly, the user then knows what to expect under specified conditions.

Table 1-2 lists a number of high-voltage applications. This compilation is not construed to be all-inclusive. A certain amount of redundancy exists in the alphabetical indexing. Nonetheless, it is evident that high-voltage supplies are in considerable demand and regulation is destined to play a greater part in virtually all of these systems and processes.

Table 1-2.
Applications and processes that involve high voltages.

Accelerator	Negative-ion generator
Cathode-ray tubes	Neon signs
Charge neutralization	Paint spraying
Corona testing	Plasma devices
Deposition of powders	Photo-etching
Dielectric testing	Photo-multipliers
Displays	Powder deposition
Electrified probes and fences	Precipitation of pollutants
Electrophoresis	Pulse generators
Electrostatic cooling (welding)	Radar
Electrostatic gyros	Scintillation
Electrostatic precipitation	Sonar
Electrostatic spraying	Strobe excitation
Fluorescent tubes	TV
High-energy capacitor charging	Vacuum systems
Ignition	Vacuum metalizing
Insulation testing	Vidicon
Ion transfer	X-ray
Lasers	Zerography
Masers	

As a first approximation, you could view the high-voltage supply as a scaled-up version of lower-voltage units. Certainly, many principles and design techniques still apply. However, before working with high-voltage supplies, you should acquire knowledge and skills in dealing with the potentially lethal aspect of high voltages and high-energy storage. Then, you must learn how to contend with x-rays, corona, ozone, ultraviolet radiation, and ionization. The deterioration of insulation by these manifestations of energy can hardly be given too much consideration. On the other hand, advances in materials, encapsulation, cables, circuitry, and devices have combined to evolve the modern high-voltage regulated supply as a compact and neatly packaged product that conveniently mates with the powered equipment. This vision contrasts dramatically with high-voltage sources from the past. These were more in the nature of power substations, often dwarfing (in size) the powered equipment.

The uninterruptible power supply or system (UPS)

Regulation is often associated with other applications besides the most obvious—where ac power from the utility line is converted into stabilized voltage, current, or power for a dc load. Speed control of motors, conditioning and regulation of ac operating power, battery charging, and *the uninterruptible power supply (UPS)* exemplify such techniques. The UPS is particularly relevant to the material discussed in this book. Such a system makes use of battery charging, inversion (generation of 60-Hz or 400-Hz power in the absence of such power from the utility line or another ac power source), and conditioning and regulation of its internally generated ac. Ideally, the load is "uninformed" that anything has happened. Although engine-driven alternators or generators often provide emergency power, the battery-electronic inverter approach to this problem is rapidly gaining popularity.

The need for uninterruptible power stems from the mutual characteristics of ac utility power and certain powered equipment. On the one hand, the ac power available from the utility system is not only plagued with noise, transients, and "brownouts," but it could disappear altogether, which would cause a "blackout." The worst feature of such behavior is its unpredictability. Even so, the occurrence of a severe brownout, or even a blackout is no more than nuisance in many situations, the chief nuisance is the stress of impatience on the part of human operators. However, in other cases, malperformance, damage, or injury can result from such loss of ac power—even though momentary. Examples of such critical loads are computing systems with volatile memories, hospital and medical applications, and certain industrial processes. Generally, the UPS automatically provides instantaneous back-up power in the event of failure or drastic malperformance of utility ac power. Most UPS systems can automatically reconnect the load to the utility line at the instant utility ac power reappears.

Figure 1-20 is a block diagram of a simple UPS. Here, the basic regulation of the internally generated ac power is provided by the characteristics of the ferroresonant-transformer itself. More refined regulation can be obtained by means of a feedback provision, in which sampled output voltage is used to vary the conduction angle of the SCRs in the controlled rectifier. Either an electromagnetic or a solid-state relay can be used to perform the automatic switching between utility and internally generated power. Various features and sophistications are found in commercial UPS equipment. In some instances, the internally generated 60 Hz is synchronized to the utility power, when present. This measure helps reduce transients upon changeover from one ac power source to the other. It is also a means of closely simulating operation from an "ideal" utility line when the UPS is used as a power conditioner. Some UPS equipment provides regulated dc output voltage in addition to, or instead of, "remanufactured" ac. Often a provision for manually overriding the automatic changeover is included. A wide range of voltage and current ratings is available, including both single- and three-phase formats.

In the basic UPS of Fig. 1-20 the hold-off diode is reverse-biased, as long as the ac utility line is operational. In the event of an outage, the hold-off diode

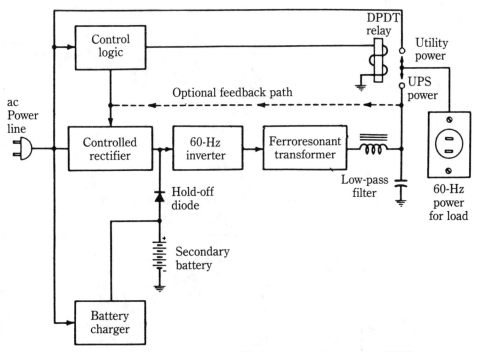

1-20 Block diagram of uninterruptible power supply or system (UPS).

becomes forward-biased and it automatically connects the standby battery to the 60-Hz inverter. Lead-acid batteries have been widely used in UPS systems, with particular emphasis on the gel-electrolyte types. Sealed or "maintenance-free" lead-acid batteries that have gained popularity in automotive use are also well-suited for such standby service in UPS installations.

Other uses for regulated power supplies

The regulated power supply is admirably suited for the control and regulation of various nonelectric quantities. This information has not been widely recognized and has not been duly publicized by the manufacturers of supplies. A case in point is the control of dc motor characteristics. A regulated power supply with remote-sense terminals readily becomes an accurate servo-system for controlling and stabilizing such motor characteristics as speed and torque. Figure 1-21 shows an adaptation for the control and stabilization of motor speed. The tachometer generator, which is mechanically ganged to the motor, provides the "error signal"; the power supply's internal operation is identical to that which prevails when the load voltage is being regulated in more conventional use.

By contrast, the same type of regulated power supply can be connected to control and stabilize the torque of a dc motor (Fig. 1-22). Here, the internal operation of the power supply is identical to that which prevails when load current is being

34 *Why use regulated power supplies?*

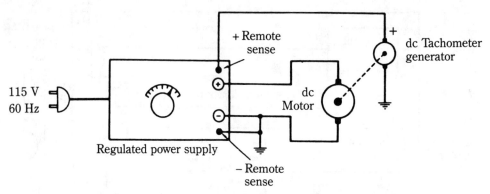

1-21 Constant-speed motor control by a regulated power supply.

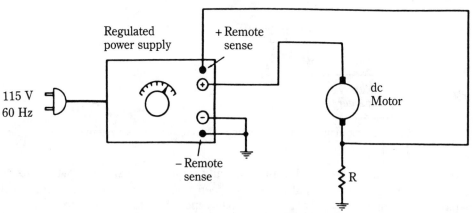

1-22 Use of a regulated supply for constant torque from a dc motor.

regulated in the constant-current operational mode. The torque of the motor is stabilized because it is a function of armature current (it can be seen that this is a more sophisticated deployment of a regulated supply than the motor-control scheme that is depicted in Fig. 1-6B).

Although the permanent-magnet motor is probably the best for both of these applications, it is interesting to contemplate that both series- and shunt-type motors can be made to work well. This is because the regulated power supply imposes its own control—the "textbook" characteristics of when the motor is no longer able to govern its operation. For example, a series motor, by itself, tends to drastically drop speed when it is mechanically loaded. However, when the series motor is connected to a regulated power supply (Fig. 1-21) its speed vs load characteristic will be flat.

Such parameter control of motors by regulated supplies can be very useful in many practical applications. The thing to watch is that the horsepower rating of the motor is not exceeded under such electronic "supervision." Often a larger motor must be selected. That would be the case if the motor were merely allowed to "do its own thing." Generally, these schemes are more amenable with linear-, rather than with switching-type supplies. However, if due regard is given to inductive kickback, back-emf, and start-up conditions, either type of regulator circuit should be feasible.

Yet another example of the regulated power supply's ability to perform servo-control functions is the temperature-stabilization system (Fig. 1-23). Here, the positive-sense lead responds to a dc voltage, which represents the temperature within the thermally insulated chamber. Interestingly, the "resistance-to-voltage converter" enclosed by the dashed lines is, itself, a voltage-regulated supply. Composing an operational amplifier, a reference voltage source, and a small transistor, its dc output is governed by the resistance of the thermistor in its operational-gain network. The nonlinearity of the thermistor does not alter the basic principle of this system, but it can enter the practical situation in the matter of dial calibration of temperature.

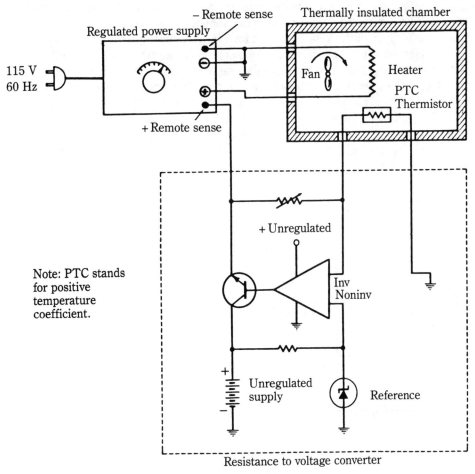

1-23 Use of a regulated power supply to stabilize temperature.

Similar control techniques can be extended to other nonelectrical parameters with the provision that the controlled quantity must be transformed to a representative voltage, current, or resistance and be appropriately applied to the sensing lead(s) of a regulated power supply. If the remote sensing lead(s) of the regulated supply are brought out as separate and independent terminals, it usually will not be necessary to make any internal modifications to the supply. In any case, numerous control and stabilization functions can be neatly and conveniently solved with the use of a regulated power supply, rather than design electronic circuitry that would be the equivalent of a regulated supply—even though this is not the designer's intent.

Using the regulated power supply to stabilize light intensity

The stabilization of light intensity from an incandescent or a gaseous discharge lamp can be achieved by the arrangement shown in Fig. 1-24. This will be recognized as a pick-up of the idea discussed for Fig. 1-13 for regulating the power delivered to a lamp by using the feedback loop of a power supply to sample light output. Actually, implementation of these kinds tend to stabilize light intensity rather than the electrical power that is delivered to the lamp. For practical purposes, however, light intensity is proportional to filament or electrode input-power, at least for small excursions. Therefore, a close description of the technique is the stabilization of power delivered to a lamp. In any event, here is a means of both adjusting and stabilizing the light intensity of a lamp via the use of a regulated power supply. Notice, also, that the use of the remote-sense terminal is equivalent to breaking into the feedback loop of a regulated supply that is not equipped with a remote-sense provision.

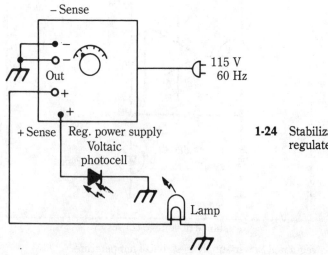

1-24 Stabilization of light intensity by a regulated dc supply.

This use of regulated power supplies can be useful in amateur TV transmission, in movie projectors, in photography, in display illumination, and in photometer techniques for measuring RF power. Of course, the optics of the arrangement have much to do with quality of performance. For example, it would be detrimental to allow the photo-detector to "see" ambient light in addition to a sample of light from the lamp to be controlled.

This scheme has an advantage for stabilizing light intensity over the apparently more-straightforward use of a regulated supply to provide either constant voltage or constant current to a lamp. The light intensity of most lamps falls off with use so that regulation of electrical input no longer maintains constant light intensity. The above described technique does, however, keep the light intensity constant regardless of the light-producing efficiency of the lamp. Probably, small lamps can be best stabilized with a linear regulator; larger lamps, such as those used in projectors, would be best served by switching regulators.

The optical output of LEDs and semiconductor lasers also diminishes with accumulated operating time. For some purposes, this stabilizing technique could enhance the performance of these solid-state "lamps."

Finally, some three-terminal voltage regulator ICs (those that are described as "adjustable") will work for this technique, even though they have no obvious remote-sense terminal. Figure 1-25 shows the basic setup for such an implementation.

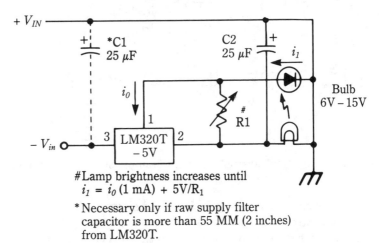

#Lamp brightness increases until
$i_1 = i_0 (1\text{ mA}) + 5V/R_1$

*Necessary only if raw supply filter capacitor is more than 55 MM (2 inches) from LM320T.

1-25 Use of a 3-terminal IC regulator to control light intensity.

Using a regulated dc supply to stabilize RF output level of a TWT

Broadband-microwave operation is one of the salient features of the *traveling-wave tube*, or TWT. However, the RF power-level vs frequency relationship can even be made more constant by a simple association with a small linear regulated power supply. Inasmuch as negligible power is consumed from the supply, the added cost of the scheme is minimal. The basic arrangement is shown in Fig. 1-26. The directional coupler contains a rectifying diode and delivers a dc voltage that is proportional to sampled microwave output power from the TWT. The control grid of the TWT controls the output power level in a way that is analogous to control-grid action in a triode or tetrode vacuum-tube.

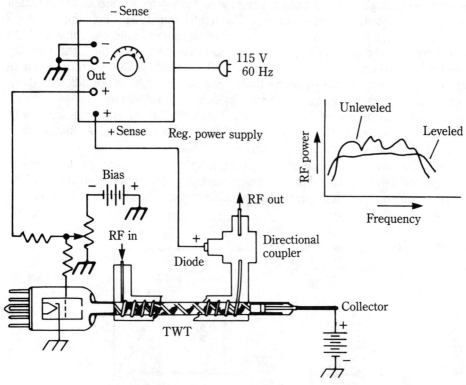

1-26 Use of a regulated dc supply to level RF output of a traveling-wave tube.

As a consequence of this previously described situation, TWT output power is decreased as the control grid is made less positive or more negative. This is where the regulated power supply with its remote-sense provision confers its stabilizing action. Specifically, any tendency for increased output power is counteracted by the control-grid bias becoming more negative. (The grid bias is the sum of the fixed bias and the bias obtained from the regulated supply.) Of course, the converse operational situation also prevails—any tendency for a reduction of microwave output power is counteracted by a change of control-grid bias in the positive direction. This basic idea can also be used for TWT amplifiers and TWT frequency multipliers. Moreover, electron tubes of the more mundane variety can be stabilized in a similar way. In most instances, amplitude stabilization of an oscillator is accompanied by enhanced frequency stability and reduced harmonic generation.

Prevention of remote-sensing problems

The remote-sensing terminals on regulated power supplies overcome the degradation in voltage regulation as a result of the resistance of the load connecting leads. They also enhance the versatility of the supply and enable it to serve purposes other than simply supplying electrical power to a load. In practice, a commonplace trouble is often experienced with such supplies, however; load damage can result when the sense lines are accidentally opened or not connected during experimentation. When the sensing circuit is open, the output voltage of the supply rises to its maximum value, and often damages the load in the process.

A simple technique for guarding against such an occurrence is shown in Fig. 1-27. Notice the polarity of the diodes connected between the output terminals and their corresponding sense terminals. As long as the overall circuit is intact, these diodes are not forward-biased and accordingly exert no effect. If, however, the sense line(s) becomes interrupted, these diodes become forward-biased and act as switches that connect the sense terminals to their respective output terminals. This action prevents the large rise in load voltage that would occur without these diodes. Why aren't such protective diodes already incorporated in the supply? They sometimes are, but you cannot naively assume that this is the case.

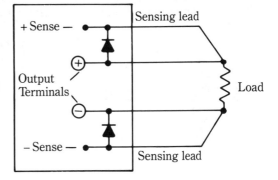

1-27 Protection against load damage from open remote-sense lines.

A possible pitfall in the use of this protective technique stems from the not infrequently encountered philosophy that the remote-sense feature will compensate for the use of otherwise inadequately sized load leads. The trouble with this notion is that the normal voltage-drop in the skimpy load-wires can then be sufficient to forward-bias the protective diodes. If this occurs, load-voltage regulation will be impaired, perhaps to the extent of negating the intended function of the regulated supply. Measurement of the voltage across the diodes during normal operation of the load will quickly reveal whether such malfunctioning is likely. Be sure to allow some safety factor for the effect of ambient temperature on the forward-conduction characteristics of the diodes, and use a good-quality digital voltmeter for such test measurements. One third of a volt or less across the protective diodes (at ambient temperature) is generally safe from unwanted diode conduction.

In the case of the remote-sensing leads, resistance is not important and small-gauge copper wire should suffice. However, the feedback-loop of the power supply is sensitive to excessive inductance in the sense leads, which might cause the operation to be unstable. Other things being equal, larger diameter sensing wires have lower inductance per unit-length than smaller wire. However, an even more practical way can be used to minimize the effective inductance of the sensing leads.

Forming the two sense-leads into a twisted pair will greatly reduce their inductance. Moreover, this technique will also guard against pickup of noise and transients; sometimes the "antenna effect" of long sensing leads can be the unexpected source of excessive noise delivered to the load. This effect is particularly noticeable in linear power supplies because of the natural expectation of clean dc at the load. Another way to reduce stray pick-up via the sense leads is to use shielded wire or miniature coaxial cable and ground the shielding braid.

Using a regulated power supply to extend transmitting-tube life

Transmitting-tube filaments are supposed to be maintained within narrow voltage limits in the interests of both performance and longevity. Traditionally, this requirement has been met by some such arrangement as depicted in Fig. 1-28A, in which the applied filament voltage is monitored at the terminals of the tube socket. Although filament emission is more exactly a function of filament current than of

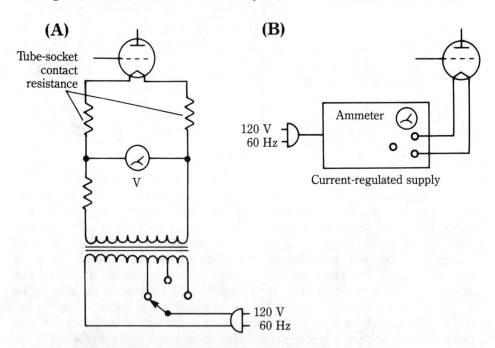

1-28 Traditional and proposed transmitting-tube filament circuits.
(A) Traditional method: manual adjustment of voltage that is monitored at tube-socket terminals.
(B) Proposed technique: filament is heated from constant-current power supply.

voltage, the voltage at the tube socket is considered a reliable indication of the current flowing through the filament. In actual practice, because of the resistance between the tube pins and the socket contacts, this assumption is not always valid. Indeed, the voltage can be considerably different if filament currents are heavy and if the contacts are corroded.

This possible mismeasurement suggests that it would be better to monitor filament current rather than applied filament voltage. Then, the actual temperature of the filament would be better controlled; less than perfect contact to the tube's filament pins would be of no consequence and the voltage measured at the terminals of the tube socket would be meaningless. You would simply adjust the variable or tapped transformer and/or rheostat to maintain a prescribed filament current.

Taking this idea one step further, why not use a constant-current power supply? Then, even though tube-socket resistance might vary, no further manual adjustment would be necessary once the regulated supply was set to deliver the desired current. Such an arrangement was not practical for many years because of the difficulty and expense of regulating heavy currents. Nowadays, this is easily accomplished with a solid-state linear supply or even better, a switching regulator. Figure 1-28B illustrates the basic idea.

Admittedly, the use of dc (rather than ac) often sparks controversy. Some feel that if dc is used, the polarity should be reversed every so many operating hours. Also, you could conceivably run into some grid-bias problems with dc filament operation. This problem could be readily overcome by connecting the grid-return lead appropriately. A nice aspect of such a constant-current set-up is that the initial current inrush to the cold filament would be automatically limited. This, alone, would be expected to extend the operating life of the tube. Notice, too, that this method not only makes filament temperature independent of socket resistance, but it also maintains its constancy in the face of ac line-voltage variations.

Other applications using regulation circuitry

It has been shown that the basic concept that underlies voltage regulation can be easily extended to the regulation, stabilization, or control of other quantities, such as current, power, temperature, speed, torque, light intensity, etc. Other less-obvious applications of regulated power-supply circuitry also accomplish useful functions.

For example, a unique AM-modulation scheme for solid-state transmitters can be implemented with a three-terminal voltage-regulator IC (AM is used in aircraft communications, and to a limited extent, in some ham bands). Interestingly, conventional AM modulators essentially vary the dc voltage that is applied to the modulated RF amplifier at an audio rate. It is only natural to suspect that the modulator, itself, could be dispensed with if some means was available to vary the output voltage of the power supply directly. This approach hasn't been practical in the past, but the three-terminal voltage regulator makes it feasible for low-power applications. On the premise that a voltage regulator would have been used even with conventional modulation circuitry, this approach merits consideration on the basis of reduced cost and part count. Also, the traditional modulation transformer is eliminated.

The basic idea of this scheme is shown in Fig. 1-29. It is best to use an adjustable regulator IC, and one with a relatively high current rating. Paralleling for greater power capability should be easier than in power-supply practice; good heat-sinking will ordinarily be needed, however. The audio signal can be introduced through a coupling capacitor of about 1000 μF. Because it will necessarily be an electrolytic type, take care to select one with low leakage. For optimum modulation percentage, some modulation will generally have to be applied to the RF driver stage, as well as to the RF power amplifier.

A more elegant AM-modulation scheme uses the pulse-width modulation circuitry of a switching-type power supply. The simplified circuit is shown in Fig. 1-30. The audio signal is superimposed on the error voltage so that the output of the supply varies at the audio rate. This technique is very efficient; unlike the previously described 3-terminal regulator used as a modulator, the power switch of the switching supply does not act as a power-wasting rheostat. Rather, it is either off or on. Both conductive states are ideally represented by zero-power dissipation.

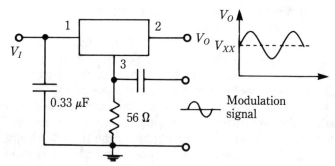

1-29 Use of a 3-terminal voltage regulator to produce AM modulation. In this interesting application, the "error signal" is the impressed audio modulation.
Lambda Semiconductors

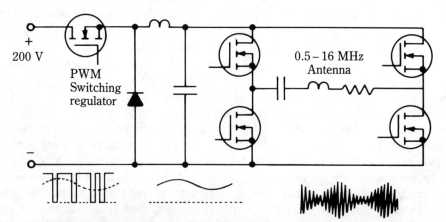

1-30 AM-modulation scheme that uses circuitry from switching-type regulator. The modulating signal is introduced at the error amplifier (not shown).
International Rectifier Corp.

A problem on the horizon

Probably, the majority of commercial power supplies have been designed to provide 5-V high-current outputs and 12- or 15-V outputs with more moderate current ratings. This availability stems from the nature of computer logic and the supportive analog electronics. Of course, a myriad of other power-supply applications are available with a wide range of voltage and current ratings for the intended loads. Both linear and switchmode supplies have been widely used. The switchmode types have gained popularity because of their inherently high operating efficiency, their compactness, and their lower manufacturing cost—especially with the higher power ratings. At the same time, linear design has prevailed where the emphasis has been on low-noise operation and very tight regulation. At power levels that exceed 20 or 30 W, heat removal can easily become a serious problem with linear regulation, unless some form of preregulator is used.

A new turn in the evolution of computer technology has evoked some new thoughts. In the continual search for devices and techniques to enable computers to be built with more memory, higher speed, and greater logic density, a new problem has been presented to power-supply technology. This problem is because theory and practice have converged on the desirability of lower-voltage computer logic, despite the success of the long enduring 5-V systems. It is quite clear that a dc operating level of 3.3 V would lead to improved speed, higher packaging density, and a relaxed thermal situation. Compatibility between existing 5-V designs and the proposed 3.3-V systems has been extensively investigated, but serious limitations exist beyond serving as a temporary expedient.

If the power-supply designer was faced with the problem of revamping say, 20-V circuitry to a new load requirement of 10 or 15 V, little effort would be expended to accommodate the change. However, going from 5 to 3.3 V is a horse of another color—once below 5-V operation, it rapidly gets into the high-dissipation area of solid-state devices. Thus, a 1-V drop in a high-current rectifying diode is taken in stride for a 12-V supply; the drop can be lived with, although undesirable, in 5-V supplies; however, it is generally unacceptable in a 3.3-V supply. To add to our concerns, computer designers have even suggested the prospect of 2-V logic. Nor can comfort be derived from the probability that regulation of sub-5-V dc sources will have to be tighter than on most 5-V systems. Some possible power-supply arrangements for the anticipated voltage requirements are shown in Fig. 1-31.

If the history of electronics technology serves as a reliable guide, it is not too far-fetched to expect some breakthrough or some presently unthinkable innovation to be on hand when the need for efficient low-voltage regulated supplies becomes strong. In the meantime, some notable developments are useful for such power supplies:

- Bipolar power transistors with inordinately low collector saturation voltage (primarily pnp types).
- Power MOSFETs with much lower drain resistance (R_D) than were available a few years ago.

- Schottky rectifying diodes with continuing improvements in such operating parameters as voltage rating, leakage current, and allowable operating temperature. A nice feature of these devices is that like types can often be directly paralleled.
- IC linear regulators capable of operating with very low voltage drop across the series-pass element.
- IC switch-mode regulators capable of efficiently converting 1 or 2 V to 3.3 V.
- IC switch-mode regulators optimally designed to convert 5 V to 3.3 V (3.3- to 5-V conversion has also received attention).
- A reconsideration of germanium devices appears merited.
- The IGBT features the high-impedance input of the power MOSFET, but the low-output voltage drop of the power bipolar transistor.
- The synchronous-rectifier circuit uses active power devices to develop voltage drops that are comparable to, or even less than, those that are readily obtained from Schottky diodes.

Whatever trend eventually evolves, it is not clear whether linear or switch-mode circuitry will dominate the power-supply scene. Indeed, some combination of the two might win out. For example, a switchmode preregulator could maintain

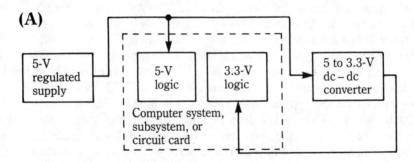

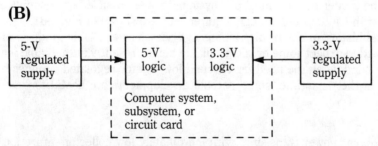

1-31   Possible power-supply schemes for computer systems in transition.
(A) Preponderance of 5-V logic circuitry.
(B) 5-V and 3.3-V logic circuitry that is about evenly divided.
(C) Preponderance of 3.3-V logic circuitry.

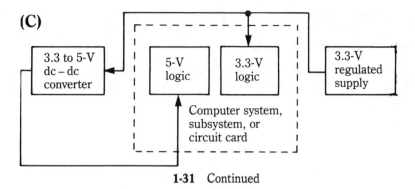

1-31 Continued

its efficiency in the face of varying dc input, and a linear postregulator could then operate with minimal voltage drop in its series-pass element. It is conceivable that the efficiency of the combined regulators might profitably exceed that of either regulator type alone—especially under worst line and load conditions.

Overall view of regulated supply features

A number of interesting applications of regulated power supplies have already been presented. Another viewpoint of the versatility of regulated supplies can be had by considering salient aspects of their behavior; in this way, you can readily utilize imagination and creativity to devise new and unusual applications.

- The family of regulated power supplies known as *dc-to-dc converters* behave as dc transformers and allow wide manipulation of dc voltage and current. Thus, you have the same application flexibility as when working directly with ac systems.
- Regulated power supplies can be used to reverse dc polarity of a dc source.
- Remote loads can have their voltage and current digitally programmed from regulated power supplies.
- Regulated power supplies can allow strapless or switchless accommodation of a wide range of ac line voltages.
- Regulated power supplies can possess sufficient "hold-up" capability to maintain load voltage and current regulation during momentary interruptions of the ac input voltage (approximately 30 milliseconds for 60-Hz power lines).
- Regulation circuitry can be modified to restore the unity power factor, which is no better than 0.6 in typical power supplies. This technique also greatly reduces harmonic energy that would otherwise be injected into the ac power line as a source of EMI.
- Regulated power supplies can be combined in parallel and in series formats to accommodate various load demands.

- The basic regulation principles can be invoked for the benefit of ac, as well as dc, loads. Thus, even regulated ac systems that include line-voltage conditioning and frequency changing are feasible.

Considering the applications and the attributes of regulated power supplies cited in this chapter, the relevant answer to "why use regulated power supplies" is "why not?"

2

Static characteristics of regulated power supplies

NOW, THE BASIC PERFORMANCE PARAMETERS THAT HELP DEFINE THE operation of regulated power supplies must be considered. Seemingly, this should be no Herculean task. The evaluation, specification, adaptation, and even design of regulated power supplies should be accomplished by following guidelines that have been standardized through extensive usage. After all, you might reason, it is only required that voltage and current (in special cases, power) must be delivered to a load that is maintained at near-constant value, despite changes in line voltage, load demand, and environmental conditions. Surely, by resorting to a mutually understandable language, maker and user can quickly agree on performance claimed for and demanded from regulated power supplies. Already, the reader must suspect that, like the iceberg, the superficial aspect of this matter is, indeed, the smaller part.

A problem in semantics

As can be attested by anyone with experience in the ordering, installation, or design of regulated power supplies, something other than common semantics often prevails. Likewise, the cause of communications and understanding has not been greatly improved by unique power-supply terminologies. The user often feels that power-supply specifications are hatched in the maker's sales department, rather than in the engineering department. Conversely, the maker has little trouble in citing case histories to show that many users apparently do not know what kind of power-supply performance they really need. Such an atmosphere of charges and counter-charges leads to the conclusion that neither maker nor user habitually spells out the true circumstances that involve the operation of regulated power supplies.

A few examples of the tyranny of words

Before dealing directly with the subject matter of this chapter, your insight will be sharpened when you can recognize the nature of the problems that can arise by using appropriate technical terms that define power-supply performance. The following examples are relevant:

- The maker might specify a power-supply current rating with the condition that the case of the supply must be limited to a certain temperature rise. The rating appears heaven-sent to a user. Following purchase, the user traumatically discovers that the mass and volume of the heatsinks that are required to prevent out-of-bounds temperature rise far exceed the packaging limitations of his system. Because this "small matter" never entered into prepurchasing discussions, it is not easy to support any legal claim by the user against the maker. Moreover, the ethics of the maker merit defense on the ground that other users, those willing to provide the necessary thermal hardware, find the maker's rating accurate, even conservative!

- The user invests heavily in a voltage-regulated power supply with very tight regulation specifications. In actual use, the effective regulation of the supply is considerably degraded by the resistance of the connecting leads between the supply and its load. Should the user have had the presence of mind to specify that long leads would be used? Or was it the maker's responsibility to remind the user that the specified regulation could be maintained only if short, heavy-gauge leads were used?

- A certain power supply is claimed to produce no more than a stipulated rms ripple voltage. The user matches this claim with his requirements and delights in what appears to be a comfortable margin of safety. Subsequently, he is distraught to find that high-amplitude spikes upset the operation of his system. On second thought, he realizes that the power-supply specifications are probably valid after all; it just happens that the rms value of very narrow, although intense, pulses can be a very modest value! Who is deceptive and who is negligent? I can only say that both parties would have benefited by discussing the actual conditions that are associated with specifications and requirements.

The purpose of this discussion has been to establish the general situation that is usually involved in specifying or in meeting the specifications of regulated power supplies. The situation humorously depicted in Fig. 2-1 is one in which parameters must be pinpointed, and all environmental conditions must be mutually discussed by the maker and the user. In the end, it still remains highly profitable to be aware of inescapable trade-offs between often incompatible performance demands. With such matters in mind, proceed with our investigation of the *static characteristics* of regulated power supplies.

A few examples of the tyranny of words   49

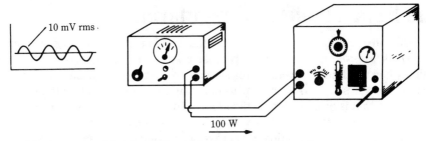

The time-pressured designer needs a regulated power supply to supply 100 watts accompanied by maximum output ripple of 10 millivolts.

(A) User's requirements in a power supply.

Zilch power supply

Extraordinary specifications:

    MTBF..............50 years ±10 years or fuse blowout, whichever comes first.
    Maximum altitude.....200 000 ft., except during peak of solar cycle.
    **Ripple..............5 mV maximum (rms, under worst conditions).
    **Power capability.......120 watts for case temperature of 25 °C.

An eager-to-comply factory rep. directs designer's attention to latest Zilch product. Mutuality is quickly achieved and no further questions or statements originate from either party.

(B) Hypothetical specifications from manufacturer.

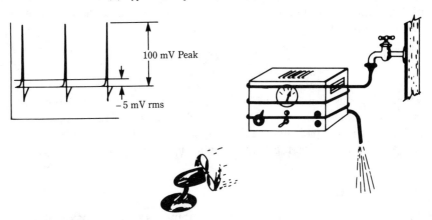

Although neither maker nor user have misrepresented the requirements or specifications (?), the above situation results in something less than cordial relationship.

(C) Realistic picture from both points of view.

2-1   Typical mishap in evaluating a power supply.

The basic concept of regulation

The variation of output voltage with respect to load in the voltage-regulated supply is known as *load-voltage regulation*. It represents the first characteristic that comes to mind when we think of a voltage-regulated power supply. The load-voltage regulation of a power supply is its ability to maintain nearly constant voltage across a variable load that is directly connected to its output terminals. Under the condition that ambient temperature and line voltage must remain constant during measurement, the load-voltage regulation is expressed as a percentage and it is defined as:

$$\text{Load-voltage regulation} = \frac{\text{zero-load voltage} - \text{full-load voltage}}{\text{zero-load voltage}} \times 100$$

Lower load-voltage regulation values indicate better stabilization.

For example, suppose that a voltage-regulated power supply is rated for a full-load current of 1.0 A. With no load, its output voltage is 33.0 V. With the 1.0-A load connected to its output terminals, the voltage drops to 32.9 V. A variable autotransformer is inserted in the line and an ac line voltmeter is observed for constancy during these two measurements. The whole measurement procedure requires no more than one or two minutes once the apparatus is hooked up. Therefore, it is logical that no change in ambient temperature has occurred during such evaluation (for extremely close regulation, and where great refinement in measurement is required, this approach to the temperature situation can be overly simplified. This is covered in more detail under *temperature effects*). The load-voltage regulation of the supply under consideration is:

$$\frac{33.0 - 32.9}{33.0} \times 100 = 0.3\%$$

More specifically, this regulation is -0.3%; the minus sign signifies a decrease of output terminal voltage as a consequence of loading. In actual practice, power-supply specifications rarely indicate whether regulation is positive or negative. Generally, it is negative for voltage-regulated supplies (see Fig. 2-2).

Beware—a trap

However, another use of plus and minus signs that merits special consideration is the *plus or minus designation* ($\pm$). For example, one manufacturer's specification claims 0.1% load-voltage regulation for their voltage-regulated power supply. When so cited, it is important to ask, "does this imply $\pm 0.05\%$, which corresponds to a total of 0.1%, or is the total regulation actually 0.2% as the consequence of $\pm 0.1\%$ regulation?" The reason for such inquiry stems from the commonplace omission of the "plus or minus" sign. Pinpoint this information prior to purchase. Otherwise, you either obtain inadequate regulation or you pay for more performance than is needed. The maker's defense will generally be that their specification method is the one recognized throughout the industry. Nor is such an assertion readily disproved. If the

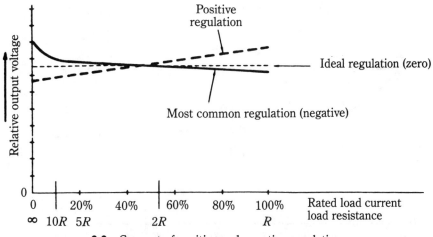

2-2 Concept of positive and negative regulation.

regulation specification was derived from an advertisement in a journal, the omission of the "plus or minus" sign might be declared a typographical error—hardly more likely to lead to compensation or satisfaction.

If misunderstanding is to be avoided, it is mandatory that a claimed or required regulation-percentage must be qualified with the word "total" or with the designation ±. Indeed, it is a good idea to use both of these qualifiers!

After the trap, a pitfall

It is not uncommon for a voltage-regulated power supply to have its load-voltage regulation specified from the slightly modified equation:

$$\text{Load-voltage regulation} = \frac{10\% \text{ load voltage} - \text{full-load voltage}}{10\% \text{ load voltage}}$$

Line voltage and ambient temperature are still assumed to be constant and this altered definition appears as a benign variation of a basic theme. You might be tempted to quickly conclude that a little mental arithmetic can closely extrapolate regulation so defined to agree with regulation defined in terms of zero and full-load voltage. Not necessarily so!

The reason for specifying regulation in terms of something other than zero load might not be incidental or innocent. Usually output voltage does not drop linearly from zero load to full load. Rather, the drop-off is between zero and, say, 7 to 15 percent or so of full load (see Fig. 2-3) is disproportionate. Therefore, the maker can present a better image of his power supply by citing its load-voltage regulation in terms of something less than its complete load range. Whether the user inadvertently acquires a poorer-than-anticipated regulation depends upon how the specification relates to his particular application. The maker might feel justified in specifying

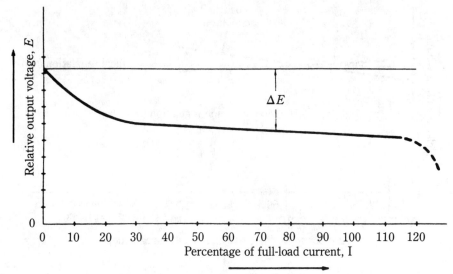

2-3 Typical load-voltage regulation curve.

load-voltage regulation in this way because zero and small-percentage load voltages are rarely important in actual applications.

An ambiguity—but whose?

When the output voltage of a voltage-regulated power supply is adjustable, we have fertile ground for naive assumption by the user and oversight by the maker. More often than not, the naivete of the former prevails with, or without, the latter's attempt to "tell it as it is." The regulation of such a supply generally varies considerably over the adjustable range of output voltage. Suppose, for example, that the output voltage of a voltage-regulated supply is controllable (from 25 to 5 V) by a front-panel knob. The load-voltage regulation is specified by the maker to be 0.1% (total) when the supply is adjusted for its maximum output voltage. So, the output voltage will be 25.0 V with no load and 24.975 V with full load. The voltage drop-off here is 0.1% × 25, or 25 mV. These figures would be cut and dried if such a specification told the whole story, but it does not!

To simplify matters, assume the unlikely coincidence that the voltage drop-off is the same for the 5-V output setting. That is, the output voltage would drop to 4.975 V when the full load is connected to the supply. Now, has load-voltage regulation remained the same? The quick answer is that it has, and this is the booby trap. A little reflection shows that the load-voltage regulation for the 5-V setting is much worse, as given by:

$$\frac{5.000 - 4.975}{5.000} = 0.5\%$$

Usually, the voltage drop-off at the lower voltage adjustments will differ from that frequency for the maximum output setting. In any event, if the maker quotes only the best regulation attainable, it is only natural for the user to infer that this holds over the whole adjustment range. The proper way for the maker to specify this sort of power supply is to state that the supply has a load-voltage regulation of 0.1% (total) or 25 mV WIG "whichever is greater". Notice that these figures are not redundant; 0.1% regulation and 25 mV are not necessarily the same thing (the essence of this discussion). Twenty-five millivolts can correspond to 0.1% regulation at best, or to 0.5% regulation at worst. The WIG notation conveys the vital information that, for any selected output-voltage, the regulation is the highest percentage from the comparison of 0.1% with the percentage regulation that is represented by 25 mV.

In the real-life situation, the regulation drop-off at the 5-V setting (or some cited intermediate setting) would probably differ from the 25 mV that correspond to the 25-V setting. The same reasoning applies; to wit, convert the voltage drop-off (resulting from application of full load) to percentage regulation and compare it to the "best-case" percentage regulation claimed by the maker (in this case, 0.1%). Then, be aware that the higher number defines the regulation for that output voltage setting.

Before leaving the subject of load-voltage regulation, consider the quantity that is often loosely referred to as *load voltage, output voltage,* or *terminal voltage.* Because of the connecting-lead resistance, it can make a considerable difference whether the dc voltage that is delivered by the voltage-regulated supply is measured at the output terminals of the supply, or at the load itself. The best way to avoid misunderstanding is to deal with the voltages that appear directly at the output terminals of the supply. The true load then consists of the actual load in series with the resistance of the connecting leads. Because the lead resistance degrades the regulation that appears at the actual load, the maker cannot be expected to specify the supply for unknown connection methods that are peculiar to the user's system (unless the special technique of remote-sensing, an extra-cost feature, is incorporated in the design).

By referring the regulation specification to the output terminals of the supply, the maker can quote the most favorable value attainable. From the user's point of view, such a regulation claim is often actually realized or, in any event, is closely approached because the actual load can be connected directly to the supply terminals, or through short leads of negligible resistance. Also, specification referred to the output terminals enables the user to readily calculate the true conditions of his actual installation. This often saves appreciable expense because nothing can be gained by ordering a closely regulated power supply, which then must be limited to much poorer regulation at the load itself.

Load-current regulation of a current-regulated power supply

The previous sections dealt with some of the important aspects of evaluating load-voltage regulation in a voltage-regulated power supply. The analogous quantity in a

54 *Static characteristics of regulated power supplies*

current-regulated power supply is *load-current regulation*. Aside from the analogy involved, switch your thoughts to the current-regulated supply. Many voltage-regulated supplies incorporate some form of automatic current control to protect both the supply and load from damage as a result of short circuits or overloads. The control manifests itself only when the load attempts to exceed a certain current. That current can be fixed at some value slightly greater than rated full-load current, or it can be controllable from the panel. The prevention of damaging current can be accomplished by limiting the current to a much lower value than that demanded by a short circuit, by causing decreasing current with increasing load demand (current foldback), or by abruptly causing the supply to transfer to a true current-regulating mode of operation. The last method is nearly universal in the better laboratory power supplies, and they are useful for certain systems applications. In the relationship:

$$\text{Load-current regulation} = \frac{\text{short-circuit current} - \text{minimum rated current}}{\text{short-circuit current}} \times 100$$

again, it is assumed that measurements are made with constant line voltage and constant ambient temperature. The line voltage is the midvalue of the specified regulating range (see Fig. 2-4).

Now, load current, rather than load voltage, is stabilized (as was the case in the voltage-regulated supply). Although this should be a very simple analogy, the equation for current regulation involves one factor that might appear strange at first inspection. The factor "minimum rated current," rather than zero current, arises

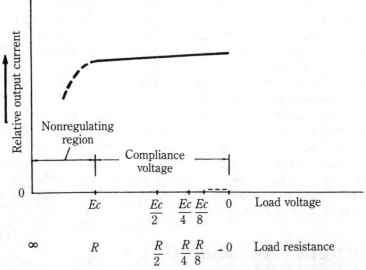

R = Highest load resistance usable for specified regulation.
Ec = Load voltage when load resistance is R (compliance voltage)

2-4 Typical regulation curve of current-regulated supply (positive regulation).

from the necessarily practical nature of the supply. The supply cannot maintain constant load current as the load resistance is increased indefinitely. To do so would, by Ohm's law, require *infinite voltage* to be developed across the output terminals. A corollary of this statement is that the current-regulated supply maintains constancy of load current despite load changes, by varying the load voltage. Obviously then, the current-regulated supply has, by its nature, extremely poor voltage regulation. *Compliance voltage* defines the load-voltage range over which load current is held constant within specification. Compliance voltage determines the smallest load (highest resistance) that can be used for a given value of constant current. Unfortunately, the user sometimes overlooks the importance of compliance voltage (it is unnatural to postulate a minimum load limitation as readily as a maximum load limitation, as in the case of the voltage-regulated supply). The failure to obtain a compliance voltage that is relevant to the load to be powered can render the current-regulated supply useless, despite its current rating!

It should not be construed that a current-regulated power supply cannot be programmed to regulate down to zero current. This can be accomplished by appropriately varying the internal reference voltage or the feedback network that is associated with the operational amplifier within the supply. The fact remains, however, that for a given constant-current adjustment, current regulation extends only to an upper limit of load resistance; this limit is accompanied by the compliance voltage.

The ideal current-regulated supply appears as an infinite resistance in series with the actual load, and it manifests this characteristic for dc and for all frequencies. The practical supply develops a high impedance over a frequency range that exceeds the needs of the powered circuitry.

Line-voltage regulation for the voltage-regulated power supply

Line-voltage regulation for the voltage-regulated power supply is the percentage change in output voltage that occurs from a specified range of line-voltage variation, provided that the measurement condition of constant output current and constant ambient temperature prevails. Thus:

$$\text{Line-voltage regulation:} \quad \frac{\text{highest output voltage} - \text{lowest output voltage}}{\text{output voltage at nominal line voltage}} \times 100$$

Nominal line voltage is generally considered to be the midpoint value between the two extremes of the specified regulatory range.

If not otherwise stated, the load current should be 50 percent of the specified full-load value. However, it is often desirable to relate the claimed line-voltage regulation to both zero and full-load condition, as well as (or instead of) the half-load condition. This relation is useful because of the likelihood that one load condition will yield an appreciably poorer line-voltage regulation than the other. Usually, the worst situation corresponds to full-load. Thus, you can be misled by assuming that line-voltage regulation at one load condition will be obtained for another. For example, the nominal ac line voltage of a voltage-regulated power supply is given as 115 V. The maker's specifications provide operation over a line-voltage range of 130 to 100 V, this being accompanied by a total output voltage change of 0.5 V, with respect to 50.0 V output with a 115-V line. The maker states that this situation is true for 50 percent of the full-rated output current:

$$\text{Line-voltage regulation} = \frac{0.5}{50.0} \times 100 = 1\% \text{ total, or } \pm 1/2\%$$

A little contemplation will show that the measurement technique must utilize an adjustable load so that the output current of the supply can be maintained at constant value as ac line voltage is varied. In this example, the line-voltage regulation would be considered positive if line and output voltage changed in the same direction. If line and output voltage changed in opposite directions, the line-voltage regulation would be considered negative. For the sake of "specmanship," a maker will sometimes indicate line-voltage regulation as so-many percent per-volt (%/V).

It is often important to know what actually happens for line voltages above and below the specified regulatory range. Several commonly encountered behavior patterns are shown in Fig. 2-5 for a voltage-regulated supply. Powered equipment can be damaged by either oscillations or by the depicted output-voltage rise (various aberrations can also occur in current-regulated supplies at low or high line voltages).

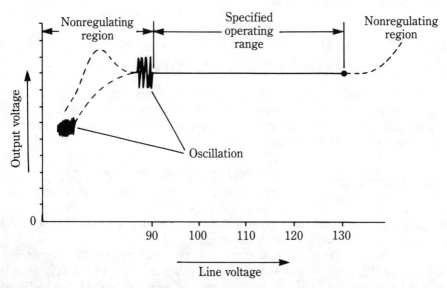

2-5 Possible behavior patterns in nonregulating regions of a voltage-regulated supply.

Line-voltage regulation for the current-regulated power supply

A similar definition of line-voltage regulation exists for the current-regulated supply:

$$\text{Line-voltage regulation} = \frac{\text{highest output current} - \text{lowest output current}}{\text{output current at nominal line voltage}} \times 100$$

This equation is valid for the measurement conditions of constant output voltage and constant ambient temperature. Again, the so-called nominal line voltage is generally considered to be midway between the highest and the lowest values of the specified regulatory range. Unless otherwise stated, the voltage across the output terminals is assumed to be constant at one-half the value of the specified compliance voltage. This statement implies that an adjustable load is used during the measurement process. As the line voltage is varied, the load is adjusted to maintain the constant output voltage.

As with the voltage-regulated power supply, the line-voltage regulation for other-than-specified loading conditions could be more relevant to the requirements of the particular application. Rather than extrapolate from one condition to another, it is much safer to obtain or measure the performance under the relevant conditions. Extrapolation assumes a linear relationship of the regulation characteristic; this does not always exist. Moreover, for both voltage- and current-regulated supplies, cessation of tight regulation occurs quite abruptly at the low end of the line-voltage range (this also occurs frequently at the high end as well).

Suppose that a current-regulated supply is specified to operate over the line-voltage range of ±20 V from the nominal 115 V. It is stated that the compliance voltage is 100 V. From this, you can deduce that the nominal load is a resistance that causes 50 V to be developed across the output terminals of the supply. Under the condition of nominal line voltage and nominal load, the output current is 1.0 A. The line voltage is varied while the load resistance is adjusted to maintain the 50 V across the output terminals. When this is done, the output current is 0.96 A for the low line voltage (95 V). Similarly, an output current of 1.04 A corresponds to the high line condition (135 V). The line voltage regulation of this current-regulated supply is:

$$\frac{1.04 - 0.96}{1.0} = 8\% \text{ total}$$

When ordering such a supply, all conditions that pertain to a performance parameter must be mentioned and mutually understood between relevant parties, such as between the systems engineer and the purchasing agent. Thus, the supply of this example would be said to have a line-voltage regulation of 8% total or ±4%

over the line voltage of 95 to 135 V when loaded to one-half specified compliance voltage. This philosophy should be followed for all regulated power supplies and for all of their performance parameters. Otherwise, confusion and disagreement are sure to arise from the different concepts of standards and references.

Temperature coefficient

The temperature coefficient of a regulated power supply is, in essence, its *thermal regulation*. The percentage change in output voltage of a voltage-regulated supply (or in the output current of a current-regulated supply) are when all parameters other than ambient temperature are maintained constant. The temperature coefficient is expressed as the percentage change in output voltage or output-current per-degree-centigrade change in ambient temperature. A casual inspection of the temperature coefficient on the specification sheet of a power supply often leads to a benign evaluation of its practical influence. However, you should appreciate the potentially overwhelming importance of the temperature coefficient.

Suppose that fairly close regulation is required and you stumble upon the specifications of a voltage-regulated supply that claim 0.005% total line-voltage regulation and 0.005% total load-voltage regulation. Combining both of these regulations, the worst-case regulation is 0.01% total. It is decided that 0.01% total combined regulation is somewhat better than the minimum needed; apparently, the power supply meets the bill and provides a reasonable safety margin as well.

But all is not well; a deeper study of the specifications turns up a temperature coefficient of 0.01% per degree centigrade. A 20-degree excursion in ambient temperature is, we learn, representative of the thermal environment where the supply must be installed. Therefore, the anticipated change in output voltage on the basis of temperature change alone is $20 \times 0.01\%$ or 0.2%. However, 0.2% completely overwhelms the 0.1% that was too quickly assumed to be the worst-case regulation. In practice, situations such as this one require a sophisticated balance between design for low temperature coefficient and stabilization of the thermal environment. Undue emphasis on either factor rapidly leads to increased expenditure.

Temperature rise and heat removal in a working environment

The temperature coefficient concept is valid for when a long period of time is allotted to thermally stabilize the supply following a change in ambient temperature. Until sufficient time passes, the temperature of zener diodes, active elements of differential amplifiers, the output sampling network, and, indeed, of all components (including the wiring itself), will be in a state of change. This results in short-time temperature coefficients that deviate considerably from the thermally stabilized value. Unfortunately, most loads have their power supplies on a continual basis during operation. Thus, you must proceed cautiously—even after applying the manufacturer's specified temperature coefficient to the overall situation.

Often, power supplies are installed so that the application and removal of heat to different sections of the supply cause relatively large temperature differences in the various sections, and slow down the thermal stabilization process following ambient-temperature changes. Thus, if you were willing and able to wait an hour, the manufacturer's claim would be vindicated; in the meantime, however, the output level of the supply could wander considerably beyond allowable limits (Fig. 2-6). Actually, a short-time temperature coefficient can exist. This coefficient can degrade regulation more than would be accounted for from the temperature coefficient claimed in the specifications.

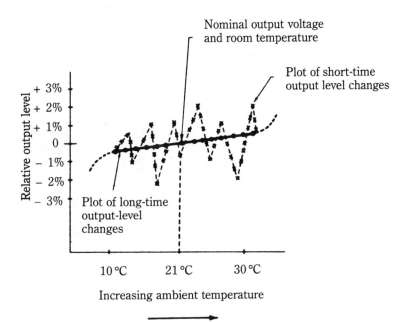

2-6 Example of temperature-effect on the output of a regulated supply.

Line and load regulation combined

For many practical purposes, especially those applications in which the ambient temperature is limited to a relatively narrow range, it is more meaningful to deal with the combined effects of line-voltage regulation and load regulation. Accordingly, the combined line and load regulation specifies the change in dc output as a result of the simultaneous effects of both line-voltage and load changes under worst conditions. *Worst conditions* are a situation where the changes produced in the dc output level are additive. The *output level* is the output-terminal voltage for the voltage-regulated power supply, and output current for the current-regulated supply. The measurement technique involves varying line voltage and load resistance to produce the maximum deviation on both sides of the nominal output level. The nominal output level is generally that which corresponds to half of the full-rated current in a voltage-regulated supply. Figure 2-7 illustrates the concept of line and load regulation combined.

60 Static characteristics of regulated power supplies

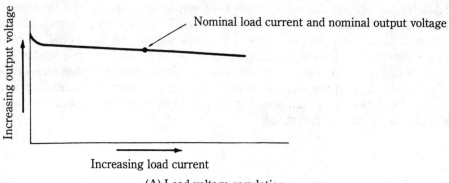

(A) Load-voltage regulation.

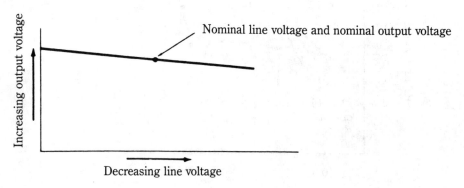

(B) Line-voltage regulation.

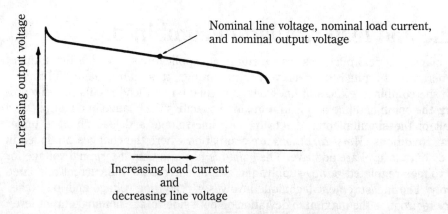

(C) Load and line regulation combined.

2-7 Load and line regulation combined.

Thus, the relationship that applies to both types of regulated power supplies is:

Line and load regulation combined =
$$\frac{\text{maximum obtainable deviation in dc output level}}{\text{nominal dc output level}} \times 100$$

As with other definitions of regulation, this relationship is valid as long as all other parameters remain constant during the measurements (ambient temperature, operating temperature, humidity, line frequency, line waveform, etc.). For most practical purposes, it is sufficient to be reasonably sure that the operating temperature of the supply remains constant while the line and load conditions are varied to produce the worst regulation. The operating temperature of the supplies can be assumed constant in most instances by allotting as short a time as possible to the measurement procedure.

Total combined regulation

We have seen that the concept of line and load regulation combined is useful in many practical installations. However, the effects of temperature and other factors that are capable of influencing regulation are ignored. As was pointed out, this is often permissible for power supplies that have moderate regulating ability and that are situated in benign environments. When very close regulation is required and a more severe environment is involved, the regulation must be defined more rigorously. It is then necessary to take into account temperature (as indicated by the temperature coefficient) and perhaps other factors, such as line frequency, altitude, mechanical vibration, humidity, etc. The total combined regulation specifies the worst regulation that can occur from the overall contribution of all factors except one: *drift*.

Drift is covered in ensuing paragraphs. The total combined regulation does provide a useful specification of *relative regulation*. Thus, if a voltage-regulated power supply with a total combined regulation of ±0.01%, with respect to a 100-V nominal output level should for some reason (say aging of the internal voltage reference) decline to an output level of only 99 V, it would then continue to have a total combined regulation of very nearly ±0.01%, relative to the new output level.

Stability

In the example just cited, a power supply was postulated to have changed its absolute output level, but to have retained its relative regulation. The change in output level over a long period of time defines the stability of a power supply. Such changes are random and indeterminate drifts. Tight regulation per se is readily and inexpensively obtainable through the use of high amplification in the feedback loop. Tight regulation, together with compatible long-time stability, or low drift, is accomplished with much more difficulty. In most cases, the inclusion of drift as a factor in total combined regulation would render the maker's regulation specifications meaningless.

Drift is so erratic and unpredictable that the manufacturer actually would not know how to do his product justice by considering drift as another factor in regulation. Fortunately, many applications require tight regulation on a relative basis, and the simultaneous constancy of the absolute output level over long periods of time is not of primary importance. It has already been pointed out in the introductory chapter, and it will be further emphasized in the succeeding chapter, that the *dynamic characteristics* of the regulated supply are generally the important operational parameters. As a matter of fact, although total combined regulation defines the static behavior of the supply, the effect of tight regulation on the output impedance of the supply (a dynamic characteristic) is usually more important than constant output level.

A treatise could be written on the causes of drift. Voltage regulator tubes can change their ionization characteristics from the effects of RF fields, light, or gradual change in gas pressure and composition. Some components might exhibit thermal hysteresis so that they do not return to original values after temperature cycling. Both tube and solid-state supplies require a warm-up period after they are first turned on. This warm-up is not always practical, and would further compound the difficulty of including drift in the regulation specification.

Although drift is not included in regulation specifications, it should not be ignored. It still remains desirable to maintain absolute output level reasonably constant. Indeed, drift is often the missing link when prices are compared. An inexpensive supply will most often be more erratic and more violent in its drift behavior than a more costly one with otherwise similar regulation specifications. The maker of quality power supplies generally provides useful information about drift, even though this factor is difficult to precisely mark.

Additional regulation criteria

Stabilization factor is a figure of merit that is derived from an early era in the evolution of voltage-regulated power supplies. It is a sort of hybrid specifying-method that incorporates elements of both line and load regulation. Perhaps this explains its worth, as judged by those who claim that it provides a quick insight into the static behavior of a regulator. Stabilization factor, S, is defined as:

$$S = \frac{E_o}{E_L} \left(\frac{\Delta E_L}{\Delta E_o} \right)$$

where:

E_O is the dc output of the regulator at half-rated load current,
E_L is the rms value of the nominally specified ac line voltage,
ΔE_L represents a small change in the line voltage,
ΔE_O is the resultant change in the dc output voltage.

The higher the value of S, the better. Later, a more rigorous figure of merit was used. This, the stability factor, S_V, is equated as:

$$S_V = \frac{\Delta V_o}{\Delta V_i}$$

where:

ΔV_i represents a small change in dc impressed upon the regulator,
ΔV_o is the resultant change in dc produced across the load.

This relationship is conditional upon load current, I_O, and the operating temperature, T, of the regulator being maintained constant during evaluation of S_V. Notice that, unlike line regulation, the stability factor does not include the shortcomings of the unregulated source of dc power.

Often, S_V will correspond to I_o at a half-rated value. However, S_V is meaningful only if the value of I_o and the prevailing thermal conditions are known. These matters should be clarified by the user in case they are not clear in the manufacturer's specifications.

In this case, the lower the value of S_V, the better the voltage regulation. The stability factor concept is most useful when applied to IC voltage regulators (IC makers now tend to designate this relationship as *line regulation*, rather than the older term, *stability factor*).

Protection techniques

Protection should not be overlooked; an otherwise satisfactory power supply can easily be the agent of its own destruction as well as that of the attached load. This can readily occur from such abnormal operating conditions as high line voltage, excessive load demand, line- or load-generated transients, inordinate temperature rise, and a number of other causes. In most applications, the so-called "abnormal" conditions can occur often enough to destroy any economy brought about by dispensing with protection techniques. Probably the component that most needs protection is the series-pass transistor in voltage-regulated supplies.

Often, even a momentary short of the output terminals of such a supply will destroy the series-pass transistor. This destruction is also readily brought about by lighter current overloads, of perhaps twice the rated output current of the supply—particularly in supplies that are rated at several hundred milliamps output or less. Such a catastrophe is generally accompanied by a chain reaction that could then destroy every solid-state device in the regulator, including the rectifiers. Moreover, this might apply either a permanent or a transient overvoltage to the load, which could destroy it. Ordinary fusing can provide only a small measure of protection because of the relatively long thermal time constant that is involved in the separation of fuse elements. During such a time interval, the destruction of transistors and diodes can easily occur by virtue of the shorter time constants inherent in many of these elements.

A simple technique to protect the series-pass transistor from overcurrent is shown in Fig. 2-8. If sufficient current is consumed from the supply, the potential drop developed across resistance R will break down the zener diode, X1. When

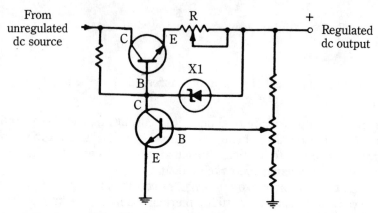

2-8 Simple current-limiting technique.

this happens, the series-pass transistor is deprived of much of its base drive and will, therefore, be nearly nonresponsive to the error signal that otherwise would cause greatly increased load current.

A somewhat more sophisticated application of this protection concept is depicted in Fig. 2-9. Here, the collector load of transistor Q1 is provided by transistor Q4. Transistor Q4 functions as a constant-current source; it, therefore, appears to the collector of Q1 as a very high load impedance. This impedance enables Q1 to develop high voltage gain. This, alone, enhances the operation of the entire regulator circuit. Yet another benefit conferred by current source, Q4, is that much less ripple current is injected into the base of Q2 than by the conventional method (resistance feed from the output of the unregulated supply).

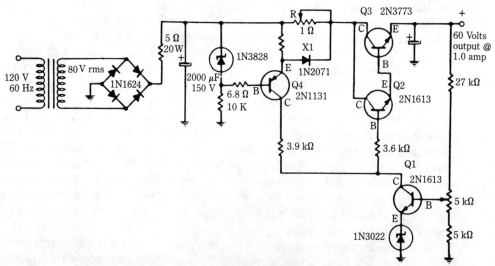

2-9 Current limiting by a constant-current control transistor.

When excessive current demand exists, the voltage drop across resistance R becomes sufficient to produce conduction in diode X1. This conduction biases transistor Q4 for negligible collector current, which thereby deprives the collector of Q1 and the base of Q2 of their nominal operating currents. In turn, the conductive state of series-pass transistor Q3 is relaxed and output current is limited.

Both this circuit and the simpler arrangement of Fig. 2-8 display automatic and instant recovery when the short circuit or overload condition is removed.

The forward-conduction diode, X1, in Fig. 2-9 requires a much smaller potential drop across resistance R than would a zener diode. Therefore, a lower value of R is sufficient and much less power needs to be dissipated in this resistance. However, the conduction transition of a zener diode is more abrupt and, as a result, the current-limiting action of the regulator is better. If a zener diode is used here, its polarity should be opposite that of the diode shown.

An additional group of various protection techniques is depicted in Fig. 2-10. In Fig. 2-10A, the diode connected across the dc output terminals protects the series-pass transistor from being subjected to reverse current. The protective diode is polarized so that it is out of the circuit under normal operating conditions when current flows out of the supply to the load, rather than vice-versa. Although this technique protects against reverse current, it does not always make the series-type regulator suitable for "active" loads. For example, a power supply that supplies negative grid bias to a class-C amplifier with vacuum tube(s) must absorb the continuous flow of rectifier grid current. Here, a shunt-type voltage-regulated supply should be used. The shunt element in this type of regulator remains polarized properly whether the current reaches it from the unregulated dc source or from the external load.

In some instances, the series-type voltage regulator can be used as a grid-bias supply with class-C amplifiers by shunting a resistive load across the output terminals with sufficient current demand to keep current always flowing from the supply. This shunting, however, can complicate operation of the class-C amplifier because additional grid bias is developed when grid current flows through this resistance. The scheme does, however, provide good protection from the back emf of dc motors and from many types of digital and inductive loads.

In Fig. 2-10B, the two clamping diodes protect the input of the differential amplifier (comparator) by limiting its input to about one volt. This protection is desirable because inadvertent load transients can momentarily produce inordinately large error signals. Aside from the tendency to damage the amplifier, such disturbances can latch up the amplifier in one or the other extremes of its conduction state. Integrated circuit amplifiers are particularly vulnerable to such operational paralysis. As a result, the power supply will be switched to either its off condition or its fully on condition, both of which are undesired departures from voltage regulation at a preset dc output level.

Methods for limiting the voltage drop across the series-pass transistors are shown in Figs. 2-10C and 2-10D. In Fig. 2-10C, the drop can be conveniently limited to about 6.2 V or even less. In Fig. 2-10D, because of the ionization potential of a neon lamp, protection cannot be reliably obtained below 65 or 70 V. However,

66 *Static characteristics of regulated power supplies*

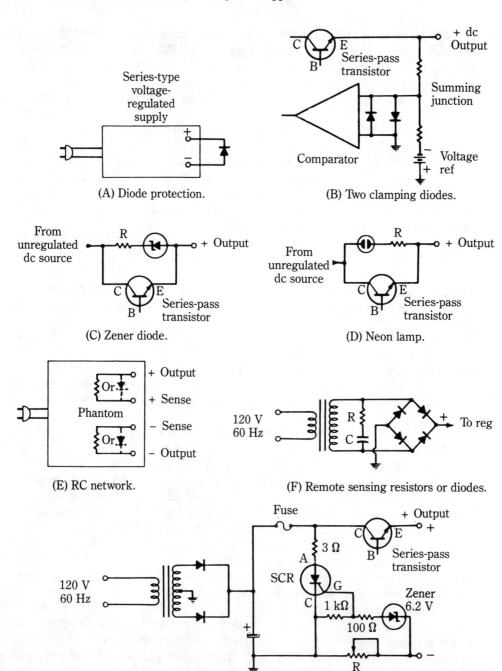

(A) Diode protection.

(B) Two clamping diodes.

(C) Zener diode.

(D) Neon lamp.

(E) RC network.

(F) Remote sensing resistors or diodes.

(G) "Crowbar" action using zener diode and SCR.

2-10 Various power-supply protection techniques.

this voltage is satisfactory for many transistors; the neon-lamp technique is generally found in high-voltage supplies. The resistance, R, which appears in both circuits, can cause degradation of protection if it is too high. However, some amount of resistance in the circuit protects the protection elements themselves if the output voltage is short-circuited.

In Fig. 2-10E, incoming transients from the ac power line are greatly attenuated by the RC network, which is connected across the secondary winding of the power transformer. Such transients can pass through the regulator and damage sensitive loads. Although the regulator filters out such transients, it might not happen if the transient is of too great an amplitude or involves extremely rapid changes.

Figure 2-10F illustrates simple, but important, protection techniques for a voltage-regulated supply, and uses the remote-sensing provision. Either resistances, or appropriately polarized diodes, can be connected to the output and sensing terminals, as shown. When so connected, either of these elements will prevent maximum output voltage (practically the entire unregulated source value) from being applied to the load if one of the sensing leads is accidentally severed. For example, if the positive sensing lead is disconnected, say at the load end, the regulator circuit interprets the situation as zero load voltage and attempts to correct it. Unfortunately, the correction involves reducing the effective resistance of the series-pass transistor or tube, and thereby impresses maximum voltage across the load. However, if either the resistances or the diodes are connected to the terminals of the supply (as shown), the feedback path of the regulator will not be broken—even though the sensing leads are interrupted somewhere between supply and load. During normal operating conditions, the diodes exert no effect. If resistances are used, they can be high enough to have a negligible effect on the regulation characteristics (if the sensing leads have appreciable inductance and the power supply has a high-gain amplifier, the use of such resistances sometimes also lessens oscillation troubles). Values of about 100 times the ohmic resistance of the sensing leads are often satisfactory.

The scheme shown in Fig. 2-10G almost qualifies as one of overprotection; here, a fuse deliberately opens in the event of excessive load current. This drastic action is greatly speeded up by the "crowbar" action of an SCR, which provides a near short circuit at the output of the unregulated dc source. In so doing, the series-pass transistor is spared the high current surge that opens the fuse. This action is caused as a consequence of the passage of load current through monitoring resistance R. When the resulting voltage drop across R becomes sufficient to cause conduction in the zener diode, the gate of the SCR is triggered and the previously described fuse-blowing action ensues.

Although the protection provided by this method can be quite good, an obvious disadvantage is the need to replace the fuse after the overcurrent condition has been remedied. A variation of this approach uses a circuit breaker in place of the fuse.

Special considerations for transient protection

Both discrete semiconductors and ICs are vulnerable to transient damage or malfunctions. Even after ample safety factors have been used in the circuit or system design, freedom from catastrophic destruction cannot be taken for granted. Anyone who has observed the 60-Hz sine wave in typical industrial (or even residential) power lines is only too well aware of the hash and spikes that come along for the ride. Often, short-interval transients of appreciable peak amplitude, because they are not synchronized to the line frequency, are not readily detected with the oscilloscope. The supposition that the energy content of such transients is too low to endanger semiconductors, or that the filter capacitor in the unregulated power supply will "iron them out" is often deceptive. It is much better to suppress these transients before they reach semiconductors. This same reasoning applies equally well to the inevitable transients that are produced when the supply is turned on or off, or when its load is switched.

Traditionally, this problem has been alleviated by connecting some type of energy-absorbing device or network across the power line, or across the secondary of the power transformer. These have included:

- Zener diodes
- Selenium thyrectors
- Silicon carbide (thyrite)
- Neon lamps
- RC networks
- Spark gaps
- Various electronic arrangements, such as "crowbar" circuits

The General Electric MOV metal-oxide varistor is now available which, for general applications, is often preferable to previous transient suppressors. This voltage-dependent resistor has a high degree of nonlinearity and symmetrical characteristics, with respect to polarity. The current-vs-voltage behavior of a typical MOV unit is shown in Fig. 2-11.

The GE MOV varistor is classified as a semiconductor. However, it contains no pn junctions in the ordinary sense. Rather, it consists of bulk polycrystalline zinc oxide sandwiched between metallized terminations. The zinc-oxide crystals are electrically insulated from one another by a thin film of bismuth oxide. At higher voltages, however, this film becomes conductive. Transient suppression occurs via I^2R; heating and the response time is in the submicrosecond range. The action of the metal-oxide varistor is described by the equation:

$$I = KV^a$$

where:

I is the current through the MOV as a result of the impressed voltage,
V is the impressed voltage,
K is a device constant that depends on material, processing, and dimensions,
a is the "figure of merit" for such a nonlinear device.

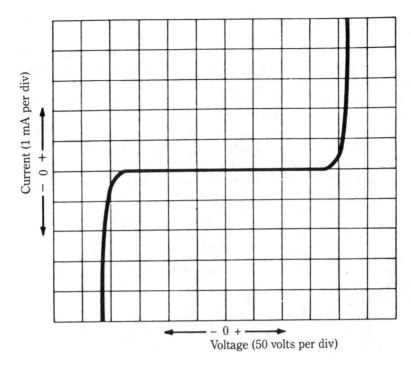

(A) Current-vs-voltage characteristics of a typical metal-oxide varistor.

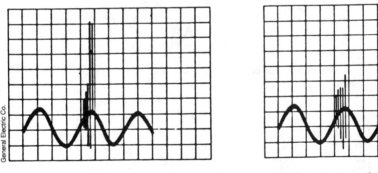

(B) Transients on the ac line without the varistor.

(C) Transients on the ac line with the varistor.

2-11 Characteristics of the General Electric metal-oxide varistor.

An ordinary resistor would be described by this equation with a being unity. On the other hand, a power zener diode could have an a value in the vicinity of 35. The GE MOV varistor generally displays a nonlinearity that corresponds to an a range of 25 to 70. Taking into consideration not only its nonlinear "crowbar" action, but also its energy and current capability, speed, size, and cost, the MOV is often the simplest and most effective means of transient suppression. Of course, a unit must be chosen with voltage and energy-absorption ratings that are appropriate to the particular application.

The situations depicted in Fig. 2-12 explain why otherwise reliable power supplies "mysteriously" fail in the field. Transients from such switching and energy-storage mechanisms can be destructive. This is particularly true of the switch that turns the power supply on and off. The transients generated as the ac line is connected and disconnected from the power transformer vary in energy content and peak amplitude. Worst conditions result from the "right" timing of switch actuation with the "right" residual magnetism in the transformer core. Such simultaneous conditions might not occur during laboratory testing, or even most of the time in the field. However, the statistical probability for such worst conditions should not be overlooked.

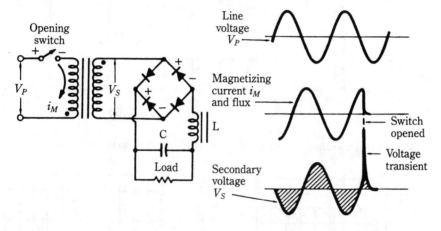

(A) Transient from opening switch in primary circuit of power transformer.

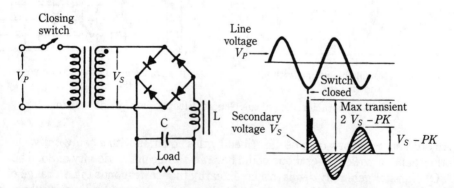

(B) Transient from closing switch in primary circuit of power transformer.

2-12 Commonly encountered transients. General Electric Co.

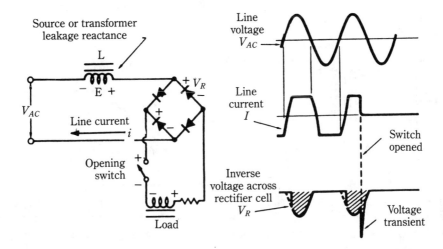

(C) Transient from switching on the dc side of the rectifier bridge.

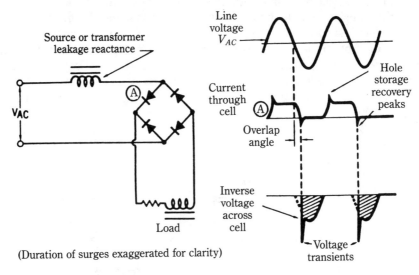

(Duration of surges exaggerated for clarity)

(D) Transient from reverse-recovery behavior of rectifier diodes.

2-12 Continued

A common application of the MOV varistor is shown in Fig. 2-13. Here, it is shunted directly across the secondary winding of the power transformer. Because of the symmetrical "breakdown" characteristics of the MOV, transient suppression can be achieved in other ways in power-supply and regulator circuits. This device can be connected directly across the ac line—indeed off-the-line switching regulators have no input transformer (in such an application, the transients that are generated by the "switcher" are largely absorbed before passing to the power

72 Static characteristics of regulated power supplies

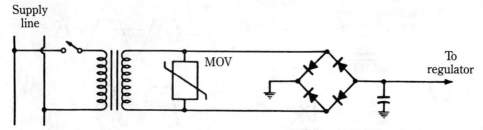

2-13 A typical circuit connection for the General Electric MOV varistor.

line). The MOV can also be connected at dc circuit junctions, as well as across rectifier diodes. Many models of the GE MOV are available. The ratings embrace rms voltages from 12 to 1000 V, energy-absorption capabilities up to 160 J, and peak current pulses from 10 to 2000 A. Although the transient phenomenon is of cyclic or recurrent nature, the average power dissipation must be considered—the available range is from 0.2 to (at least) 15 W.

Figure 2-14 shows the use of an MOV transient suppressor to protect a switching-regulator transistor from a commonly encountered failure mode—which penetrates the transistor's SOA by the initial output-capacitor charging current when the supply is first turned on. Such failure often has an aura of mystery because the supply might otherwise be electrically rugged in the face of heavy load-current, high temperatures, and prolonged operation. However, initial capacitor surge current might only be tolerated by the transistor for a relatively few times before catastrophic destruction occurs. Notice the initially high current carried by the transistor (Fig. 2-14A).

In Fig. 2-14B, the MOV enables a quick charge build-up of the capacitor. By the time the transistor turns on, it "sees" a fully charged capacitor and consequently

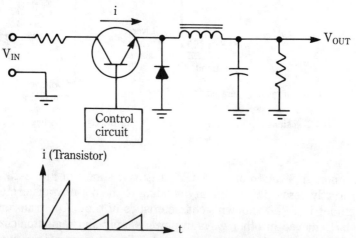

(A) Initially discharged capacitor causes initial current surge.

2-14 Scheme to prevent the turn-on transient through the switching transistor.

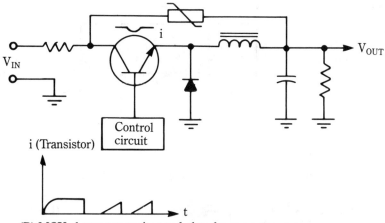

(B) MOV charges capacitor and absorbs current surge.

2-14 Continued

does not have to endure a heavy surge current. On the other hand, the MOV effectively drops out of the circuit because of the relatively low voltage that is impressed across it once the switching transistor becomes active. This voltage is approximately $V_{in} - V_{out}$ and is sufficient to produce only a negligible current in the MOV.

Special IC for overvoltage protection

The primary purpose of the previously discussed protection techniques has been the protection of the regulator circuit and/or the input power supply. Generally, such techniques also protect the load considerably. However, it is often desirable to incorporate a fast-acting "crowbar" circuit directly across the output of the voltage regulator. Then, if something goes wrong with the regulator circuitry, any tendency for the output voltage to soar is immediately followed by a short circuit. If the regulator has current limiting or current foldover, these operating modes are automatically brought into action. The net result is zero voltage across the load and a restricted current in the power supply. If current restrictive modes are not available or if they become inoperative, then the crowbar circuit must be allowed to open a fuse or circuit breaker. In any event, the load is well protected against damage from overvoltage.

A common provision for such "crowbar" protection is to simply use an SCR with a zener diode in its gate circuit (Fig. 2-15). Resistances R_1 and R_2 are selected so that the zener diode will conduct if the output voltage of the regulated power supply increases to, say, 125% of its set value. Resistor R3 is not critical, but it is used to keep the gate of the SCR from "floating." Generally, R_3 can be several times the combined series resistance of R_1 and R_2. Such an arrangement is simple enough, but it is not always easy to implement because of the triggering safety margin that is needed to avoid inadvertent SCR triggering by noise pulses. Of course, you can use a sharp-knee zener diode in conjunction with RC networks. A more reliable approach is to use the MC3423/3523 overvoltage-sensing circuit,

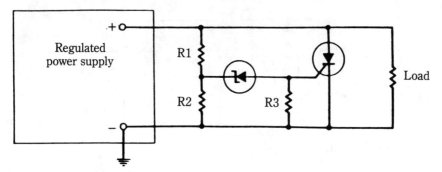

2-15 Simple crowbar scheme for protecting a load from overvoltage.
Motorola Semiconductor Products, Inc.

which was designed specially for this purpose. Not only does this IC provide an adjustable threshold for triggering an external SCR, but it can also be programmed for minimum duration of overvoltage before triggering.

A block diagram of the internal circuitry of this IC is shown in Fig. 2-16. Its connection to the output of a power supply is shown in Fig. 2-17. A study of the associated waveforms reveals that a minimum duration of the overvoltage condition is necessary to trigger the external SCR. With capacitance C equal to 0.1 μF, the time delay (t_d) in triggering is one millisecond. A delay of 0.1 ms is obtained when C is 0.01 μF. Notice that triggering occurs when the voltage accumulated across capacitor C reaches V_{ref}. Noise immunity is achieved because short duration noise pulses cannot readily charge capacitor C.

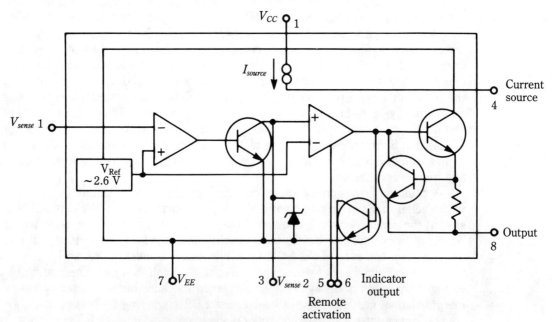

2-16 Block diagram of the MC3423/3523 overvoltage-sensing circuit.
Motorola Semiconductor Products, Inc.

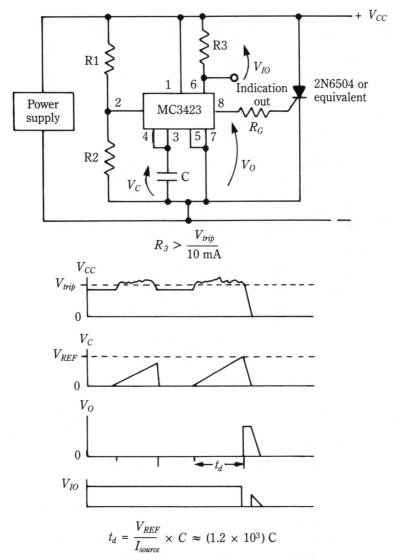

2-17 Implementation of the overvoltage-sensing IC and relevant waveforms. Motorola Semiconductor Products, Inc.

Using the crowbar technique when supply and load are far apart

Whether you use a primitive crowbar protection circuit or a sophisticated IC, a special consideration merits attention when considerable physical separation is between supply and load. In Fig. 2-18A, the overvoltage crowbar senses the voltage at the output of the supply—the voltage at the distant load can be appreciably

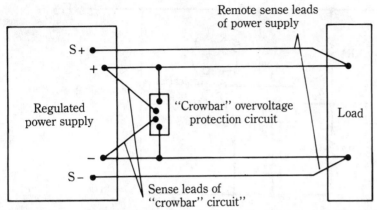

(A) Conventional method serves practical considerations especially when crowbar circuit is integral part of the power supply.

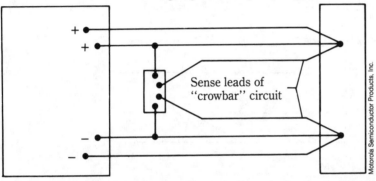

(B) Bringing crowbar circuit sense leads to *load* terminals can result in smaller differential triggering voltage and/or greater reliability from false triggering of crowbar circuit.

2-18 A "right" way and a better way to connect the crowbar circuit.

higher because of the voltage drop in the connecting wires, which transfer the current from supply to load. The load voltage is, however, tightly regulated because sensing leads from the voltage comparator in the supply are not connected to the output terminals of the supply; they are connected to the load itself. A similar technique must be used with the crowbar circuit to provide maximum overvoltage protection for the load (Fig. 2-18B). Considering that our prime objective is to protect the load from overvoltage, this approach is logical.

Using the scheme depicted in Fig. 2-18B, a much smaller triggering differential can be reliably maintained, which is especially true in low-voltage, high-current systems. A little contemplation will reveal that, to a first approximation, neither the dc voltage drop nor the transients in the main power supply leads are likely to influence the triggering voltage differential of the crowbar circuit when its sensing leads are connected to the load terminals. Particularly good results can be obtained when the crowbar-circuit sensing leads are twisted pair, or, even better, if shielded wire (such as mini-coax) is used.

Even when power-supply sensing leads are not brought out to the load, it remains desirable to bring sensing leads from the overvoltage protector to the load terminals, rather than to the output terminals of the supply. Generally, the shorting element of the overvoltage protector (usually an SCR) can be connected to the supply output terminals. In some instances, however, it can be more conveniently connected to the load terminals (for example, when the protection circuitry is adjacent to the load). This situation dispenses with the crowbar circuit sensing leads. A possible drawback could be difficulty in producing an effective short circuit across the supply.

The improvement that results from using the connection scheme of Fig. 2-18B diminishes as the distance between supply and load increases. This increase has to do with the remote-sensing technique of the power supply. At long separation distances between supply and load, it remains easy enough to preserve tight static regulation of load voltage. However, this is not necessarily true for dynamic regulation—and especially as this concept applies to transients. The reason is that the power-supply sensing leads contain inductance and stray capacitance. Because these leads are directly in the error-signal feedback path of the power supply, the presence of reactance degrades response time (this same reactance also degrades the stability of the supply and can cause oscillation, even though the intrinsic design of the supply might incorporate a large safety-margin against such instability).

Comments on current-limiting modes

Of all the protection techniques that have been discussed, the provisions for *constant-current* and *current-foldback operation* (Fig. 2-19) are probably used the most. Regardless of the various other protection methods used, you are very likely to also

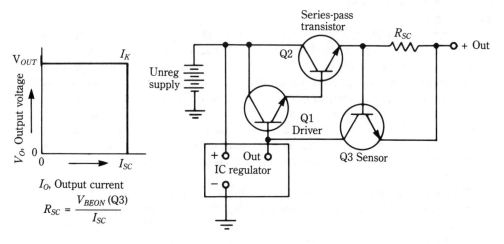

(A) Constant current.

2-19 Typical implementaton of current-limiting modes for protection.

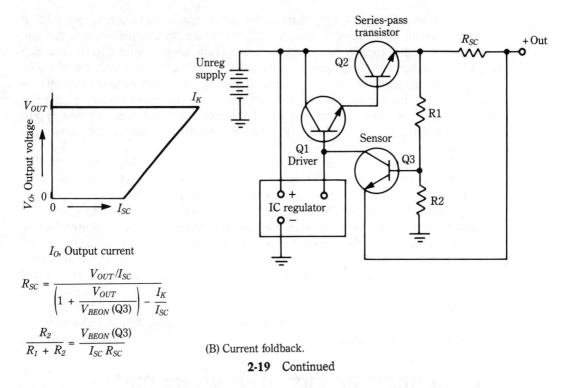

$$R_{SC} = \frac{V_{OUT}/I_{SC}}{\left(1 + \dfrac{V_{OUT}}{V_{BEON}(Q3)}\right) - \dfrac{I_K}{I_{SC}}}$$

$$\frac{R_2}{R_1 + R_2} = \frac{V_{BEON}(Q3)}{I_{SC} R_{SC}}$$

(B) Current foldback.

2-19 Continued

find either the constant-current or the current-foldback mode. Obviously, a short-circuited regulator operates under safe conditions with either constant current or current foldback. Because the easier-to-implement constant-current mode is sufficient to prevent catastrophic destruction of the series-pass transistor, the rectifiers, and, perhaps, the load, the current-foldback mode has a feature that might not be apparent.

Figure 2-20 illustrates the SOA (safe-operating area) of a popular series-pass transistor, the 2N3055. The load line (any combination of collector voltage and collector current) must always be inside of the enclosed area. Otherwise, the transistor is subject to permanent damage. Suppose that you wish to have your regulator automatically go into its constant-current mode when the load current reaches I_{SC} (Fig. 2-21) and that the collector-emitter voltage can approach V_{IN}. Although the enclosed rectangle is safely within the SOA, much of the current capability of the transistor is not available. Or, stated from another viewpoint, a larger and more expensive transistor must be chosen than would otherwise be justified by considering I_{SC} alone.

The situation that involves current foldback is shown in Fig. 2-22. As in the previous case, V_{IN} is the same, but now a much higher current (I_K) is available for the load. Thus, the operation of the transistor is equally safe in both modes, but current foldback allows more effective usage of the transistor. Do not surmise that the thermal conditions are the same for both modes. For the operation depicted in Fig. 2-22, you must provide for greater heat removal. Generally, a larger heatsink

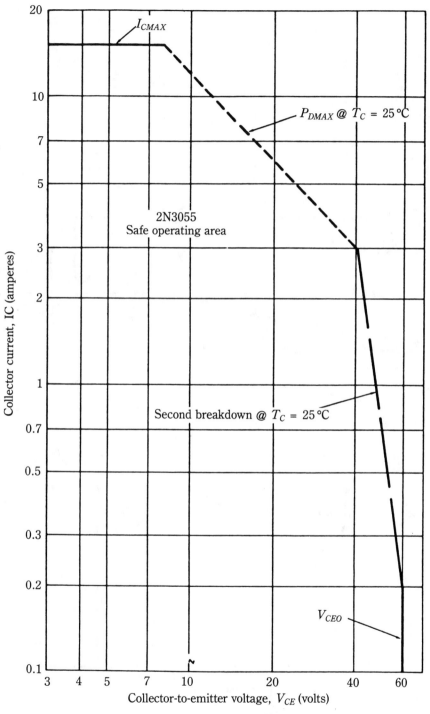

2-20 Safe operating area for the 2N3055 power transistor.

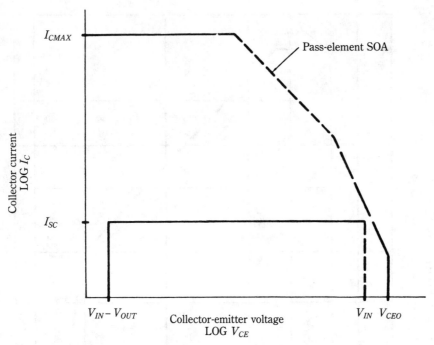

2-21 Allowable operation for constant-current mode with limit at I_{SC}.

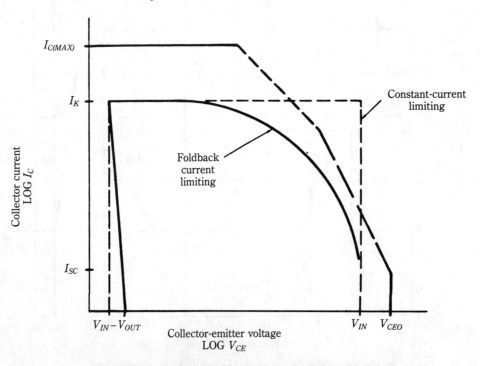

2-22 Allowable operation for current-foldback mode.

would be used for such an application. Notice that, despite the availability of current I_K in normal operation, the regulator only needs to supply current I_{SC} when the load voltage is short-circuited.

Protection techniques for the three-terminal IC regulator

Three-terminal IC regulators have been extensively publicized for their simplicity of implementation. This is essentially true, but do not believe that protection techniques are never needed or desired. Although protection circuitry is seldom shown, it is, nevertheless, true that these devices are quite vulnerable to catastrophic destruction under certain fault conditions. This destruction can occur despite the fact that considerable internal self-protection in the form of current limiting, overload limiting, thermal shut-down, and safe-area protection. The culprit, however, is often the energy storage in the capacitors that are associated with the regulator—the internal protection circuitry is useless under some fault conditions, which "dump" the capacitor energy into the regulator.

Figure 2-23 shows a three-terminal adjustable voltage-regulator with protection diodes. Diode D1 protects against discharge of output capacitor C1 into the regulator, which could occur if the input of the regulator was inadvertently shorted. Diode D2 protects against discharge of bypass capacitor C2 into the adjust terminal, which could occur if either the input or output to the regulator were shorted. Both D1 and D2 can usually be the 1N4002 (or similar) diode. Whether or not these protective diodes need to be incorporated in a final design depends much on the capacitors and the voltages. As a rough rule of thumb, 10-μF 10-V capacitors can discharge sufficient surge current to damage the regulator IC under the described fault conditions. Perhaps five volts and several microfarad might be "safe," but it is difficult to be certain. A low-value capacitor with inordinately low esr (internal resistance) could be as dangerous as a large capacitor with higher esr. Also, the adjust terminal is particularly vulnerable to damage from the discharge of C2. In any event, it is probably wise to use these diodes during experimentation.

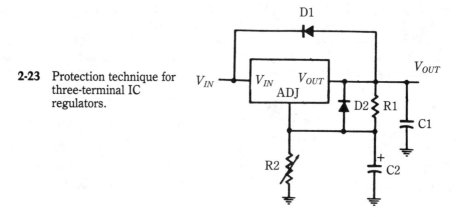

2-23 Protection technique for three-terminal IC regulators.

They do not affect ordinary operation of the IC regulator (in fixed-voltage three-terminal regulators, the middle terminal is generally grounded so that bypass capacitor C1 is not used. Thus, diode D1 is unnecessary).

Evaluation and selection of power transistors

Selecting power transistors often depends upon the priorities that are associated with the particular application. For example, the older, but still viable, germanium transistors merit consideration for low-voltage high-current linear regulators. If your options include additional space and weight for heat removal, it is possible to achieve worthwhile cost savings over more sophisticated silicon transistors. On the other hand, the expenses of thermal hardware and packaging could eliminate any overall cost savings. Also, other trade-offs must be considered.

The unique application of a regulated power supply might dictate its cost, packaging factors, and other considerations. In most instances, however, the name of the game is to balance one feature against another without jeopardizing the operating reliability of the power transistors. Realistic ratings require operation within the manufacturer's SOA curves under worst conditions. In this situation, the highest line-voltage, greatest load, and maximum ambient temperature that are deemed possible usually occur simultaneously. It is also wise to make additional allowances for line or load transients, even though it is thought that these might occur on rare occasions or that their energy content might be too low to justify concern.

The information included in Table 2-1 will provide some guidance in the evaluation and selection of power transistors. Transistors that are capable of dissipating 1 to 10 W are considered *low power*. *Medium power* includes devices with power dissipations in the 10- to 100-W range. *High-power devices* are capable of dissipating more than 100 W. With regard to current capability, it is practical to view transistors with a rating of less than 3 A as *low-current devices*. From 3 to 10 A are considered the realm of *medium-current transistors*. Current capabilities that exceed 10 A qualify as *high-current transistors*. Voltage ratings to 90 V represent *low-voltage transistors*. Ratings within the 90- to 300-V range are *medium-voltage devices*. *High-voltage transistors* have voltage capabilities beyond 300 V (much of the recent progress in transistors has been with the development of high-voltage transistors).

Power transistors that are capable of switching up to 5 kHz without incurring appreciable rise-and-fall time losses are, for the purposes of regulated power supplies, generally classified as low-frequency devices. The middle range is from 5 to 25 kHz, and a device selected for higher switching rates must have high-frequency capabilities. Thus, the cutoff frequency (f_T) should generally be well into the megahertz range. Notice that the f_T of the once-popular germanium power transistors (e.g., 2N178) was only about 10 kHz.

Although this breakdown involves somewhat arbitrary classifications, it is nevertheless based on actual rating statistics. Although it is true that evolving technology changes the present concepts of "high" and "low," the basic guidance provided by Table 2-1 is likely to retain general validity.

Table 2-1. Guide to application of power transistors.

Germanium (pnp)	Silicon single diffusion (npn)	Silicon epitaxial base (npn/pnp)	Silicon double diffusion (npn/pnp)
Features Low cost. High current capability. Low collector-emitter saturation voltage.	Electrically rugged. Superior thermal characteristics often make this silicon device a better choice than germanium.	Practical combination of ruggedness, frequency capability, and voltage rating. A useful workhorse for many power-supply applications.	Best bipolar process for high-frequency capability and high switching speed. Current rating is good.
Shortcomings Limited frequency capability. Effective heat removal needed (junction temperature is limited to vicinity of 110 °C). Collector voltage is generally rated below 100 V. No npn.	Limited frequency capability. Moderate voltage ratings. Not commonly available in pnp.	Possibly marginal performance in high-frequency switching circuits.	If too much emphasis is placed on frequency capability; electrical ruggedness suffers. Generally, the SOA is low.
Uses Series pass elements in linear regulators. Inverters, converters, and switching regulators operating at low and medium frequencies.	General-purpose power device in linear regulators and in switching applications up to medium frequencies.	General-purpose device with parameter blend which is often easier to apply than those of the single diffusion transistor.	High-speed switching applications utilizing good design. Protective techniques are often used.

Silicon triple diffusion (npn)	Silicon Darlington (npn/pnp)	Power MOSFET (n-channel)	Synthesized transistor (simulated npn)
Features Second-best bipolar device for frequency capability and switching speed. Has best voltage rating.	High current gain. High input impedance. Favorable cost and production aspects. Parameter combo is suited for many uses. Electrically rugged.	No thermal runaway. No secondary breakdown. Very high switching speed. Very high input impedance. Internally protected gate. Enhancement mode. Easy paralleling.	Complete overload protection. Base drive as high as 40 volts with no damage. 0.5 μsec switching time. 3 μA base current. Interfaces CMOS or TTL. Low cost.
Shortcomings Current rating is generally inferior to other silicon transistors. Cost tends to exceed "workhorse" types.	Some loss of circuit flexibility. Trade-offs in frequency, voltage, and current capabilities must be carefully evaluated. Cannot be driven to saturation. Therefore this device has a high voltage drop.	In some circuits, the internal (body) diode can cause trouble. ON resistance might cause excessive voltage loss and degradation of efficiency.	Limited current and voltage ratings. (Approx. 2A and 40V.) No pnp version is presently available. Vulnerable to self-oscillation if care is not exercised in layout and by-passing.
Uses High-voltage regulators, inverters, and converters. Generally used in switching applications where both high frequency and high voltage exist.	Series-pass elements. Drivers. Moderate speed switching. Useful in reducing parts count and manufacturing costs. Motor control.	Appears optimally suited for linear and switching supplies involving wide power-range. Merits consideration for high switching rates. Excellent for motor control.	High-reliability, low-power output element in both switching and linear supplies. Especially useful in cost-effective designs because protective circuitry is not needed.

Power MOSFET (p-channel)	Depletion-Mode power-MOSFET	SENSEFET power MOSFET	Insulated Gate Bipolar Transistor (IGBT)
Features Many are marketed with very similar characteristics to popular n-channel types. Simplifies circuit topography in certain applications.	Device normally ON. Gate bias can be applied to either cause harder turn-ON or to turn the device OFF. Development has focused on n-channel types, p-channel types are possible.	Has extra element that provides a tiny fraction of actual load current for control purposes.	Exhibits high input impedance for a power MOSFET and the low voltage drop of a bipolar transistor can have good reverse-blocking ability. No body-diode effect.

Table 2-1. Continued.

Power MOSFET (p-channel)	Depletion-Mode power-MOSFET	SENSEFET power MOSFET	Insulated Gate Bipolar Transistor (IGBT)
Shortcomings			
Availability not as extensive as n-channel MOSFETS and cost tends to be higher for similar ratings.	Not widely available compared to enhancement-mode MOSFETS. Commonly used circuit techniques are geared to enhancement devices. Limited current and voltage ratings.	Not widely available in extensive range of ratings.	This is a high-voltage and high-current device that is best-suited for low and moderate frequencies. Best IGBTs are limited to 5-kHz, 20-kHz and 50-kHz types might be useful.
Uses			
Most applications are similar to those in which n-channel types have been used. These included switch-mode power supplies and motor control. The p-channel MOSFET is particularly well-suited for use in bridge-type synchronous rectifiers.	Well-suited for certain constant-current circuits. Can be used in linear regulators. Simulates normally ON mode of certain relays.	Excellent for use in current-mode switching regulators. When so deployed, power-wasting sampling resistance is not needed. (The resistance that is required has negligible power dissipation.)	Excellent for switch-mode supplies operating at 600 volts (and higher) and at tens of amperes, but at 20 kHz or preferably lower rates. Device enables very efficient motor control.

Motorola

The last three columns of Table 2-1 contain information about devices of relatively recent development. To the experimentally and progressively minded, these devices appear to offer many interesting features. Because of its electrical ruggedness, the silicon Darlington transistor has been used considerably in regulated power supplies, especially the linear type.

In contrast to earlier hybrid structures, presently-available Darlingtons are of monolithic fabrication and display surprisingly good parameters in categories that are relevant to uses in regulators. Of particular significance, modern power Darlingtons can often perform efficiently in 20-kHz switching regulators. The power MOSFET has become the star performer of designers, although some still identify the FET as a device that is inherently limited to small-signal power levels. In the category of synthesized transistors, the LM195, is actually a specialized IC. Not only is this device inexpensive, but its cost effectiveness, when used in regulator circuits, merits consideration. This is because the device incorporates its own protective techniques, thereby dispensing with the need for external components. Other interesting ICs are appearing on the horizon. Included is the RCA HC2000H power op-amp, with ratings of 7 A (peak), 30-kHz bandwidth, and up to 100-W-rms output power. IGBTs hold much promise for efficient switching-regulators below 30 kHz or so. Power MOSFETs seem destined to monopolize high-frequency designs up to several MHz.

Interconnecting regulated power supplies

A frequently encountered question with regard to regulated power supplies is whether increased ratings can be attained by connecting the outputs of two or more units. The greatest interest concerns the possible paralleling of voltage-regulated power supplies in order to provide greater load current.

A quick, but valid, reply would be to consult the manufacturer's application notes. If such an answer smacks of vagueness, it is not without intention, for it is neither easy to recommend nor to advise against the practice. Paralleling power supplies without the manufacturer's blessing or without certain precautionary techniques is almost a sure invitation for trouble. Difficulties likely to arise are unequal current division, undesired operation of one or more protective circuits, loss of regulation, loss of control or programming functions, large circulating currents between units, and destruction of one or more series-pass transistors.

On the other hand, successful paralleling can be an effective means to make a system conveniently expandable and to meet various performance, cost, and space requirements. Many technical people are attracted to modular techniques wherein "beefed-up" operation can be quickly and neatly achieved by simply plugging in the required number of duplicate power supplies.

At first, it would appear that you should be able to parallel two or more voltage-regulated supplies by merely connecting their identical-voltage outputs together. This would indeed be practical if their output voltages could be adjusted and maintained at, or very nearly at, the same value. Consider, however, that the effective output impedance of a good regulator might near $0.01\ \Omega$. Then, just one tenth of a volt difference in the output voltages of two such regulators would produce a local circulating current of 10 A! It is true that a dc path for such a circulating current does not always exist. This, however, does not necessarily make paralleling feasible, for it will generally be found that one supply tends to hog the load. Fortunately, some techniques are available to properly divide the load current among the participating power supplies.

Various methods for interconnecting power supplies require access to the internal circuitry of the supplies and often involve minor changes. Thus, it is desirable to have circuit-strapping options on the back panel of the supplies so that such alterations can be conveniently made. Certain manufacturers incorporate these construction techniques in their products. The simplified schematic diagram in Fig. 2-24 illustrates such a technique.

Parallel operation by means of separate pass-element circuits

A simple, but effective, method of paralleling voltage-regulator outputs is shown in Fig. 2-25. In this arrangement, the *slave unit* has been divested of all of its control circuitry and is no longer a regulated supply in its own right. However, when connected with the *master supply* (Fig. 2-25), both supplies are forced to develop the same output voltage and therefore participate equally in delivering current to the load. Whether more than one slave unit can be accommodated in this fashion depends upon the base-drive capability of the master power supply. The power supplies interconnected in this way are usually (but not always) identical units. It is generally only essential that each supply has sufficient load-current capability.

Parallel operation by parallel programming

A similar, but somewhat more sophisticated method of forcing voltage-regulated power supplies to share load current is shown in Fig. 2-26. Here, the error amplifi-

86 *Static characteristics of regulated power supplies*

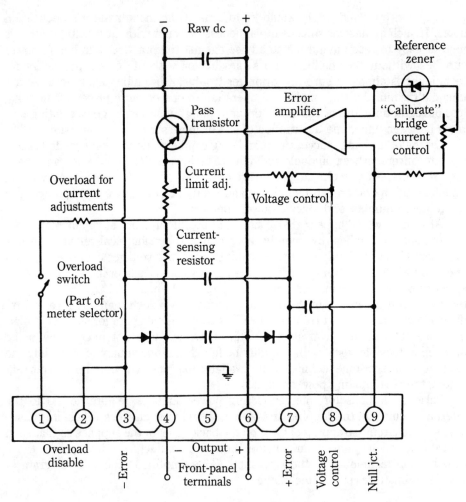

2-24 Regulated supply with terminals for circuit-strapping options.

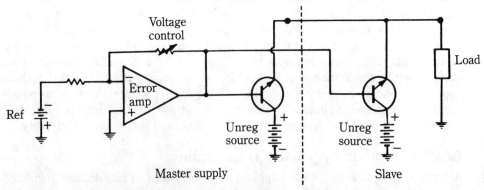

2-25 Parallel operation via separate pass-element circuits.

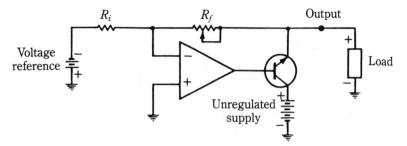

(A) Simplified circuit of typical series-pass, voltage-regulated power supply.

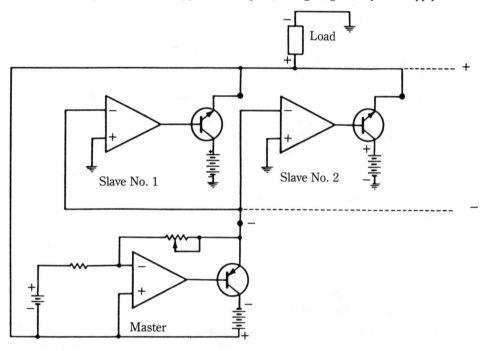

(B) Parallel programming system with master power supply and two slave units.

2-26 Basic technique of parallel-programming of power supplies.

ers in the slave units are retained and the master supply functions as the voltage reference for the entire system. In order to clarify this scheme, the basic building block is shown in Fig. 2-26A. For best results, the slave units should be alike. The master supply in this application can be a relatively small unit, however, it must be a floating type if it is utilized in the manner indicated.

Automatic parallel operation of two voltage-regulated power supplies

A unique master/slave system of parallel power-supply loading is shown in Fig. 2-27. The master supply is essentially similar to the typical voltage-regulated sup-

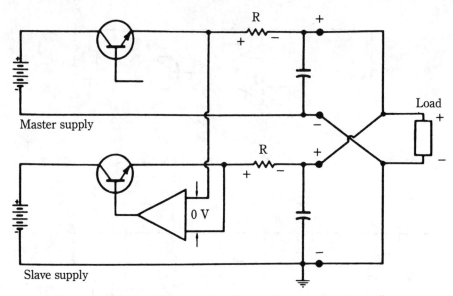

2-27 Automatic parallel operation of two voltage-regulated supplies.

ply (Fig. 2-26A). Both supplies in Fig. 2-24 have small current-monitoring resistances (R) in their output leads. The error amplifier in the slave unit compares the voltage drops developed across these resistances and always controls the current output of the slave unit to produce a voltage null. In this way, the two supplies always develop equal output voltages and share equally in supplying load current. In order to avoid needless dissipation, the value of the monitoring resistances (R) should be as low as the gain of the error amplifiers will permit.

Automatic tracking operation of two voltage-regulated power supplies

The scheme depicted in Fig. 2-28 is not a paralleling technique. However, it is of relevant interest because it enables the control of load currents that are delivered by two voltage-regulated supplies. In this arrangement, the load voltage produced by the master supply is necessarily greater than that produced by the slave unit. The simplified schematic of the master supply can be assumed to be essentially similar to the somewhat-more-detailed regulator circuit (Fig. 2-26A). The master supply not only supplies its own load, but it also acts as the voltage reference for the slave unit. This method of load-current control is useful in systems where the ratio of dc voltages is more important than their individual values. Analog logic circuits and certain operational amplifiers are examples of such a requirement.

Series-connected voltage regulators

Several techniques have been presented for paralleling voltage-regulated power supplies. You can deduce from the various schemes that it is not normal for voltage regulators to cooperate in the partitioning of load current—something had to be

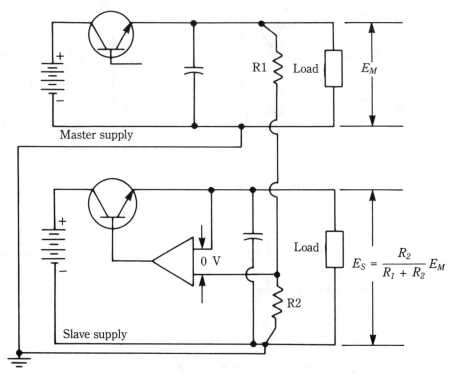

2-28 Automatic tracking operation of two voltage-regulated supplies.

specially arranged for desired operation. On the contrary, series-connected power supplies are relatively easy to place into operation. It is only necessary to connect the supplies in series and add their individually regulated voltages. A series cascade, which involves three identical voltage-regulated supplies, is shown in Fig. 2-29A. It is preferable to use identical units, but this is not a requisite. If dissimilar power supplies are connected in series, the maximum available current will be that of the supply with the lowest current rating. Additionally, the unit with the lowest isolation voltage rating should be placed closest to ground (Fig. 2-29B). Included in the series arrangements of Fig. 2-29 are protective diodes across the output terminals of each supply. These diodes conduct the reverse current that could result if that particular power supply was not turned on along with the other supplies of the cascade. The diodes protect the filter capacitors and the series-pass transistors. The ac-power-line connections are shown in Fig. 2-29 in order to emphasize the desirability of having ac power applied and removed simultaneously for all of the power supplies in the series cascade.

Automatic series operation of two voltage-regulated power supplies

A somewhat more sophisticated arrangement for connecting voltage-regulated power supplies in series is shown in Fig. 2-30. A master/slave relationship exists

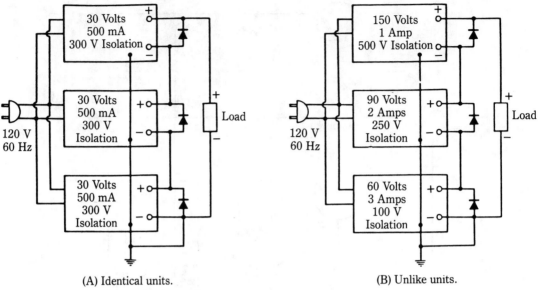

(A) Identical units. (B) Unlike units.

2-29 Series connections for voltage-regulated supplies.

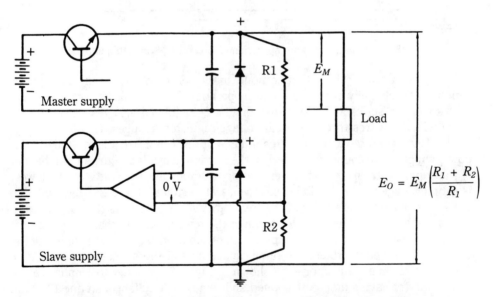

2-30 Automatic series operation of two voltage-regulated supplies.

between the two supplies because of their control circuitry. Thus, the master unit not only participates in delivering power to the load, but it also functions as the reference voltage for the comparator circuit in the slave unit. For the sake of simplicity, the control circuitry of the master supply is not shown, but it is assumed to be representative in that a comparator and a voltage reference are involved.

Dual-output tracking operation of two voltage-regulated power supplies

The arrangement shown in Fig. 2-31 is very useful because of the large number of IC modules, which require dual voltages of opposite polarity. Actually, this scheme is a simple modification of the power-supply system that is depicted in Fig. 2-30. A split load has been substituted for the single load and ground has been moved to the center tap of the dual supply. Although the positive and negative output voltages (with respect to ground) can have a wide range of values, in most applications of this technique the two voltages are equal. These equal voltages are easily caused by making resistors R1 and R2 equal in value.

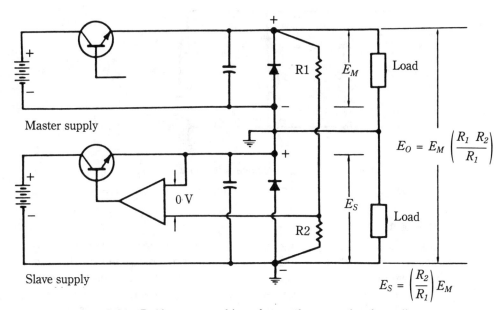

2-31 Dual-output tracking of two voltage-regulated supplies.

Parallel operation of current-regulated power supplies

Current-regulated power supplies work well in parallel operation. Indeed, if all participating units have the same compliance-voltage rating, any number can be simply connected in parallel (Fig. 2-32A). If the compliance-voltage ratings differ among the units, the diode isolation technique (Fig. 2-32B) should be employed. This technique protects units with low compliance-voltage ratings from having excessive voltage impressed across their terminals. Such a situation would otherwise exist in the event that the load was removed. In normal operation, all of the diodes are forward biased and conduct current to the load. The power supplies can either be current regulated or crossover types that operate in the constant-current mode.

92 Static characteristics of regulated power supplies

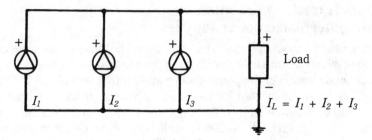

(A) Units with identical compliance-voltage ratings.

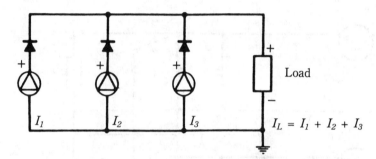

(B) Units with different compliance-voltage ratings.

2-32 Parallel operation of current-regulated power supplies.

Other situations

Although the most useful interconnections have been discussed, other combinations could merit consideration. For example, you might desire to combine the load currents from two or more voltage-regulated supplies, which incorporate current limiting, but do not operate in the constant-current mode. In general, it is difficult to operate such supplies in parallel, although some degree of success can be obtained if one supply operates as a voltage source and the others operate as current sources. As might be expected, much depends on the nature of the current-limiting characteristic.

Voltage-regulated switching supplies can sometimes be successfully paralleled because their output impedances could be relatively high. However, all voltage-regulated supplies can be operated in parallel by the scheme shown in Fig. 2-33. Inasmuch as load regulation is necessarily degraded, this method must be considered a last resort and is likely to be of value only with a constant load or where close regulation is not important.

Basic aspects of heat removal

This book, as with others involved with various electronics themes, is primarily concerned with circuits, devices, and components. It would, however, be amiss to

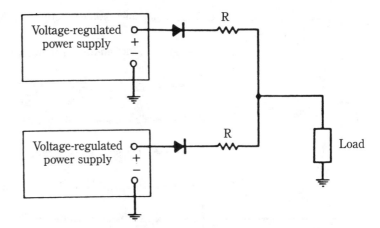

2-33 Paralleling scheme when close regulation is not important.

exclude at least a basic discussion of *heat removal*. Although the competent designer ensures that the generation of heat is minimal, that which cannot economically or technologically be further reduced must be conveyed to the external environment. Otherwise, life-limiting or destructive temperatures will develop in vulnerable devices and components. Even before this stage is reached, circuit malperformance is likely. Heat systems, like electrical circuits, exhibit both static and dynamic behavior. This section covers static (steady-state) behavior.

Static behavior is a basic phenomenon that often tends to be overlooked. Simply, entrapped heat-energy manifests itself as temperature rise. Consider, for example, a one-watt resistor operating under rated conditions. Such a resistor is not likely to be more than lukewarm to the touch. However, that is only because the nominally one watt or so of heat dissipation escapes from the resistor via conduction, convection, and radiation. It might seem implausible, but if these three "leakage paths" could be blocked, the resistor would rather quickly attain charring temperature. Here, the manufacturer imposed the one-watt rating because he knew that in ordinary situations and with conventional mounting methods, such heat retention would not occur. However, with semiconductors, capacitors, and larger resistors (higher power rating), a more marginal heat-removal situation prevails. With such components and devices, damaging internal temperatures very readily accrue because the generated heat cannot leave rapidly enough.

The quantity that impedes the flow of heat energy is, appropriately enough, called *thermal resistance* or *thermal impedance*. This concept is very important and much effort is expended to reduce thermal resistance. This term is commonly ascribed as the total opposition to heat transfer in the face of the three transfer mechanisms, conduction, convection, and radiation. Heat problems deal with the net consequences of heat transfer, as if a thermal current passes from a higher to a lower temperature region.

Effective heat removal is caused by an appropriate resort to an "Ohms law" of the thermal circuit. This enables us to circumvent much of thermodynamic theory and the physics of heat transfer. An "equivalent" circuit, such as that depicted for a power transistor mounted on a heatsink, is shown in Fig. 2-34. At the outset, realize

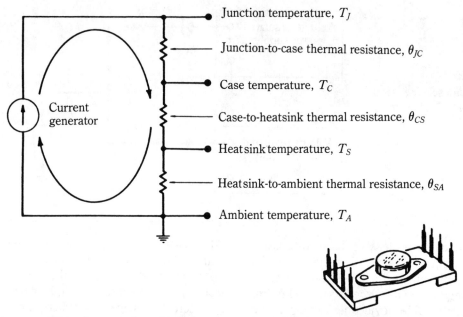

2-34 An equivalent circuit for the development of a thermal Ohm's law.

that the analogy between thermal and electrical parameters is not perfect on a one-to-one basis. Nonetheless, for practical purposes, this concept is very useful.

Thermal resistance corresponds to electrical resistance. And thermal "pressure" or temperature corresponds to electrical "pressure" (voltage). Accordingly, a voltage can be measured with respect to ground potential at certain points. Remember that a voltmeter readily provides us with the voltage drop between any of the indicated points and ground. The voltmeter's thermal counterpart is a thermometer. The thermometer does not indicate the temperature of one of these points relative to "ground" or ambient temperature. In order to obtain the various temperatures that are relative to the ambient, a subtraction must be made. Thus, the junction temperature relative to ambient is the difference between the measured junction temperature and the ambient temperature: $T_J - T_A$. Similarly, the case temperature relative to ambient temperature is: $T_C - T_A$. Finally, the heatsink temperature relative to ambient temperature is: $T_S - T_A$. The basic idea here is that voltage drop and temperature drop are considered analogous quantities. However, "ground" is at zero electrical potential, whereas, the ambient temperature of ground is not zero. Ambient temperature is commonly assumed to be 25 °C.

So far, this is an "equivalent" circuit: the temperature drop represents voltage drop and thermal resistance represents electrical resistance. The current postulated as flowing through the resistances is driven by a current generator. Here the equivalency between thermal and electrical quantities apparently falters. For the thermal analogy, electrical current is taken to be *power*, the electrical power dissipated in the device. It might initially appear that nothing is accomplished by such an assumption and that you have deviated from your goal of correlating appropriate thermal and electrical parameters. The assumption is thermodynamically valid, however, and is extremely useful from a practical standpoint.

The apparent lack of perfection in this scheme derives from the missing "thermal current." Indeed, such a quantity would be dealt with if the problem was attacked purely with the procedures used in thermodynamics and the physics of heat transfer. These rigorous approaches are, however, much more complex than this engineering shortcut. The measured electrical power dissipation, although an electrical quantity, represents the missing thermal current and leads to a simple Ohm's law solution of the thermal "circuit." Fortunately, electrical power dissipation can be an easily measured quantity in practice (in the foregoing, ΔT usually represents a temperature drop or difference. Thus, ΔT might be $T_J - T_A$, $T_C - T_S$, etc. Because ambient temperature is not referred to a lower temperature, it is simply T_A).

Because of the preceding considerations, this thermal circuit is governed by three equations that are directly similar to the three forms of Ohm's law, which relate voltage, current, and resistance in electrical circuits. These thermal counterparts of Ohm's law are:

1. $\Theta = \dfrac{\Delta T}{P}$, that is, thermal-resistance is equal to temperature relative to the ambient divided by electrical power dissipation. The unit of thermal resistance is, appropriately, degrees Celsius per watt. As previously pointed out, this relationship is analogous to $\dfrac{E}{I}$ in the electrical case.

   Admittedly, our thermal form of the equation uses two thermal parameters and one electrical parameter. The electrical parameter, *power*, accurately represents thermal current. You might have an imperfection of form, but not of substance. Again, $\Theta = \dfrac{\Delta T}{P}$ is similar to $R = \dfrac{E}{I}$.

2. $\Delta T = P\Theta$. This algebraic transposition of (1) enables you to calculate the temperature relative to ambient if we have power and thermal resistance given. This form of the equation is useful because P is readily measured or assumed and Θ is often available from specifications and from tables. You can then investigate whether a critical point, such as a transistor junction, will develop an excessively high temperature rise. $\Delta T = P\Theta$ is similar to $E = IR$.

3. $P = \dfrac{\Delta T}{\Theta}$. This algebraic transposition of (1) enables you to calculate the electrical power dissipation, which corresponds to a known temperature rise and a known thermal resistance. It leads to the solution of such problems as, "how much power dissipation is allowable in a transistor in order that the maximum permissible junction temperature is not exceeded?" Here, you must know the thermal resistances in the path of the heat transfer or thermal current. Often, these thermal resistances are found in the specifications of transistors and heatsinks. $P = \dfrac{\Delta T}{\Theta}$ is similar to $I = \dfrac{E}{R}$.

Strictly speaking, your thermal circuit might be expected to involve only conduction as a means of transporting the thermal current. This analogy would seem to be closer to the electric circuit. In actual practice, however, the thermal resistance is not only considered a function of the material and its dimensions, but also of the operation of convection and radiation on that material. Convection and radiation generally make the thermal resistance lower than would be obtained from conduction alone. Thus, the use of fins on a heatsink increases the area of heat-conducting material that is exposed to the air, which thereby enhances convection. By making the fins a dull black color, heat radiation is optimized. The thermal resistance between the heatsink and the surrounding air will, accordingly, be determined by the combined influences of conduction, convection, and radiation.

The term *convection cooling* usually pertains to still air. Forced-air cooling, as imparted by a fan or blower, exposes more air molecules per unit time to the hot surfaces, thereby carrying away more heat.

Radiation cooling is actually an electromagnetic phenomenon that involves the infrared region of the spectrum. As such, a pure vacuum is sufficient for the passage of the heat energy into surrounding space. The air plays a secondary role here because some of the radiation raises the temperature of the air molecules. This, in principle, degrades the effectiveness of the air as a convecting agent. In practice, overall heat removal is generally enhanced by the mutual operation of convection and radiation. Only in the stagnant atmosphere of an undersized enclosure would the two mechanisms interfere with one another.

A practical example of the use of the thermal circuit

Thermal circuit theory is necessary to avoid "working in the dark." Practical insight into the considerations and computations involved in the design and selection of heat-removal hardware can now be attained via a step-by-step example. This example uses an appropriate heatsink for the series-pass power transistor of a regulated power supply. This problem has numerous other ramifications; an understanding of its solution will enable other thermal situations to be worked out via the same basic procedure (Fig. 2-34). The Motorola 2N6544 silicon npn transistor, under worst operating conditions, will dissipate 30 W of power. Under such conditions, the transistor handles 1 A and drops 30 V. The required thermal data for this transistor is listed in Table 2-2.

Step 1. Assume an operating junction-temperature, T_J. Experience and judgment generally dictate a margin of safety, with respect to the maximum-allowable temperature of 200 °C. Semiconductor reliability is a strong function of temperature—the lower the temperature, the greater the reliability. Also, thermal-circuit calculations are, at best, reasonable approximations in practice, because of the difficulty of precisely accounting for variables. For example, "nonlinearities," where thermal resistance might not be the same at different temperatures, can often occur. The combined effects of conduction, convection, and radiation do not exert the same effect in different hardware arrangements or packaging methods. Slight variations in surface-to-surface contacts can have relatively great effects on thermal operation. Accordingly, T_J should be limited to 135 °C.

Table 2-2. Typical thermal data that is found in transistor specifications.

Rating (Maximum)*	Symbol	2N6544	2N6545	Unit
Collector-emitter voltage	$V_{CEO(sus)}$	300	400	Vdc
Collector-emitter voltage	$V_{CEX(sus)}$	350	450	Vdc
Collector-emitter voltage	V_{CEV}	650	850	Vdc
Emitter base voltage	V_{EB}		9.0	Vdc
Collector current—continuous —peak (1)	I_C I_{CM}		8.0 16	Adc
Base current—continuous —peak (1)	I_B I_{BM}		8.0 16	Adc
Emitter current—continuous —peak (1)	I_E I_{EM}		16 32	Adc
Total power dissipation @ $T_C = 25°C$ @ $T_C = 100°C$ Derate above 25 °C	P_D		125 71.5 0.714	Watts W/°C
Operating and storage-junction temperature range	T_J, T_{stg}		−65 to +200	°C
Thermal characteristic	**Symbol**		**Max**	**Unit**
Thermal resistance, junction to case	$R_{\theta JC}$		1.4	°C/W
Maximum lead temperature for soldering purposes: 1/8" from case for 5 seconds	T_L		275	°C

*Indicates JEDEC registered data
(1) pulse test: pulse width = 5 ms, duty cycle ≤ 10%.

Courtesy Motorola Semiconductor Products, Inc.

Step 2. Ambient temperature, T_A, is the "ground" temperature, the environmental temperature to which heat removal is directed. In many situations, this temperature is the nominal room temperature. For a simple cooling system where the heatsink is exposed to free air, such an assumption is essentially valid (a common and often costly oversight is to apply this assumption to where the air contacting the heatsink is stagnant, or even worse, is at an elevated temperature from exposure to other equipment). In this situation, T_A is assumed to be 30°C.

Step 3. $\Delta T = T_J - T_A \, 135°C - 30°C = 105°C$. This temperature drives the 30 W of thermal current through the complete thermal circuit.

Step 4. Total thermal resistance:

$$\Theta_{JA} = \frac{\Delta T}{P} = \frac{105}{30} = \frac{3.5\,°C}{W}$$

This is the summation of the three thermal resistances depicted in the thermal circuit of Fig. 2-34. Thus,

$$\Theta_{JA} = \frac{3.5\,°C}{W} = \Theta_{JC} + \Theta_{CS} + \Theta_{SA}.$$

Step 5. Obtain the case to heatsink thermal resistance, Θ_{CS}.

Refer to Table 2-2, assuming that 3-mil mica will be used as an electrical insulator between the transistor and the heatsink. Because the 2N6544 is packaged in a TO-3 container, $\Theta_{CS} = \frac{0.4\,°C}{W}$. Remember that this desirably low thermal resistance will be obtained only if a suitable heatsink compound is applied to the mating surfaces.

Step 6. Now, the thermal resistance, Θ_{SA} of the heatsink can be evaluated. From Table 2-1, the junction-to-case thermal resistance, Θ_{JC} of the transistor is $\frac{1.4\,°C}{W}$. From Table 2-3, the thermal resistance from case to sink is $\frac{0.4\,°C}{W}$. From

Table 2-3. Typical case-to-sink thermal resistances.

			θCS	
		Metal-to-metal	Using an insulator	
		With heatsink	With heatsink	
Case	Dry	compound	compound	Type
TO-3			0.4 °C/W	3 mil mica
TO-3	0.2 °C/W	0.1 °C/W	0.35 °C/W	Anodized aluminum
TO-66	1.5 °C/W	0.5 °C/W	2.3 °C/W	2 mil mica

Courtesy Motorola Semiconductor Products, Inc.

step 4, the total thermal resistance of the circuit is $\frac{3.5\,°C}{W}$. Because we are dealing with a series-connected array of resistances, any one resistance must be equal to the total resistance minus the sum of the remaining resistances. Accordingly:

$$\Theta_{SA} = \Theta_{JA} - [\,\Theta_{JC} + \Theta_{CS}\,], \text{ or } \Theta_{SA} = 3.5 - [1.4 + 0.4\,]$$

Finally:

$$\Theta_{SA} = 3.5 - 1.8 = \frac{1.7\,°C}{W}$$

Step 7. Referring to Table 2-4, you can now select the heatsink. The basic idea is to choose one that has a thermal resistance that is equal to, or lower than, the Θ_{SA} value of $\frac{1.7\,°C}{W}$, which is obtained in step 6. Appropriate Delco heatsinks have a thermal resistance of $\frac{1.6\,°C}{W}$, such as the models 7281361 or the 7281364.

From the above, you can also determine the operating case temperature of the

Table 2-4. A family of heatsinks for power transistors.

Heatsink model	7281366 7281369*	7281361 7281364*	7300811 7270606* 7270725 7281352	7281353* 7281355	7281351 7281354*	7276040 7277151 7278482*	7281357 7281360*
θ_{SA} in °C/watt	1.45	1.6	2.3	2.3	2.8	3.0	5.0

*Unmachined and unpainted blanks.
Equivalent finishing is required to obtain the appropriate thermal resistance.

Courtesy Delco Electronics Division of General Motors Corp.

transistor. This temperature is important because transistor specifications always involve thermal derating factors (the 125-W rating of this particular transistor is a "book value"—it only pertains to the nonpractical situation in which the case temperature is held to 25 °C). The temperature drop from junction to case, that is, T_{JC}, is $P \times \theta_{JC} = 30 \times 1.4 = 52$ °C. Therefore, case temperature, $J_C = T_J - T_{JC} = 135 - 52 = 83$ °C. The transistor will safely operate at this case temperature while dissipating 30 W. This figure can be determined from the 0.714 derating factor in Table 2-2. Also, the combination of voltage and current involved is within the safe-operating area, but this combination will not receive further attention here, because the objective has been to show how the thermal circuit can be dealt with.

The effect of forced air cooling

If air is set in motion by a fan or blower and directed at the heatsink, the effective thermal resistance of the heatsink is lowered (Fig. 2-35). Here, a Delco 7281352 heatsink, with a nominal thermal resistance of $\frac{2.3 \, °C}{W}$ in still air, is subjected to airflow. As might be expected, the greatest percentage-wise improvement occurs in the region of low and moderate air-flow velocities. Thereafter, with increasing velocity, the improvement continues, but with diminishing returns. In actual practice, the designer often finds that some empirical investigation is necessary in order to evaluate the effectiveness of forced-air cooling. This evaluation is primarily necessary because the actual air velocity in the vicinity of the heatsink is not usually accurately known and whether the flow is *laminar* or *turbulent* makes a difference. Another factor complicating prediction is the temperature of the moving air; although it is generally assumed to be at ambient, it might not actually be so. In many practical situations, the awareness that moving air lowers the thermal resistance of a heatsink to ambient interface is sufficient when numeric data is not easily obtained.

The effect of heat radiation

Radiation also lowers the effective thermal resistance of the heatsink. A commonly encountered fallacy is that polished surfaces are good radiators. The converse is true, however. Although polished surfaces are excellent reflectors, they

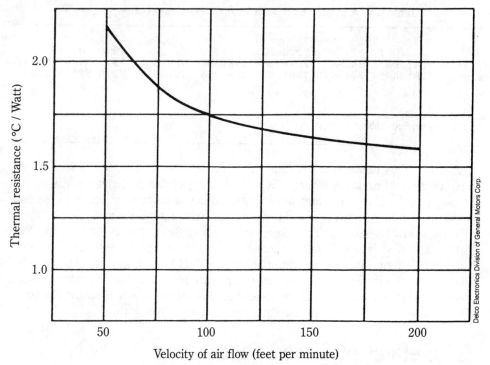

2-35 Typical effect of forced-air cooling on a heatsink.

are the poorest radiating surfaces (Table 2-5). Here, a perfect heat radiator would have an emissivity of 1.00 and a surface incapable of radiating its contained heat would have an emissivity of zero. Interestingly, aluminum almost embraces this gamut, depending on whether its surface is polished or anodized (the anodization used usually has a dull-black finish). Sometimes metal surfaces are sandblasted or otherwise roughened to increase their emissivity (i.e., their ability to radiate contained heat). Remember that the very properties that impart high emissivity

Table 2-5. Emissivities of surfaces in thermal hardware.

Surface	Emissivity, ϵ
Aluminum, anodized	0.7 – 0.9
Alodine on aluminum	0.15
Aluminum, polished	0.05
Copper, polished	0.07
Copper, oxidized	0.70
Rolled sheet steel	0.66
Air-drying enamel (any color)	0.85 – 0.91
Oil paints (any color)	0.92 – 0.96
Varnish	0.89 – 0.93

Courtesy Motorola Semiconductor Products, Inc.

to a surface also make it a good absorber of heat. Thus, a heatsink can absorb heat energy from a nearby hot object—even though the intervening air is at, or near, ambient temperature.

The all-important matter of thermal conductivity

Very important in the matter of heat removal are the thermal conductivity of the heatsink and/or the intervening heat-flow paths between the region to be cooled and the heatsink. Obviously, the heat path between a semiconductor junction and the case of the device is not accessible. Also, the functional heatsink might not be a discrete piece of hardware, but it could be the device itself, or more likely, PC board copper foil or portions of the aluminum chassis. Nonetheless, it is usually important to know something about the thermal conductivity of those heat-flow paths so that you can specify the nature and dimensions of the materials used.

In this regard, it is useful to clear an oft-encountered controversy regarding the best metal for heatsinks. Some believe that aluminum is best; usually as a result of the popularity of aluminum for this purpose. Others are adamant that copper has about twice the thermal conductivity of aluminum and should, therefore, be the selected metal for heatsinks. It is only natural to ponder which is correct, for successful heat removal is the bottom line in the design and operation of power semiconductors.

Actually, some nice things can be said for both parties to this controversy. Referring to Table 2-6, copper does, indeed, exhibit about twice the thermal conductivity

**Table 2-6.
Thermal conductivities of metals and common materials.**

Material	Conductivity
Air	0.000054
Aluminum	0.49
Brass	0.26
Cement	0.0007
Copper	0.91
Cork	0.0001
Cotton	0.0005
Glass	0.002
Ice	0.005
Iron	0.15
Lead	0.08
Platinum	0.17
Silk	0.0002
Silver	0.99
Slate	0.005
Steel	0.11

of the more commonly used aluminums. Incidentally, the thermal conductivities listed are in the metric system. These can be converted to English-system equivalents by multiplying by 7×10^4 (i.e., by 70,000). Although such a conversion gives us a new list of numbers for thermal conductivities, remember that percentage relationships between the thermal conductivities of the various materials remain unchanged. English-system notation was favored by the older physics texts and they remain the system of choice in some newer books, as well.

Interestingly, the sequence of thermal conductivities from best to worst parallels that of electrical conductivity. For example, copper is usually recognized as a superior electrical conductor to aluminum. It so happens, however, that you could play the devil's advocate and insist with validity that the reverse is true! This brings us, surprisingly, to the alluded controversy regarding the best heatsink material.

When it is said that copper is a superior electrical conductor to aluminum, it is implied (whether knowingly or not) that the comparison is made on the basis of equal cross-sectional areas of the compared materials. However, is this the only practical way to make a comparison? Comparison can also be made on a weight basis. Consider the electrical conductivity of materials based upon equal weights. A given length of aluminum is a better electrical conductor than the same length of copper of the same weight. A little contemplation reveals that such an aluminum conductor will have a greater cross-sectional area than the copper conductor. So, aluminum is the preferred material (lower I^2R dissipative losses than the copper conductor). In order to use the aluminum conductor, you must first ascertain that space is no problem. Then, you must investigate such practical matters as strength and cost. In the real world, a practical way of utilizing the better conductivity of aluminum (on a weight basis) for electrical transmission of power is to use aluminum-clad steel-core wire.

For the sake of heat removal, a massive aluminum element can provide a great deal of thermal conductivity. It can, moreover, satisfy this need economically, and with ease of manufacturing intricate shapes. There are complicating factors because of the roles played by convection and radiation in a large piece of hardware broken up into sections or fins. Regardless, if space is available, the aluminum heatsink can perform its heat-removal function very well. At the same time, remember that the same heatsink made of copper would transport heat even more effectively. Such a heatsink might be quite costly, however. Translated into more practical terms, it is generally true that a smaller and simpler copper heatsink can remove heat just as effectively as a relatively-large and complex aluminum heatsink (assuming that the comparative heatsinks are subjected to similar convection environments and that their surfaces are treated so as to produce similar thermal emissivities). Without detailing the fact that a precise comparison is difficult, copper should be investigated when conventional aluminum heatsinks are inadequate.

Nonmetallic substances also are of importance in the quest for the efficient removal of heat via thermal conduction. These materials are usually also *electrical insulators*. Beryllium oxide (beryllia) is often used to conduct heat away from semiconductor junctions to the case of the device. This material is also used as insulating spacers and washers for mounting power devices on heatsinks. Beryllium oxide is classified as a ceramic.

Warning Be cautious when working with devices or hardware made with beryllium oxide. In its powdered or pulverized form, ingestion or absorption into the body is hazardous. Therefore, do not attempt to cut, machine, drill, or abrade it. It should not be incinerated and should not be allowed to fall into the hands of unaware persons. Disposal by burial is advocated by some manufacturers.

Despite this warning, do not be fearful of toxicity from the mere handling of the solid material as you might reasonably expect in most electronics work. Those who handle beryllium oxide on a workday basis should follow safety procedures, including wearing rubber gloves and properly ventilating the work area. Beryllium oxide is also known as *beryllia*.

Diamond is another nonmetallic substance with excellent thermal conductivity. It, too, is an electrical insulator, although gem-quality material is not necessary. Nonetheless, cost is an inhibiting factor, and its use tends to be limited to military, space, and research projects where performance, rather than expense, merits priority. Another drawback of diamond is that it is not readily available in large pieces, whether mined or man-made.

It is interesting to compare the thermal conductivities of Beryllium oxide and diamonds to that of the metals. Surprisingly, diamond is about 1.5 times better than copper in this respect! It is also approximately three times as effective a heat conductor as aluminum. More practical comparisons can be made with beryllium oxide because of its availability in relatively large pieces and diverse shapes. Beryllium oxide has about 90% of the thermal conductivity of aluminum; this conductivity, of course, makes it superior to such common metals as brass and steel.

Do not be dismayed that thermal-conductivity tables in different texts reflect different dimensional units used in their calculation. In practical electronics applications, what is important is the comparison in thermal conductivity between various materials. Thus, regardless of units, all tables will show copper to be about twice as effective as aluminum in conducting heat. Also, regardless of units, these tables base their results on a cross-sectional area basis. This is familiar because electrical-conductivity tables are also presented in this manner.

Representative heatsinks are listed in Table 2-7. Those shown for TO-220 and TO-3 packages are especially relevant to power-supply applications. Most are either cast or extruded aluminum, but a few are probably copper. To the manufacturer, aluminum exhibits advantages of cost and ease of production. Copper is used more for heatsinks that are intended for use in larger power supplies—especially those designed around high-current SCRs and GTOs. Notice the listing of the addresses of the four main heatsink makers: Staver, IERC, Thermalloy, and Wakefield.

Because safe thermal operation of power devices is one of the important evidences of good design, you should be aware of possible pitfalls in heat removal. Remember that it is not enough to extract heat from a power device via a good thermal path if no place is available to "dump" the heat energy. Usually enough free convection is required to keep the heatsink at a relatively-low temperature;

Table 2-7. Representative heatsinks.

θSA^*(°C/W)	TO-3 & TO-66 Manufacturer/Series or Part Number
0.3 – 1.0	Thermalloy—6441, 6443, 6450, 6470, 6560, 6590, 6660, 6690
1.0 – 3.0	Wakefield—641 Thermalloy—6123, 6135, 6169, 6306, 6401, 6403, 6421, 6423, 6427, 6442, 6463, 6500
3.0 – 5.0	Wakefield—621, 623 Thermalloy—6606, 6129, 6141, 6303 IERC—HP Staver—V3-3-2
5.0 – 7.0	Wakefield—690 Thermalloy—6002, 6003, 6004, 6005, 6052, 6053, 6054, 6176, 6301 IERC—LB Staver—V3-5-2
7.0 – 10.0	Wakefield—672 Thermalloy—6001, 6016, 6051, 6105, 6601 IERC—LA, uP Staver—V1-3, V1-5, V3-3, V3-5, V3-7
10.0 – 25.0	Thermalloy—6013, 6014, 6015, 6103, 6104, 6105, 6117

Courtesy Motorola Semiconductor Products, Inc.

sometimes this can only be accomplished by forced-air cooling. No matter how heat is removed from the heatsink, the heat must ultimately be disposed of in the environment outside of the equipment—heat that is trapped within a cabinet or box will cause a temperature rise as the equipment is operated.

Another pitfall stems from a too-hasty interpretation of the table of thermal conductivities of materials. For example, many aluminum alloys have no more than 66% of the thermal conductivity of pure aluminum. Similarly, it might be construed that the various brasses, because they are essentially copper alloys and are commonly encountered in electronics work, might readily substitute for copper. Here again, it is all too easy to wind up with about 66% of the thermal conductivity of copper. Remember, too, that although metals tend to lose a few percent of their room-temperature thermal conductivity at 100°C, beryllium oxide suffers an almost 30% reduction over a similar temperature-range.

In the practical implementation of heat-removal hardware, too little attention is often given to the interface between the power device and the heatsink. If the mating surfaces were perfectly smooth and flat, the junction they form would not add thermal resistance to the flow of heat. A reliable approach to this ideal condition is not, however, likely to be achieved in actual practice. Imperfect mating of these surfaces is equivalent to the insertion of an air gap in the path of heat flow. Because of the very low thermal conductance of air, the efficiency of heat removal is degraded, even though we might have what appears to be a high-integrity butt joint from the mechanical viewpoint. The application of more pressure between the surfaces invariably improves the situation, but it is difficult to overcome the cumulative effect of micro air bubbles trapped in the tiny valleys of imperfectly smooth and flat mating surfaces. The application of various greases or fluids to the interfacing surfaces improves the thermal conductance of the joint. These coatings, although their thermal conductance is not high, still provide considerably less resistance to heat flow than does air.

Because most greases, oils, and the like have objectional physical and chemical properties, it is much better to use thin coatings of one of the various thermal pastes that are specifically made for such use. These provide relatively high thermal conductance, and do not deteriorate with temperature change or time. Such thermal paste should be applied to the surfaces of both metal-to-metal junctions and metal-to-insulator junctions. Remember, however, that the objective is to displace very thin films and bubbles of air. Therefore, it is not a case of the more the better; an excessively thick application of the thermal paste will likely produce less-than-optimum results. Table 2-8 lists several manufacturers of thermal-paste products. Best results will be obtained from reasonably smooth and flat mating surfaces, thin coatings of a commercial thermal paste, and high pressure between the joined parts. Thermal paste will not, however, remedy the deleterious effects of grossly rough and nonflat mating surfaces.

Table 2-8. Commercial preparations for improving thermal bonding.

Thermal paste or compound	Manufacturer
120 Compound	Wakefield
Thermalcoate	Thermalloy
640 Compound	General Electric
340 Compound	Dow Corning
980 Compound	Aham

The use of mica or silicone rubber washers or spacers to provide electrical insulation between the power device and the heatsink is sometimes considered thermal resistance in the heat-flow path and they should be circumvented, if possible. It is often better to insulate the heatsink, rather than the power device. Conflict between thermal and electrical requirements can be best resolved by appropriately using optoisolators or transformers within the power-supply circuit itself. Pay heed to these matters early in the overall design philosophy; it seems to be a law of nature that heat removal in power supplies is less effective than hoped for. It is certainly best to err on the conservative side for no other reason than that the thermal circuit involves many parameters that are not easy to precisely pin down.

3
Dynamic characteristics of regulated power supplies

THE DYNAMIC CHARACTERISTICS OF REGULATED POWER SUPPLIES INVOLVE their behavior as a result of nonsteady loads, transients, and, in general, alternating currents that are superimposed on the dc output level. One of these dynamic characteristics, the output impedance, has already been discussed. This chapter looks at important aspects of dynamic supply response from as many viewpoints as possible.

A look at dynamic output impedance

Circuitry operation often becomes clearer as you look at it first one way, and then in a slightly different way. With this in mind, it would be appropriate to commence our discussion with the concept of dynamic output impedance. This section is confined to the dynamic output impedance of a series-type voltage-regulated supply. This is probably the most common regulated power supply in use. By extrapolation, reciprocal relationship, or analogy, the general ideas developed can be readily applied to other types of regulated power supplies.

An appropriate example of the application of the dynamic characteristics alone is shown in Fig. 3-1. Such so-called "electronic filters" are often used in TV receivers to provide clean, but unstabilized dc. Massive chokes and inordinately large filter capacitors are thereby eliminated. Notice the absence of the reference voltage necessary to stabilize the dc output-voltage level.

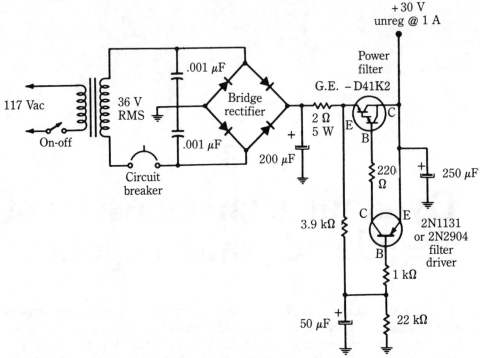

3-1 Example of a regulator circuit that is used as an electronic filter.

Operation of the voltage regulator

Figure 3-2 is a simplified block diagram of a series-type voltage-regulated power supply. It would be difficult to explain its operating principle without using the word *change* or one of its many synonyms, such as *vary, fluctuate, differ*, etc. Such words imply an unsteady condition; we are already in the realm of dynamic operation, even though we might be concerned primarily with the maintenance of a constant dc output level. The arrangement in Fig. 3-2 is essentially a negative-feedback amplifier, in which the error signal is the difference between a fixed reference voltage and a sampled portion of the output voltage. In operation, this error-signal voltage tends to extinguish itself. That is, the voltage derived from the sampling network changes in direction and magnitude in order to equal the reference voltage, E_R. The "mixing" of the sampling and reference voltages occurs in the comparator. The comparator might be a differential amplifier, or a single active element in which these two voltages are introduced so that a third voltage, E_e, represents their differences.

Any tendency for the output voltage, E_o, to change is counteracted by a change in the effective resistance of the series-pass element. The series-pass element is controlled by the amplified error signal. The counter action of the output voltage will be great enough to drive the error signal back toward zero. With the error signal ideally at zero, equilibrium is attained and the output voltage is restored to its

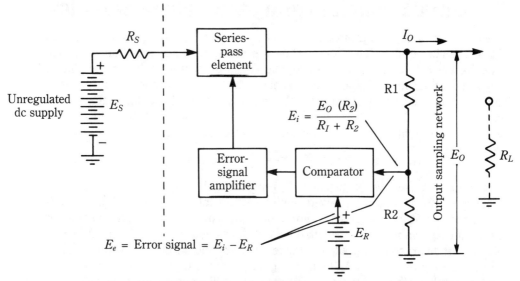

3-2 Simplified block diagram of a series-type voltage-regulated supply.

regulated level. This explanation is simplified in that you will not at this time concern yourself with the never-ceasing "hunting" action that prevails in a practical supply. Nor should you involve yourself with the fact that the practical supply falls short of the ideal 100% restoration. Rather, your objective is now to visualize the general sequence of events that accounts for the static regulation of the dc output level. However, this objective will lead to the consideration of the dynamic characteristics of power supplies.

The salient features of the arrangement depicted in Fig. 3-2 are:

- The series-pass element functions as a voltage- or current-controlled resistance that, in turn, controls the dc voltage that is impressed across the sampling network (and load). Notice that it accomplishes this by actually controlling the output current, I_o.

- The comparator functions as a difference amplifier and one of its two inputs is a fixed voltage.

- The feedback that results from impressing the generated error signal, E_e, at the input of the error-signal amplifier is negative; a nullifying action results as a consequence of any output voltage change. This action leads to restoration of the output voltage. The greater the disturbance, the greater the restoring action. Moreover, the circuit seeks to restore deviant output voltage, whether the disturbance has been brought about by a change in the supply voltage, E_S, or by a change in the load R_L. Actually, the error signal never attains zero level, as with mechanical servo systems. However, when the overall gain of the amplifier stage is high, the nulling action forces the error signal to a very low value. The error signal is constantly "hunting" for zero and is therefore an ac voltage.

Dynamic behavior from static characteristics

The output-voltage and output-current changes that were involved in the previous explanation of regulator-circuit operation were assumed to be small ones. If such disturbances are too great or too fast, the active elements might not be able to respond with sufficient magnitude or speed. Thus, the restoration of output voltage would be ineffective and the supply would not be well regulated against the disturbances. If, then, you are concerned with slow changes in output voltage and current and stipulate these to be of small magnitude, these quantities are ΔE_o and ΔI_o. Thus, the output impedance, Z_o, is: $\dfrac{\Delta E_o}{\Delta I_o}$. Under appropriate measurement conditions that use low frequencies, Z_o represents the static output impedance. Because small changes are involved, such derivation also represents dynamic output impedance at low frequencies. For many practical purposes, Z_o from static- or low-frequency evaluation is reasonably valid throughout regions A and B in Fig. 3-3. In particular, static output resistance and dynamic output impedance closely coincide throughout region A. Thus, if region C and the higher portion of region B are not involved in the load requirements, dynamic output impedance can be obtained from the specifications on static behavior. In region A, Z_o is governed mainly by amplifier characteristics. Regions A and B overlap considerably, but the output capacitor now exerts increasing influence with rising frequency. From region C on, the inductance of output leads eventually results in a rise in output impedance.

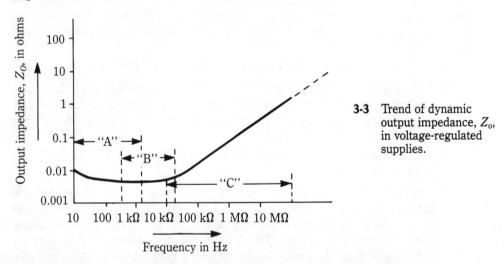

3-3 Trend of dynamic output impedance, Z_o, in voltage-regulated supplies.

Regulator as an ac feedback amplifier

Happily, static and dynamic output impedance merge at the lower frequencies. Pursuing the matter further leads to another way of looking at the regulator circuit. Consider the simple regulator circuit shown in Fig. 3-4A. The orientation of

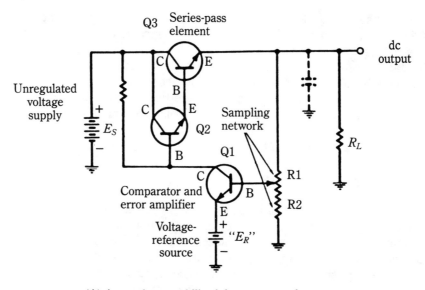

(A) As a voltage-stabilized dc power supply.

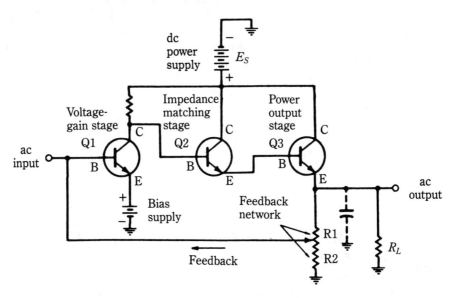

(B) As an ac amplifier with negative feedback loop.

3-4 Two ways to view a regulator circuit.

the circuit symbols, as well as the names that are assigned to the stages, emphasize the overall function of a dc voltage stabilizer. We become conditioned to think of this configuration as a voltage-controlled resistance (the series-pass element), which is associated with a servolike circuit; a keen eye on the voltmeter and a trained hand on the rheostat knob are necessary.

Now inspect Fig. 3-4B. The circuit connections remain the same (as in Fig. 3-4A), but the components have been shifted to resemble the general layout of a three-stage ac feedback amplifier. Indeed, this is just such an amplifier circuit. The stages are direct-coupled so that response is immediately known to go down to zero frequency (dc). The "transformation" has been aided by renaming stages and components to comply with the nomenclature that pertains to amplifier circuits. This is an equivalent circuit in a truer sense than is generally encountered in circuit analysis. The significance of this duality in functional roles should not be taken lightly. This viewpoint of operation provides useful computations and approximations of dynamic regulator behavior. The load resistance, R_L, is retained as part of the amplifier circuit because the parameters of the "power output stage," Q3, are affected by the dc collector-emitter current.

The schematic diagram of Fig. 3-5 essentially repeats the configuration of Fig. 3-4B, but several test techniques have been added to facilitate the measurement of dynamic characteristics. Switch S1 is inserted in the feedback path, which enables evaluation of ac performance with and without feedback. Blocking capacitor C2 makes the dc operating bias applied to the base of transistor Q1 independent of the position of S1. Resistance R_x reestablishes dc operating bias, which otherwise would have been lost as a result of the insertion of S1 and C2. Blocking capacitor C1 prevents disturbance of this bias by the ac signal generator, which might have a dc path across its terminals. Variable resistance, R_T, and its associated blocking capacitor C3 are test-switched in and out of the load circuit by means of switch S2. Capacitors C1, C2, and C3 should be large enough that ac operation will be unaffected down to 10 or 20 Hz.

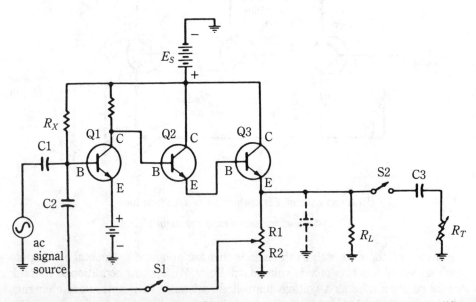

3-5 Determination of regulator output impedance, when viewed as an ac-feedback amplifier.

In making the ac measurements, an oscilloscope monitors the peak input and output voltages. If, during the measurement procedure, the top or bottom of the output wave is clipped, the indication is that R_x is not properly adjusted to permit linear operation of the amplifier stages. The smallest signal that can be used practically is applied to the input of the amplifier.

The foregoing is for the purpose of utilizing the equation:

$$Z_o = \frac{R_o}{1 - AK}$$

where:

Z_o is the dynamic output impedance for the frequency being investigated and for the value of R_L chosen,

R_o is the output impedance without feedback—that is, with switch S1 in its open position,

A is the total voltage amplification without feedback (because of polarity inversion, A is negative. Therefore, $1 - AK$ is positive),

K is equal to $\frac{R_2}{(R_1 + R_2)}$ = percentage of output voltage fed back to the input.

The evaluation technique is straightforward. First, measurements are best made at a low frequency, 100 Hz, for example. The value of A is simply the ratio of output to input voltage with S1 and S2 open. That is, $A = \frac{E_{out}}{E_{in}}$.

Next, S2 is closed and R_T is adjusted to cause the peak-to-peak output voltage to become 50% of the value that was observed with S2 open. Now you have obtained sufficient data to satisfy the needs of the equation for dynamic output impedance. K and A are already known. Also, sufficient information is available to compute R_o. The law of parallel resistances can be used to calculate R_o, thus: $R_o = R_T R_L - R_T$. Although this is an approximation by virtue of the fact that we have not included the effect of paralleled branch R1-R2, the error will be small if R_L is chosen to load the "amplifier" corresponding to at least 30% of the fully rated dc output current of the actual power supply.

Finally, we insert the appropriate numerical values into the equation,

$$Z_o = \frac{R_o}{(1 - AK)}$$

Reiterating, Z_o is the dynamic output impedance with feedback and is representative of dc power-supply operation, where regulated voltage is being applied to a load that corresponds to R_L. The measurements can be made for a wide frequency range in order to plot a curve of dynamic output impedance vs frequency (Fig. 3-3). The usefulness of this test method is because Z_o is often a very small fraction of an ohm and is difficult to determine by more direct measurement techniques.

Some comments on the output capacitor

The described theoretical and practical approach for determining dynamic output impedance assumes that the large output capacitor generally connected across the output terminals of voltage-regulated power supplies remains intact. Therefore, ac amplifier behavior is investigated under the same conditions that prevail in actual power-supply operation. This capacitor severely limits the high-frequency response of the amplifier. If further insight is desired regarding amplifier response capability, the capacitor can be disconnected. This disconnection is not always feasible, however, because oscillation can occur if sufficient loop gain is present. Often a compromise situation can be attained by substituting a much smaller capacitor; one that is large enough to stabilize the phase-gain characteristic of the amplifier, but small enough to permit only a relatively small narrowing of the bandwidth. A better approach is to dispense with the output capacitor altogether and attempt to suppress oscillation by connecting a small capacitor from the collector to the base of a voltage-gain stage, such as Q1 in the example.

The foregoing technique is feasible for resistive loads. It is often surprising to find that the output capacitor can be greatly reduced in value or even eliminated, as long as the load provides no inductive component. This change not only enables extended investigation of the intrinsic characteristics of the regulator, but the greater gain-bandwidth product then allows a higher slewing rate (i.e., higher programming speed). Naturally, the manufacturer, above all else, wants his supply to be stable for all load impedances. A large output capacitor is the easiest way to ensure a safe gain-phase relationship. Such a capacitor tends to impose its 6-dB-per-octave rolloff, regardless of the load. For many applications, programming is either not involved or is not required at a high rate. Even if the large capacitor limits the bandwidth of the amplifier, it compensates for this action by providing low impedance at higher frequencies, where the amplifier response is down (region B of Fig. 3-3).

Another feedback arrangement for the voltage regulator

A somewhat different arrangement of circuitry components from the configuration thus far considered is shown in Figs. 3-6A and 3-6B. These two block diagrams appear at first glance also to differ greatly from one another. However, a little study reveals that these two configurations are identical. The bridge layout of Fig. 3-6A suggests the concept of balance. Assume that the voltage source, E_{REF}, causes a common current through R_i and R_f. This assumption is made on the premise that no direct current can flow into the input of the amplifier. Other than this, the action of the amplifier drives its input voltage to zero. It accomplishes this by varying the effective resistance of the series transistor to balance the bridge. Bridge balance exists for the condition that $\frac{E_{ed}}{E_{ad}} = \frac{E_{eb}}{E_{ab}}$, where the small-letter subscripts define the four bridge arms.

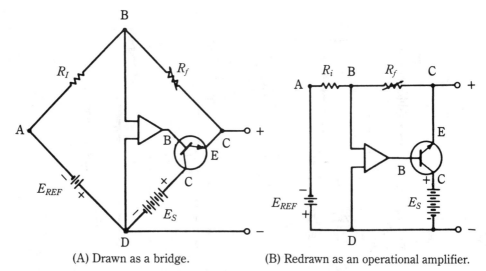

(A) Drawn as a bridge.   (B) Redrawn as an operational amplifier.

3-6 Other feedback arrangements of the voltage regulator.

Transconductance and the regulator circuit

Why is the concept of transconductance tailor-made for analyzing the voltage-regulated power supply? Notice that the tube transconductance equation is valid as long as the plate voltage is constant. The reason for this is that the basic idea is to show a simple cause-effect relationship. The change in plate current per unit change in grid voltage is a figure of merit because it shows how much effect can result from a known small cause, e_g. If the plate voltage was not kept constant, the effect, i_b, would be influenced by grid voltage and plate voltage; we would not then have a simple relationship that would be suitable for assessing the "activeness" of the tube. Thus, it is fortunate that the output voltage of the voltage-regulated supply is, by its intended nature, also maintained constant. Moreover, this output voltage is achieved automatically (for test purposes, the unregulated supply, E, can be assumed to remain reasonably constant because the load, R_L, will not be changed).

A second reason why the regulator circuit can be readily analyzed in terms of transconductance is that the basic quantity that really produces the stabilized output voltage is a change in plate (or collector) current. Thus, for the whole regulator circuit, the transconductance concept can be applied, just as it would be for a single vacuum tube (this statement applies to field-effect devices in the same manner. Simply substitute drain, gate, and source for plate, grid, and cathode, respectively. Indeed, many designers prefer to extend the transconductance concept to bipolar transistors and ICs).

The transconductance viewpoint applied to tube and transistor circuits

Figure 3-7 shows I_o as the basic output quantity in a simple vacuum-tube voltage regulator. Here, a small input signal voltage, E_o, is converted to an incremental out-

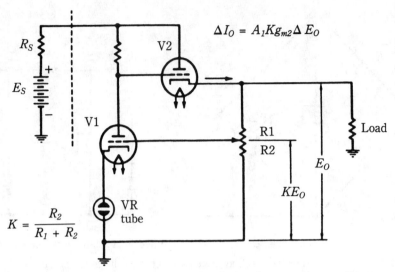

3-7 Transconductance concept applied to the electron-tube regulator.

put current, ΔI_o, via the factor $A_1 K g_{m2}$. A_1 represents the voltage amplification developed by tube V1. This being the case, the entire regulator circuit can be viewed as a single active element with transconductance $g_m = A_1 K g_{m2}$ where g_{m2} represents the transconductance of tube V2. The factor K accounts for the attenuating effect of the voltage divider in the feedback loop and is equal to $\frac{R_2}{(R_1 + R_2)}$. In so many words, it can be stated that the transconductance of tube V2 is increased by the total effective voltage amplification that precedes it.

The output impedance, Z_o, is then the reciprocal of this enhanced transconductance (i.e., $Z_o = \frac{1}{A_1 K g_{m2}}$). This evaluation for region A of Fig. 3-3 is valid (region A is generally where manufacturers' specifications designate dynamic output impedance, except when explicitly stated otherwise).

Voltage gain for triodes

The voltage gain of stages that precede the series-pass tube can be evaluated by well-known methods. For triodes,

$$A = -\frac{\mu R_L}{R_L + r_p + R_k(\mu + 1)}$$

where:

μ is the voltage amplification factor of the tube,
R_L is the effective load resistance seen by the plate,
r_p is the dynamic plate resistance of the tube,
R_k is the unbypassed resistance in the cathode lead.

R_k is often zero, whereupon the equation simplifies to

$$A = -\frac{\mu R_L}{(R_L + r_p)}$$

Gain for pentodes

The voltage gain for pentode tubes is given by the formula:

$$A = -\frac{g_m R_L}{1 + g_m R_k\left(\frac{1 + I_{g2}}{I_p}\right)}$$

where:

I_{g2} is the average screen-grid current,
I_p is the average plate current,
g_m is the tube transconductance,
R_k is the unbypassed resistance in the cathode lead.

In place of this complex equation, the simpler approximation $A = -g_m R_L$ is usually sufficient. Indeed, satisfactory results are often obtained by using this formula for triode stages as well. The basic idea is to facilitate design or to reasonably evaluate the performance of a regulator. Rigorous calculation is generally not required; the objective is usually to achieve a dynamic output impedance that is several times lower than an estimated minimum allowable value. This objective can be easily reached without precisely pinpointing the output impedance.

Transconductance of transistors

The situation for the transistor regulator is shown in Fig. 3-8. Again, the transconductance is enhanced, but in a somewhat different manner from the tube circuit. Here, the transconductance of the first stage (the comparator) is multiplied by the subsequent current amplification. The factor K has the same significance as in the tube circuit. We can think of K as modifying the voltage gain of the first tube (V1) in the tube circuit and as modifying the transconductance of the first transistor (Q1) in the transistor circuit. In both circuits, ΔE_o appears across the output terminals and $K\Delta E_o$ is applied to the actual input of the first stage (the comparator).

The entire transistor regulator can be viewed as a single active stage with transconductance $g_m = \beta_3 \beta_2 K g_{m1}$, where g_{m1} represents the transconductance of transistor stage Q1.

The output impedance is again the reciprocal of the enhanced transconductance, but this time it is the first stage, rather than the last, that has been operated on. With this in mind, output impedance $Z_o = \frac{1}{\beta_3 \beta_2 K g_{m1}}$.

The different technique stems from the fact that transistors are specified by their current gain rather than by any analogy of the voltage-gain factor, μ, for vac-

118 *Dynamic characteristics of regulated power supplies*

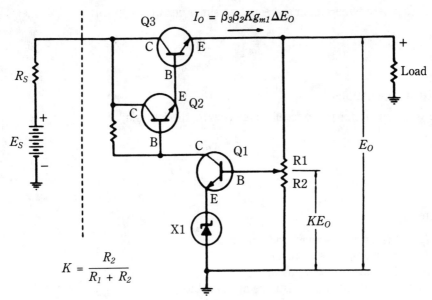

3-8 Transconductance concept applied to the transistor regulator.

uum tubes. Transistor data sheets generally provide the hybrid parameters of the common-base configuration, such as h_{21} and h_{11}. Often, the corresponding parameters of the more-often encountered common-emitter configuration, h_{fe} and h_{ie}, are given. In any event, an adequate approximation of the transconductance, g_m, can be computed from the following relationships which yield g_m in mhos:

$$g_m \approx \frac{h_{21}}{h_{11}}$$

Because h_{21} is very close to unity in modern transistors, we have:

$$g_m \approx \frac{1}{h_{11}}$$

Also,

$$g_m = \frac{h_{fe}}{h_{fe}} = \frac{\text{beta}}{\text{input impedance of common-emitter circuit}}$$

h_{21} represents alpha, whereas h_{fe} represents beta. Additionally, an often satisfactory estimate of transistor transconductance can be obtained from $\frac{I_E}{26}$, where I_E is the dc emitter current in milliamperes, and the transconductance, g_m, is again derived in mhos.

A look in retrospect at the voltage regulator

Having considered the voltage regulator from several viewpoints, it is now worthwhile to summarize quickly some of the salient features that are associated with voltage stabilization. For this purpose, refer to the representative regulator (Fig. 3-9). Although this section covers a regulator constructed around an operational amplifier, these points generally apply to other feedback arrangements as well.

The error voltage, E_e, in Fig. 3-9 is very small. This error voltage results from the basic principle of the regulator in which a null is established at the input of the amplifier. Indeed, as the intrinsic (open-loop) gain of the amplifier approaches infinity, the error signal approaches zero. Aside from ideal performance, the error signal of practical regulators can be a very small fraction of a volt. Now, this seemingly infers a low resistance between the two input terminals of the amplifier. In other words, you might suppose that the current I_{REF} would complete its path from E_{REF} through R_i, and thence through the amplifier input circuit to ground (arbitrarily, the flow of electrons is considered to emanate from the negative terminal of a source of emf). Such is not the case!

The current I_{REF} is governed by the Ohm's law relationship, $\dfrac{E_{REF}}{R_i}$, but it does not flow into the amplifier terminals. Moreover, I_{REF} flows through feedback resistance R_f, but it is not affected by the value of R_f. Although this phenomenon is startling when encountered for the first time, it is "old-hat" to those acquainted with operational-amplifier techniques. It is said that a "virtual ground" exists at input 1, the inverting terminal, of the amplifier. Ideally, the amplifier input behaves as a short circuit, except that it does not allow the passage of current. Although this sounds contradictory from the component point of view, it is a very useful concept for gaining insight concerning the operation of this type of circuitry. The input terminal is also referred to as the *summing junction*, for it is here that current I_{REF} is

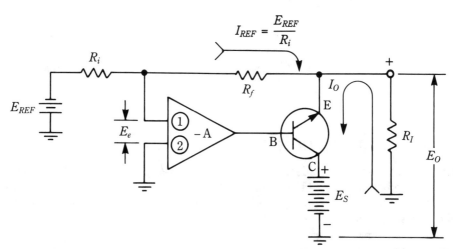

3-9 Regulator circuit that is constructed around an operational amplifier.

flowing to it; flowing away from it, it is algebraically summed. Because this summation is zero, such terminology infers that no current flows into the amplifier, and that voltage E_e across the amplifier input terminals is zero. A corollary of the operation thus far described is that output voltage $E_o = I_{REF} R_f$. Because $I_{REF} = \dfrac{E_{REF}}{R_i}$, $E_o = \dfrac{R_f E_{REF}}{R_i}$. The classic relationship for the operational amplifier is that the output voltage is equal to the input voltage multiplied by the "operational gain." The output voltage can be programmed by making either E_{REF} or R_f variable. Programming can also be accomplished by making R_i variable, but then a reciprocal relationship exists between R_i and E_o. Yet another programming technique consists of the substitution of a variable dc source for R_f.

A closer look at the error signal

The relationships indicated in the preceding paragraph are mathematically rigorous for the assumption that the error signal is zero, which is necessary for assuming that the open-loop gain, $-A$, of the amplifier is infinite. In a practical power supply, a small error signal does exist. Its value is the ratio of the output voltage to the open-loop gain (i.e., $E_e = \dfrac{E_o}{A}$). Also, now that the ideal conditions are not ideal, the equation for output voltage becomes:

$$E_o = \frac{R_f E_{REF}}{R_i} + E_e$$

From this a second equation for error signal is obtained:

$$E_e = E_o - \frac{R_f E_{REF}}{R_i}$$

This can be written: $E_e = E_o - G E_{REF}$, where G is defined as the operational gain of the amplifier, being given by: $\dfrac{R_f}{R_i}$.

A still closer look at the error signal

Having first dealt with an ideal amplifier with infinite open-loop gain, the error signal vanished altogether—a beautiful situation! In a more practical mood, it is conceded that real-life amplifiers have high, but finite, gain, in which case the error signal is small, but not zero. Now, in the spirit of nit-picking, another nonideal characteristic can be attributed to real-life amplifiers; to wit, they all have some amount of voltage and current offset. That is, if operated in a simple nonfeedback circuit, an output would exist with no input applied to the amplifier. It is as if an input voltage of E_{io} and an input current of I_{io} were associated with the input terminal of the amplifier.

Taking these "false" error signals into account, a more rigorous expression for output voltage is:

$$E_o = GE_{REF} + E_e \pm E_{io}(1+G) \pm I_{io}R_f$$

Here, $GE_{REF} + E_e$ equates the output voltage in an amplifier with zero offset. The subsequent terms represent contributions as a result of the presence of voltage and current offset and can be responsible for much of the fluctuation in static stability (drift) of the regulated supply. Thus, it is not sufficient to have a near-perfect reference voltage to ensure low drift. An amplifier with excessive offset can produce unacceptable deviations of the absolute output level because offset tends to be temperature dependent and inconsistent.

Ripple

The term *ripple* at one time implied the residue of ac delivered to the load as the consequence of imperfect rectification and filtering. In simple nonregulated supplies, the concept of ripple voltage was hardly controversial. Thus, for full-wave rectification from a 60-Hz ac line, you could expect to find some 120-Hz voltage superimposed on the dc output level. As a result of the imbalance of the rectifying elements or of the center-tapped power transformer, a smaller percentage of 60-Hz voltage could also be detected in the dc output. Finally, harmonics, predominantly odd, were invariably present because of rectification process nonlinearity and saturation effects in the transformer. With half-wave rectification, both odd and even harmonics were present in relatively good strength. In any event, all frequency components that were superimposed on the dc output level were synchronized to the ac line frequency. Moreover, the waveshape of the ripple sufficiently resembled a sine wave so that you could often specify ripple amplitude either in rms or peak-to-peak units without the danger of invoking gross errors when converting measurements.

A new look at ripple

In the regulated supply, superimposed ac on the dc output level presents itself somewhat differently. The regulatory process also confers electronic filtering because ac in the output is acted upon just like any other amplitude deviation. The electronic filtering action is often likened to that of a very large filter capacitor. This is because the electronic filtering retains full effectiveness right down to zero frequency, that is, to dc. The output capacitor in regulated supplies is used to prevent oscillation and to provide low impedance to the higher frequencies where the amplifier gain has fallen. Such frequencies are much higher than the first several harmonics of the ac line frequency. The output capacitor is generally too small to provide any appreciable attenuation of the ripple so far discussed. Therefore, what we have previously termed *ripple* is of much smaller magnitude than generally exists in the output of the nonregulated supply. Also, its waveshape is often considerably distorted.

The main difficulty in dealing with ripple at the output of the regulated supply is that other ac components can be of comparable, or greater, amplitude and might not be synchronous with the ac line frequency. Often, the on-off conduction transition of semiconductor rectifiers contributes high-amplitude spikes of low energy content (i.e., with very high peak-to-rms ratios). Thermal and semiconductor noise (other than switching) constitutes, for practical purposes, *white noise*—the electrical disturbance that is caused by the random movement of free electrons in a conductor. Another source of ac stems from the null-seeking action of the supply as it continuously tries to minimize its error signal. The high amplification that is generally involved in regulated supplies renders the supply susceptible to all manner of electromagnetic interference.

All noise sources shock-excite the null-seeking action so that instead of the simple ripple situation of the nonregulated supply, it is a conglomeration of transients and line-synchronized disturbances. How then do you speak of "ripple" in the more primitive sense of the term? Most certainly, many pitfalls exist. For example, the rms value of narrow, very high amplitude spikes is deceptively low. It would be misleading to use the sine-wave peak-to-rms factor of 1.41 for conversion. Actually, the ratio of peak to rms amplitude of such spikes could easily be 5 to 50 or more. To compound this confusion, some manufacturers have excluded rectification spikes from their ripple specifications (this, of course, does not prevent the supply from destroying the logic elements in a digital system).

Pard

Sometimes the acronym *PARD* (periodic and random deviation) is encountered. Pard is measured under conditions of nominal line voltage and 50% load current with constant operating temperature (in practice, this merely implies the thermal equilibrium attained after an adequate warmup period). Pard is particularly meaningful when it is measured both by an rms-responsive instrument and by an oscilloscope for worst-case peak-to-peak value.

Excluding drift and line and load transients, pard sums the ac effects that are internally generated (Fig. 3-10). Drift is too slow and too unpredictable to be meaningfully included. Line and load transients cannot be consistently tied to the operating behavior of the supply. Also, on similar premises, the effects of electromagnetic interference obviously must be excluded. In pard specifications, the line and load are assumed to be free of transients, drift is too slow to be considered "dynamic" in nature, and no noise from electromagnetic interference is present.

A practical consideration where clean dc is of importance

Although the definition of pard excludes the effects of line transients, in practical applications, the supply can be powered by an ac line that contains considerable

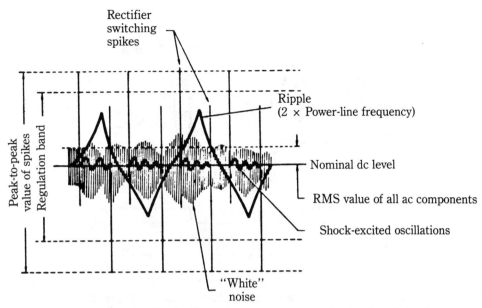

3-10 Concept of *Pard*—all ac components are superimposed on dc level.

electrical noise. Remember when ordering or specifying a regulated power supply that tight line regulation can be very helpful in preventing noise riding on the ac line from reaching the dc output supply terminals.

Transient responses

If a voltage-regulated power supply is to be used with circuitry that generates pulses or spikes, it is not enough to specify low output impedance over an appreciable sine-wave frequency range. Switching transients can prevent realization of the output impedance, which prevails for a passive, "clean" load. Regulated power supplies require a relatively long time to recover from the effects of abrupt load or line changes. Although specification criteria differs among manufacturers, the underlying philosophy is essentially similar. When differences are encountered, they generally involve the percentage deviations that are represented by the considered disturbance.

Referring to Fig. 3-11, a voltage band composes the limits of static voltage excursion that are called out in the specifications. An abrupt change in load current provokes an oscillatory response that is generally characterized by overshoot and undershoot through the limits of the static-regulation band. The severity of the overshoot and undershoot, as well as the number of oscillations that are required to return to the regulation band, varies with different designs. On the one hand, we can have a highly damped response that is characterized by minimal overshoot and undershoot, and by return to the regulation band within a half cycle or less of the shocked oscillation. On the other hand, an underdamped system can exhibit con-

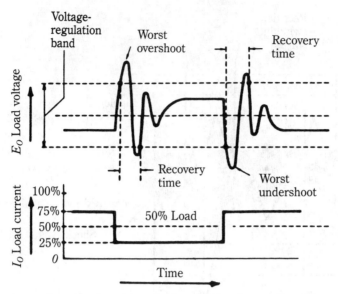

3-11 Concept of transient load regulation for a voltage-regulated supply.

siderable overshoot and undershoot and many cycles of shocked oscillation before the response returns to the regulation band. Damping is achieved at the expense of gain and/or bandwidth. Most power supplies achieve a practical compromise so that some overshoot and/or undershoot occurs and is followed by a half cycle to several cycles of damped oscillation.

In Fig. 3-11, the abrupt changes in load current represent 50% of the rated full-load current. This is where the differences occur among manufacturers' specifying methods. Some use more, some use less than the 50% change that is selected here for purposes of illustration. Obviously, the user should attempt a mental adjustment when comparing the transient response of different supplies in which the load-current disturbances are specified differently. Also, it is always assumed that the line voltage is maintained at its nominal value and that all other variables are held constant.

Figure 3-12 illustrates the concept of transient line regulation. Here, $\pm E$ represents abrupt variations from nominal line voltage. The supply is loaded to 50% of its rated full-load current. All other variables are held constant. Similar, although not necessarily identical, results to those in Fig. 3-11 are produced. Oscillatory characteristics are considerably influenced by the filter capacitor that follows the rectifier and by the output capacitor that follows the regulator. In both load and line transient response, the important parameters are the worst amplitudes of overshoot or undershoot, and the recovery time. Notice that recovery time is not necessarily the time that is required to end the oscillatory response. Rather, it is the time that the output voltage is outside the specified voltage-regulation band.

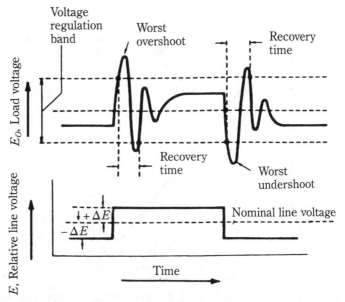

3-12 Concept of transient line regulation for a voltage-regulated supply.

Dynamics of the constant-current supply

By means of analogies, equivalent circuits, reasonable assumptions, and the emphasis of different viewpoints, the objective of this section has been to develop a deeper insight of the dynamic characteristics of the voltage regulator. Most of it has been focused on dynamic output impedance because this parameter is the most important, with respect to the ac behavior of the supply. The same is true of the constant-current power supply, except that in this case, the supply should simulate a very high dynamic output impedance. The advantage of such simulation is quite dramatic when compared to the same result that is achieved by the brute force method of an actual physical impedance. Either it would require an impractically large inductor or a very high resistance that would be backed by an impractically high voltage source.

Figure 3-13 shows the typical output-impedance plot of a current regulator (compare it with Fig. 3-3). The current regulator does not utilize the large output capacitor that is generally associated with voltage regulators. Here, it maintains an output impedance that is as high as possible with increasing frequency. Thus, the other methods, such as lag networks and local feedback paths, must be used to stabilize the phase-gain characteristic of the amplifier. Also, if you desire to retard the fall-off of output impedance with rising frequency, a series inductance can be inserted in an output lead (this is opposite to the function of the shunt capacitor in the voltage regulator).

Figure 3-14 shows how a voltage regulator, such as the one depicted in Fig.

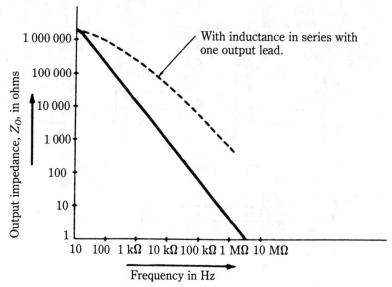

3-13 Dynamic output impedance, Z_o, in current-regulated power supplies.

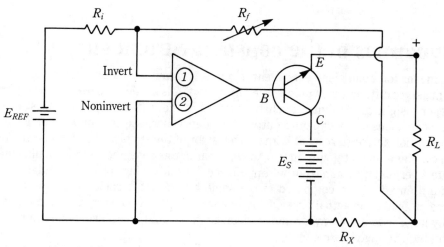

3-14 Feedback technique for obtaining constant load current.

3-9, is modified to achieve constant current through the load, R_L. The modification consists of the addition of current-sampling resistance R_x and the connection for the "sensing" end of feedback resistance R_f. To understand the new mode of operation, remember that R_x is very small compared to R_L. The basic action is still that of a voltage regulator, but now the voltage is stabilized across R_x instead of R_L. The voltage across R_x is proportional to the current through it. Because this current is the same as that in the load, R_L, the regulation of the voltage drop across sampling resistance R_x implies constancy of current through R_L. A change in R_L cannot be accompanied by a change in current through R_L because that would tend to change the voltage drop developed across R_x, the "regulated" parameter (curiously, both voltage and current regulators accomplish their mission by controlling the current that is permitted to flow through the series-pass element; likewise, both regulators accomplish their mission by sensing a voltage. Thus, both types of regulators achieve their respective results by indirect action).

In the current regulator, the voltage across the load is no longer regulated and therefore varies considerably as the load resistance is changed. Indeed, the constant-current supply exhibits poor voltage regulation. This is generally of no consequence because current-activated devices (such as solenoids) are not voltage sensitive. In some instances, protection is required from the effects of rising output voltage and the current-regulated supply must utilize some such technique (as described for Fig. 1-9 of chapter 1) in order to limit output voltage.

Two ways of specifying programming speed (slew rate)

A common way to specify programming speed is by stating the maximum rate that the output voltage can be changed as a consequence of applying the appropriate programming signal. Dimensionally, this rate is expressed as voltage per unit time, $\frac{dv}{dt}$. Usage favors volts-per-microsecond, but other units can be used, such as kilovolts-per-second. The rate thus indicated is said to be in the time domain. This concept is illustrated in Fig. 3-15. A second method of indicating programming speed is by expressing it in the frequency domain, where a sine wave is described at a certain frequency and a certain peak amplitude. This approach is more practical for many applications because the frequency of the output-voltage variation produced by programming is a function of the amplitude excursion that we wish to impart to the output voltage. Small excursions can be made at higher frequencies better than large excursions. This can be seen from the formula which equates the time and frequency domains:

$$\frac{dv}{dt} = 2\pi f E_p$$

where $\frac{dv}{dt}$ is identical to $\frac{\Delta E}{\Delta T}$ in Fig. 3-15.

For example, suppose that a programmable power supply can have its output voltage changed at a rate of 0.5 V per microsecond. If you are interested in slewing the output voltage through a total (peak-to-peak) excursion of 10 V, how fast can this be done with a sinusoidal programming signal?

$$\frac{dv}{dt} = \frac{0.5}{10^{-6}} = 2\pi f(5), \text{ or } f = \frac{0.5 \times 10^6}{10\pi} = 15,915 \text{ Hz}$$

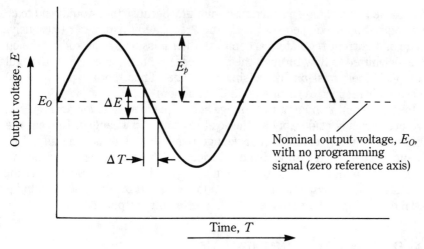

3-15 Programming speed, $\Delta E/\Delta T$, as the slope of a sine wave crossing the zero axis.

If twice the amplitude excursion is desired, the allowable programming frequency is reduced to one-half the above value (7958 Hz).

Conversely, the formula can be solved for E_p. To determine how great an amplitude excursion must be accommodated for a certain sinusoidal programming frequency, f, when $\frac{dv}{dt}$ is known, the algebraic transposition yields:

$$E_p = \frac{\frac{dv}{dt}}{2\pi f}$$

Suppose that $\frac{dv}{dt}$ is specified as 1000 kV per second. It is desired to process a signal at 10 kHz (sine wave). What is the amplitude (E_p) of the 10-kHz variation that is superimposed upon the quiescent dc output level?

Substituting the appropriate values, and solving for E_p:

$$E_p = \frac{1000 \times 1000}{2\pi \times 10 \times 1000} = \frac{100}{2\pi} = 15.9 \text{ V}$$

The safest way to make such calculations is to express $\frac{dv}{dt} = \frac{\Delta E}{\Delta T}$ as so many *volts-per-second*. Then, frequency f will be expressed or derived in Hz.

In the preceding examples, the nominal output voltages of the supplies would necessarily have to be high enough to accommodate the indicated "modulations." In other words, the lower alternation of the sine wave of Fig. 3-15 must not reduce the output voltage below zero as the ultimate limit. Other matters being equal, a power supply with a nominal 100-V output can accommodate a larger excursion of its output level than a supply with a nominal output voltage of 60 V, for example. On the other hand, when programming is expressed as a *rate*, $\frac{dv}{dt}$, such specification is valid without regard to the nominal output voltage of the supply.

Specifications of regulated power supplies

It is now instructive to inspect a few specification sheets that pertain to regulated power supplies. Tables 3-1 through 3-6 provide technical descriptions of randomly selected manufacturers' supplies; remember to establish communications with the maker before arriving at conclusions. For example, three of the makers specify ripple in both rms and peak-to-peak values, and only three of the makers include noise or spikes with ripple. Four makers do not specify dynamic output impedance.

Notice the temperature coefficient specifications—none of them are preceded by a plus or minus sign. In one instance, temperature coefficient is expressed in so many parts per million per degree C. The temperature coefficient for four of these supplies is expressed in so many percent per degree C. Naturally, both methods of

Table 3-1. Boschert regulated power supplies.

	Model			
	OL60	OL80	OL150	OL300
Output power	60 W	80 W	150 W	300 W
Input	100 – 130 Vac 40 – 440 Hz (220 V option available)			
Line regulation	±0.05%		±0.2%	
Load regulation	±2% typical	±0.2% for one output, ±2% for other outputs		
Ripple and spikes	40 mV RMS 100 mV P-P			
Transient response	0.5 mS			
Temp coeff	0.04%/°C		0.01%/°C	
Protection	Fold-back current-limited, short-circuit proof with automatic recovery, reverse-polarity protection, and overvoltage protection.			
Temp rating	– 20 °C to + 55 °C convection cooled 70 °C with forced air.			
Efficiency	70%		75%	
Hold up	Supply will maintain regulation for 16 ms after input power removal at full load.			
Weight	1.5 lb	2 lb	4 lb	4.5 lb
Outputs	4 output standard up to 300 V.			

Courtesy Boschert Associates

Table 3-2. Datel regulated power supplies.

	Modular-line operated	
	Single output Model no. UPM-5/350*	Dual output Model no. BPM-15/60A*
Outputs		
Output voltage	5 Vdc	±15 Vdc
Output voltage accuracy	±0.5%	±0.5%
Output current	350 mA	±60 mA
Current limiting	Yes	Yes
Output capacitor (internal)	20 μF	6.8 μF
Inputs		
Input voltage	115 Vac	115 Vac
Input voltage range	±10 Vac	±10 Vac
Frequency range	50–420 Hz	50–420 Hz
Maximum input current	NA	NA
Isolation		
Resistive coupling	100 MΩ	100 MΩ
Capacitive coupling	100 pF	100 pF
Breakdown voltage	300 Vdc	300 Vdc
Regulation		
Line regulation	0.05%	0.05%
Load regulation	0.05%	0.05%
Temperature coefficient	0.005%/°C	0.005%/°C
Noise and ripple (RMS)	2 mV	1 mV
Output impedance		
@10 kHz	150 mΩ	150 mΩ
Transient response	50 μs max	50 μs max
Thermal		
Operating temperature	0 to 70°C	0 to 70°C
Storage temperature	−55 to 85°C	−55 to 85°C
Heatsinking requirements	NA	NA
Physical		
Case material	Diallyl phthalate/epoxy	Diallyl phthalate/epoxy
Case/pin configuration	G2	12
Weight	1.5 oz	1.5 oz
Mating socket	DILS-1 or -2	DILS-1 or -2

*Note: (1) Above 35°C (95°F) mounting surface temperature, derate 1.3 mA/°C

Engineering note: Although heatsinking is not required to meet specifications over the ambient temperature range, good engineering practice should include heatsink provision to avoid "hot spots" within the final equipment.

Courtesy Datel Systems, Inc.

Table 3-3. Fluke high-voltage regulated power supplies.

	Model			
	408B	410B	412B	415B
Output voltage	0 to ±6000 V	0 to ±10,000 V	0 to ±2100 V	0 to ±3100 V
Calibration accuracy 90 days, 25°C ±5% C	±0.25% of setting or 250 mV, whichever is greater		±0.25% of setting or 100 mV, whichever is greater	

Specifications of regulated power supplies

	Model			
	408B	410B	412B	415B
Output current, max Line (±10% line change)	20 mA	10 mA ±0.001% or 2 mV, whichever is greater	30 mA	30 mA ±0.005% or 2 mV, whichever is greater
Regulation Load (full load change)	±0.001% or 5 mV, whichever is greater			±0.005% or 2 mV whichever is greater
Stability Hour Day	±0.005% ±0.02%			±0.002% ±0.01%
Ripple RMS Peak-to-peak	1 mV 5 mV		0.5 mV 1 mV	0.1 mV 1 mV
Resolution	5 mV	5 mV	5 mV	5 mV
Resetability	±0.05% of setting or 50 mV, whichever is greater			
Overcurrent trip (Internally adjustable)	5 to 25 mA	5 to 15 mA	5 to 40 mA	5 to 40 mA
General specifications				
Output polarity	+ or − grounded via front-panel switch			
Temperature coefficient	< 20 ppm/°C, 10°C to 40°C			
Operating temperature	0°C to +50°C			
Storage temperature	−20° to +70°C			
Humidity	up to 80% RH			
Altitude Operating Nonoperating	0 to 3048 M (0 to 10,000 ft) 0 to 15,240 M (0 to 50,000 ft)			
Shock and vibration	MIL-T-28800 Class 5 Style-E Equipment			
Input power	115/230 Vac ± 10% 50–60 Hz; 300 VA at full load			
Output mating connector, One supplied	Front and rear, MS3102A-18-16S		Front and rear, UG931/U	
Dimensions	cm in	cm in	cm in	cm in
Height	22.2 8.75	22.2 8.75	8.9 3.5	8.9 3.5
Width	48.2 10	48.2 19	48.2 19	48.2 19
Depth	38.1 15	38.1 15	38.1 15	38.1 15
	kg lbs	kg lbs	kg lbs	kg lbs
Weight	26.76 59	26.76 59	12.70 28	13.64 30

Courtesy John Fluke Mfg. Co., Inc.

Table 3-4. KEPCO regulated power supplies.

Model	dc output range Volts	dc output range mA	Output impedance Voltage mode dc ohms + series L	Output impedance Current mode dc ohms + shunt C	Ship wt lbs.	Ship wt kg.
APH 500M	0–500	0–40	0.625 Ω + 10 μH	125 MΩ + 6.0 μF	27	12.3
APH 1000 M	0–1000	0–20	2.5 Ω + 20 μH	500 MΩ + 3.0 μF	27	12.3
APH 2000M	0–2000	0–10	10 Ω + 40 μH	2000 MΩ + 1.1 μF	27	12.3

Influence quantity		Output effects Voltage mode*	Output effects Current mode (internal sensing)	Uncommitted amplifier offsets ΔE_{io}	Uncommitted amplifier offsets ΔI_{io}	References ±6.2 V (1 mA max)
Source: 105–125 Vac		<0.001%	<0.005% or 0.2 μA‡	<5 μV	<1 nA	<0.0005%
Load: No load—full load		<0.005%	<0.01% of I_o max	—	—	—
Time: 8-hour [drift]		<0.01%	<0.01% of I_o max	<20 μV	<1 nA	<0.005%
Temperature: Per °C		<0.01%	<0.01% of I_o max	<20 μV	<2 nA	<0.005%
Unprogrammed**	rms	<1.0 mV	<40 μA	—	—	—
Output Deviation: (Ripple and noise)	p-p†	<5.0 mV	<200 μA	—	—	—

*Specifications are expressed as a percent-of-setting for the output range 10% to 100%. Below 10% output, the specification limit is the rated percentage of the 10% output setting.

**One terminal grounded or connected so that the common-mode current does not flow through the load or (in current mode) through a sensing resistor.

†20 Hz to 10 MHz.

‡Whichever is greater.

Courtesy KEPCO, Inc.

Table 3-5. Lambda 100-volt regulated power supplies.

Model	Regulation (line, load)	Ripple (RMS)	Max amps at ambient of 40°C	Max amps at ambient of 50°C	Max amps at ambient of 60°C	Max amps at ambient of 71°C	Pkg size	Dimensions (inches)
LCS-A-100	0.01% + 1 mV, 0.01% + 1 mV	250 μV	0.18	0.18	0.18	0.18	A	3³/₁₆ × 3³/₄ × 6¹/₂
LCS-B-100	0.01% + 1 mV, 0.01% + 1 mV	250 μV	0.46	0.46	0.46	0.34	B	3³/₁₆ × 4¹⁵/₁₆ × 6¹/₂
LCS-C-100	0.01% + 1 mV, 0.01% + 1 mV	250 μV	0.65	0.65	0.65	0.65	C	3³/₁₆ × 4¹⁵/₁₆ × 9³/₈
LM-D-100	0.05% + 4 mV, 0.03% + 3 mV	1 mV	1.70	1.50	1.30	1.10	D	4¹⁵/₁₆ × 7¹/₂ × 9³/₈

Courtesy Lambda Electronics Corp.

Table 3-6. Power/Mate regulated power supplies.

Model with MM, MD, or MT Prefix

Input voltage: 115 ± 10 Vac, 50–400 Hz.

Output voltage: Up to 28 Vdc.

Output current: Up to 2 Amps.

Line regulation: ±0.05% up to 6 Vdc. ±0.02% 10 Vdc and above.

Load regulation: ±0.05% up to 6 Vdc, ±0.02% Vdc and above.

Ripple: Less than 500 microvolts RMS up to 6 Vdc. Less than 1.0 millivolt RMS 10 to 18 Vdc. Less than 3 millivolts RMS 20–28 Vdc.

Polarity: Positive or negative with respect to ground, or "floating" up to 300 volts.

Overshoot: No voltage overshoot on turn-on, turn-off, or power failure.

Temperature coefficient: 0.02%/°C Max.

Ambient operating temperature: 0–71 °C.

Storage temperature: −25 °C to +85 °C.

Output current vs. temperature: Each output is rated for full current at temperatures between −20 °C and +40 °C and is linearly derated from +40° to 60% of the full output at 71 °C.

Overload and short-circuit protection: Built-in. The output can be shorted without causing damage to the power supply. However, prolonged operation with the supply overloaded or shorted is not recommended.

Precision output adjust: The output voltage of each supply is factory set to within ±2% of the specified voltage. The TRIM connection, where provided, allows a means of externally adjusting the output voltage by placing a resistor between the TRIM pin and the positive output to lower the voltage, or between the TRIM pin and the negative output to raise it.

Courtesy Power/Mate Corp.

specification are associated with operation at rated output, but exceptions do exist. For example, makers often avoid specifications that pertain to extremely light loads. Therefore, it does no harm to ask about the temperature range that the temperature-coefficient specification pertains to. Notice that one maker does not cite the temperature coefficient.

Three of the manufacturers make direct mention of transient behavior. Two makers cite stability in terms of *output drift*. Here, one specifies maximum drift for one hour and for one day, whereas the other sets a limit for eight-hour operation. Experience shows that it is unwise for the user to extrapolate such data if his drift requirements are other than the specified time period.

Two of the manufacturers specify line and load regulation as a percentage that is preceded by a plus or minus sign. Another manufacturer specifies regulation as a plus or minus percentage or as a fixed number of millivolts—"whichever is greater." The line and load regulation for one of the supplies is given as a percentage plus a fixed number of millivolts. One manufacturer specifies regulation only as a percentage, but the specifications for one of the power supplies does not include regulation.

For one product, the manufacturer explicitly informs that, although no heat-sinking is required to meet specifications over the ambient temperature range, it nonetheless constitutes "good engineering practice." Another maker informs that no cooling is required over a certain ambient temperature range, which is quite extensive and is likely to encompass the majority of installations. With a third product, the options of using convection or forced air depends on the operating temperature range. For the remaining products, you should inquire further to ascertain what, if any, special techniques of heat removal are required for a given application.

The family of regulated supplies from one maker is intended for use in either the voltage-stabilized or the current-stabilized mode. For the latter mode of operation, it is refreshing to see the output impedance specified as so many hundreds of megohms plus so many microfarads. It is just as important for the dynamic impedance of the current regulator to be high as it is for the dynamic impedance to be low in voltage regulators. Unfortunately, this same maker neglects to state the compliance voltage that pertains to current regulation (the compliance voltage can probably be inferred from the output voltage of a power supply when operating as a voltage regulator, but it is preferable to provide this information directly).

These six regulated power supplies are of diverse design and characteristics. It is irrelevant to compare their specifications, because their application areas involve different philosophies of merit. Their specifications have been presented in order to emphasize the need for clarity and understanding in all cases. All of these supplies are quality products—it is primarily up to the user to determine which type is most economical and applicable to his or her particular needs.

EMI and RFI

EMI is electromagnetic interference and *RFI* abbreviates radio-frequency interference. Previously, these electrical noise propagations were considered to involve similar manifestations. The tendency now is toward the use of the term EMI. With regard to power supplies, EMI is generally inferred to include all self-generated noise that "escapes" from the power supply and pollutes the electrical environment. Some controversy exists as to whether this should also include output ripple. Inasmuch as higher harmonics of the ripple frequency often penetrate the load equipment via electromagnetic fields, you could consider EMI as any ac signal or transient that is impressed on the load by the regulated power supply.

The inherent EMI of the switching-type power supply has been long overemphasized and the possibilities for electrical noise generation in linear supplies has been underemphasized. Such attitudes stem, on one hand, from early construction techniques for switching power supplies, where it often appeared hopeless to significantly reduce the injection of transients and harmonics into the environment. On the other hand, many have succumbed to the naive notion that all functional blocks in the linear supply operate in a smooth manner.

The linear power supply can be the source of troublesome switching transients. A clue to such behavior is in the pard diagram of Fig. 3-10, where rectifier switching spikes contribute to the pollution of the dc output. The pn diodes display properties other than the intended rectifying action, where alternate halves of the impressed ac wave are passed and blocked. The common picture of pn diodes performing rectification is suitable simplification for many, but not all, purposes. Such diodes store electric charge in their junctions and bulk material during forward conduction. As a result, the diode conducts reverse current for a brief, but not insignificant, time after the impressed voltage wave has crossed its zero axis. Discharge of this stored energy impairs rectifying action and generates switching transients.

The behavior of pn diodes in rectifying circuits is shown in Fig. 3-16. Notice that various reverse-recovery characteristics are possible. The area and shape of

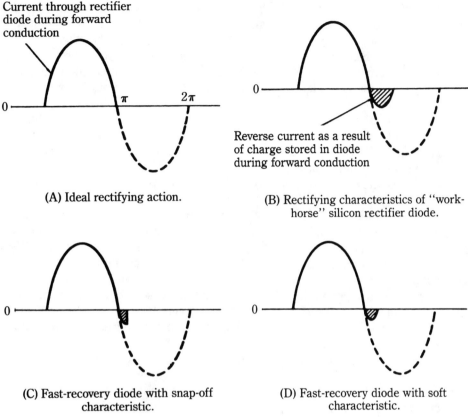

3-16 Charge-storage effects in pn rectifying diodes.

the recovery region are governed by the way the manufacturer processes the semiconductor material. For 60-Hz circuits, not much degradation of rectifier efficiency results; however, at 20 kHz, a "workhorse" diode might run hot, produce excessive ripple current, and adversely affect the operation of the regulator circuitry. Although fast recovery is needed for high-switching rates, it is not always wise to push this too far because the snap-recovery characteristic often results in inordinately high production of radio-frequency components.

One way in which silicon pn diodes are processed for control of reverse-recovery characteristics is by *gold-doping* the semiconductor material. However, certain trade-offs must be contended with. For example, the recovery speedup thereby attained is at the expense of increased forward voltage drop, as well as the higher leakage current.

For 60-Hz operation, you can simply select the soft characteristic in a fast-recovery diode. Here, you are concerned primarily with considerations of noise generation. At 20 kHz, however, both rectification efficiency and the spectrum of the rectification noise are involved. It is often helpful to experiment with different diodes in order to avoid unfavorable excitation of offensive noise frequencies. This will vary for different power supplies because of differences in both circuitry and in construction techniques. Sometimes different diodes of the same type will display significantly different noise spectrums—this might or might not be correlated by visual inspection of the recovery waveform. However, modem construction techniques that feature compact layout, small components, and printed-circuit boards are vastly superior to older equipment, where long interconnections often functioned too effectively as "tank" circuits (and antennas), thereby enhancing the EMI from pn rectifying diodes. In general, the smaller the area that is encompassed by the reverse-recovery characteristic, the less energy is available for the production of EMI from rectifier switching transients. Thus, the fast-recovery diodes are often advantageously used—even in 60-Hz service.

Where feasible, the use of Schottky hot-carrier diodes can help avoid such rectifier switching transients. This type of semiconductor rectifier does not involve the storage of minority charges and, therefore, switches cleanly up to very high frequencies. Indeed, the high-frequency limitation is ultimately imposed only by the internal capacitance of these diodes; for all practical purposes, rectification is both "clean" and efficient up to at least 100 kHz in power circuits. These diodes, however, do not have the voltage capability of the pn junction type. They have attained considerable popularity in 5- to 15-V systems where they also outperform silicon pn diodes in the matter of low forward-voltage drop (in this respect, they combine advantages of present silicon and formerly popular germanium rectifying diodes).

Although the gold-doped silicon diode has proved to be a good performer in 20-kHz regulators, trade-off compromises sometimes have been troublesome in the parameters for avalanche voltage, forward voltage drop, and leakage current. New semiconductor processing techniques are now coming into prominence, and they hold considerable promise for desirable reverse-recovery characteristics with relatively little compromise in other important rectifying features. Included in these sophisticated production methods are ion implantation, and electron and gamma irradiation. At this writing, such diodes are already available in 100-A ratings and with recovery speeds in the neighborhood of one-tenth of a microsecond. Unlike the gold-doped diode, these recently developed units actually exhibit much lower forward-voltage drops than do "workhorse" diodes.

Packaging the power supply in a metal case also keeps the supply from polluting the environment with electromagnetic radiation. Many open-frame supplies have appeared on the market. The radiated electrical noise from such equipment cannot be meaningfully specified. To be sure, the manufacturer usually implements reasonable measures, such as compact structure and the individual shielding of some components. You, however, must be alert to the vulnerability of the load to noise and should be prepared to exercise caution. Metal partitions and load shields are recommended, and the supply and load should be physically separated.

The ac power line can both conductively transfer and electromagnetically radiate electrical noise. The effects of conductive transfer can be reduced or eliminated by inserting an appropriate low-pass filter in the incoming ac line at the load. However, the appearance of switching transients on the ac power line often results in reradiation and this problem is not so readily alleviated. Then, RFI scatters "all over the place" and even renders difficult the interpretation and operation of measuring instruments, such as oscilloscopes and digital voltmeters. It is obviously good to prevent switching noise from entering the power line. Quality switching power supplies have built-in low-pass filters in their power-line circuits. The inductor is usually a bifilar-wound choke with a ferrite core. Thus, a series inductor is in each side of the incoming power line. The bifilar arrangement provides additional features, such as preventing core saturation by the 60-Hz current and providing greater effective inductance per turn than would be the case with separate chokes. Because these filters are often designed to attenuate a switching frequency of 20 kHz and higher, the capacitors are electrically and physically small compared to the filter capacitors in a 60-Hz power supply.

These line filters provide increasing attenuation up to 5 or 10 MHz. Higher radio frequencies often find sufficient stray capacitance to bypass the filter and make their way to the power line. If these higher frequencies are harmonics of relatively "clean" switching, their strength might not be of consequence. If, on the other hand, they are generated by the snap-off action that is caused by the charge-recovery characteristics of diodes, a surprising amount of high-frequency energy can appear on the power line (sometimes switching transistors are the culprits—if so, it will often be found that they also have snap-off charge-recovery characteristics). Where a slight loss in operating efficiency can be tolerated, it is sometimes beneficial (from the standpoint of noise reduction) to slow down the switching waveform so that it is nearly a trapezoidal shape. Unfortunately, you cannot go very far in this direction before the most important feature of the switching process, *negligible dissipation*, is lost. In addition to the described line filter, it is often beneficial to install small RF chokes in each lead of the ac line. These chokes attenuate the higher frequency energy that manages to bypass the larger filter network.

Although this section focuses on rectifier circuits because this function block is common to both linear and switching-type supplies, the free-wheeling diode in the switching regulator is also a notorious source of EMI. This diode should have a faster recovery than the turn-off time of the switching transistor. Otherwise, the reverse current of this diode will be superimposed on the current that is demanded from the switching transistor. This degradation of overall efficiency tends to increase EMI production. Even worse, a slow-recovery diode in this circuit is often the cause of "mysterious" failures of the switching transistor. Of course, the mystery is dispelled by the realization that the switching transistor is burdened with a higher peak current than it was rated for.

Figure 3-17 depicts some of the routes whereby EMI can be transferred from the power supply to load. Ground loops and chassis currents (A) are very much a function of the way that common connections are made within the supply, the load equipment, and in external grounding. Minimizing this coupling medium generally requires experimentation during breadboarding, final design, and installation. Although it is generally best to strive for single-point grounding techniques, some important exceptions exist. One of these involves the free-wheeling diode in a series-type switching regulator. Usually, the "ground" connection for this component should be physically associated with that of the input filter capacitor, and not with that of the output filter capacitor.

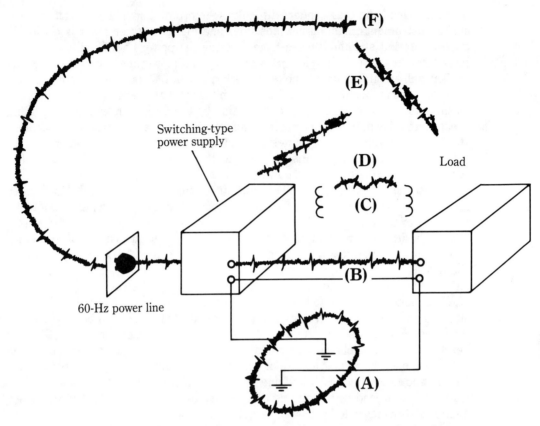

3-17 Transference of electrical interference from supply to load.

The transfer of transients via superimposition on the output leads (B) represents the ripple or pard characteristics of the supply. As previously stated, many loads will not be adversely affected by this source of electrical noise. Filtering is directly applicable here, but you must remember that an LC filter that is not included within the feedback loop tends to degrade regulation. In many practical situations, adequate filtering is provided by filters that do not affect regulation. Included are feedthrough types and small capacitors (0.1 to 1.0 μF). Such capacitors, usually mylar or ceramic types, attenuate high-frequency components that are not affected by the large electrolytic capacitors because of their internal resistance and inductance.

Sometimes inductive coupling of switching currents is between the supply and the load (C). Inductive coupling is best minimized during the design stage by selecting toroidal cores, or other core configurations with relatively low external magnetic fields, for transformers and inductors. Ground-loop and chassis currents also transfer by inductive coupling as well as by direct conductive paths. When inductive effects are severe, it should be helpful to enclose the power supply in a steel box, rather than in aluminum. Frequently, a practical compromise must be achieved with thermal requirements because excessively large ventilation holes or slots can enable electrical radiation to "escape."

Yet another source of interference might be in the rectifying circuits. It is common knowledge that the half-wave current pulses in any rectifying diode contain many harmonics of the power-line frequency. These harmonics are relatively low in energy content and, except for very large power supplies, interference at high radio frequencies might not be incurred. However, under conditions that are now prevalent with regulated power supplies, interference in the audio frequency range, as well as in the ultrasonic and low radio frequencies, is a distinct possibility.

Figure 3-18A is the common depiction of current in a rectifier diode. This assumes a simple resistive load, one rarely encountered in practical power supplies. Before the advent of regulated power supplies, choke input filters (Fig. 3-18B) were more common at 60 Hz than is now the case. In theory, the rectifying diode that is connected to a load with an infinite inductance should experience a unidirectional current with a rectangular waveform. Such current waveforms are very rich in high-energy harmonics over an extended frequency range. In practice, harmonic generation was usually not troublesome from this source because of the finite inductance and inherent resistance that is always associated with any practical choke. The diode current waveform is actually a rounded trapezoid (Fig. 3-18C) relatively free of harmonics—especially with respect to high frequencies.

The modern trend in unregulated power supplies has been to use bridge rectification in conjunction with a large single filter capacitor (Fig. 3-18D). At first, such circuits were implemented with a small resistance to protect the diodes from the capacitor charging current. In the relentless quest for greater operating efficiency, this protective technique is now used to a lesser degree than before. This usage is feasible because of the excellent peak-current and surge ratings of modern rectifier diodes. However, the waveform of Fig. 3-18D becomes narrower and more peaked. Increased harmonic generation and more EMI might be the price paid for such simplified circuitry.

An interesting aspect of switching-type power supplies is that their relatively noisy dc outputs are not a problem with many types of loads. Although it is generally unwise to use a switching-type supply to provide operating power for a communications receiver or for inordinately sensitive instrumentation purposes, the most common applications (computer and logic circuits) tend to tolerate a reasonable combination of ripple and switching transients riding on the dc output bus. Such loads are themselves the source of considerable switching noise. So, a certain amount of immunity to pulses on the power-supply lines is designed into digital loads. How, then, is this reconciled with the well-known fact that logic circuits often are subject to false-triggering when a switching regulator is associated with the system?

This phenomenon is usually caused by the effects of EMI reradiated from the power line, and by conduction and induction of EMI via ground loops. When EMI enters logic circuits by these routes, flip-flops and other digital circuits can be more vulnerable to malfunction than they are from the presence of spikes on the dc supply lines. More than anything else, the antenna effect of transient-laden ac power lines has tarnished the reputation of the switching-type regulator. Once

140 *Dynamic characteristics of regulated power supplies*

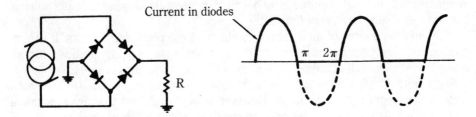

(A) Half sine waves from full-wave rectifier circuit feeding into resistive load.

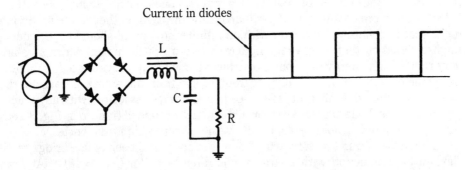

(B) Infinite-inductance choke input filter produces harmonic-rich square waves.

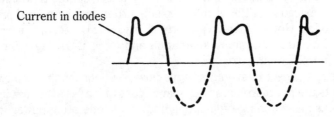

(C) Choke input filter with practical inductance produces less-offensive waveform than (B).

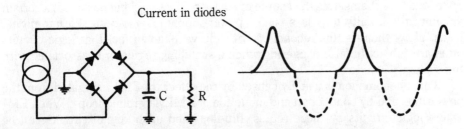

(D) Capacitor input filter can produce narrow current pulses and many strong harmonics.

3-18 Diode current waveforms in various rectifier circuits.

appreciable energy, in the form of switching transients or radio frequencies, has contaminated the ac feeder lines, the battle against EMI suffers a serious setback. On the other hand, if suitable precautions have been taken to prevent or reduce EMI on the ac line, it is surprising how "quiet" the operation of a switching-type supply can ultimately be made. Other EMI routes are then readily blockaded by such remedies as shielding, filtering, bypassing, etc.

The output capacitor as a contributor to EMI

The output filter of a typical switching regulator resembles, in configuration at least, filters that are associated with simple unregulated supplies. Thus, you might suppose that excessive ripple or noise could be drastically reduced by merely making either the inductor or the capacitor larger, or perhaps both. However, design constraints having to do with the switching frequency, peak current, and magnetic saturation restrict freedom in how high an inductance can be accommodated. The size of the output capacitor is also tied to the switching rate, but because no electrostatic counterpart of magnetic saturation exists, it could appear allowable to "fudge" a bit by piling on additional capacitance in order to achieve greater attenuation of unwanted frequencies and noise. In regulating circuits that are driven, rather than being self-oscillatory, this can sometimes be done to a degree. However, you soon will encounter another obstacle—the nature of the electrolytic capacitor at high frequencies.

First, these capacitors do not maintain their 60- or 120-Hz capacitance at high frequencies. At the popular 20-kHz switching rate, the effective capacitance has already dropped considerably from lower-frequency values. At 100 kHz and beyond, you can easily deceive yourself in estimating the amount of capacitance that is "seen" by the higher frequency components of noise and switching rate. Often, a remedy to this situation is to parallel a relatively small nonelectrolytic capacitor. Paralleling additional electrolytic units via a "brute-force" approach is generally not satisfactory because of the *effective series resistance (ESR)* and *effective series inductance (ESI)* of the capacitors. When paralleled filter capacitors are seen in a commercial switching supply, the intent of the designer usually has not been to simply increase the capacitance, but rather to reduce the ESR and often the ESI. In so doing, whatever the effective capacitance thereby achieved, it is better able to attenuate ripple and noise when completing its circuit path through lower ESR and ESI. This technique is often more cost-effective than ordering capacitors with inordinately low ESR and ESI parameters (paralleling is also used to get more current-carrying capability for ripple current).

Figure 3-19 shows the equivalent circuit of an electrolytic capacitor, together with commonly encountered impedance characteristics. Up to approximately 10 kHz, the behavior of the 1600-μF capacitor is ideal in the sense that its capacitive reactance is a straight line with constant slope (a plot of $X_c = \dfrac{1}{2\pi fC}$ on log-log graph coordinates conveniently assumes this form). Notice, however, that the continuance of the plot does not follow along the dashed line. Rather, a "V" pattern is

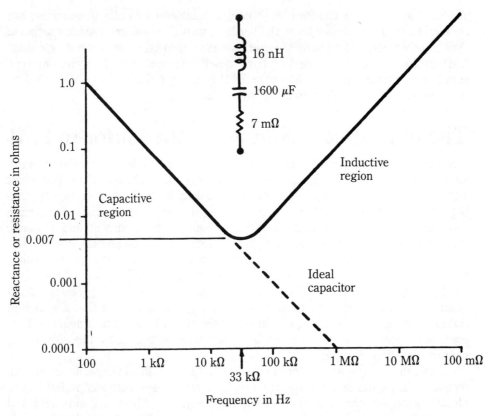

3-19 Electrolytic-capacitor equivalent circuit and impedance characteristic.

traced out. This is not surprising considering that the equivalent circuit is that of a low-Q series-resonant network. The equivalent inductance is caused by the physical construction of the capacitor, including, of course, its leads. The equivalent series resistance is primarily derived from the nature of the electrolytic dielectric. At the resonant frequency (in this case, approximately 33 kHz), the effective impedance is neither capacitive nor inductive, but is resistive. Resonance corresponds to the horizontal portion of the curve and it is here that ESR is specified. As can be seen for this capacitor, it is about 0.007 Ω.

From Fig. 3-19, you can see that the attenuation of noise frequencies above 33 kHz will not be as good as would be the case for an ideal capacitor, or at least for one with zero inductance. The fundamental ripple frequency would be of lower amplitude if the ESR could be made lower. Because these "stray" capacitor parameters plague all real-world electrolytic capacitors, design is often both an art and a compromise. The designer, unfortunately, cannot sample feedback voltage directly across the equivalent series resistance, so it is difficult to appraise the phase-gain characteristic of the feedback loop. It is often found empirically that the amount of error-voltage amplification that would be desirable in order to reduce ripple and noise cannot be incorporated because of instability and oscillation. And, it is very difficult to reduce the stray inductance beyond a certain point. Remember that cheap capacitors and sloppy wiring or connection techniques can aggravate both ripple and noise in the switching supply. In practice, low filter-capacitor ESR might cause loop instability. A proper design and a good layout should, in the interest of noise reduction, accommodate arbitrarily low values of ESR; high ESR treats the symptom, not the cause of instability.

Noise benefits from synchronization of regulator switching rate

The switching regulator necessarily generates greater output ripple than does the linear regulator (often about an order of magnitude more). The problem might arise with certain sensitive equipment—even after all ordinary precautions have been taken via filtering, shielding, and isolation. This need not necessarily doom the switcher, for sometimes immunity to the effects of ripple can be obtained by synchronizing the switching rate to one of the important operating frequencies within the critical equipment. For example, in a video display system, the beat frequency and noise problems, which otherwise might exist, tend to disappear if the horizontal deflection frequency is used to synchronize the switching rate of the regulator.

This is readily accomplished by adjusting the free-running switching rate to be somewhat lower (perhaps 15%) than the horizontal deflection frequency. Then, a sampled signal from the deflection circuit is applied to the RC network of the oscillator in the regulator. The oscillator frequency can, by this technique, be locked in step with, or synchronized to, the deflection frequency. In many instances, good results will also be obtained when the synchronizing signal frequency is approximately two or three times that of the oscillator. That is, the regulator oscillator or clock can be stably locked to 50% or 33% of the synchronizing frequency.

From a noise or interference standpoint, any of such locking arrangements might be better than allowing just a random oscillator frequency. Experimentation is generally needed to determine the best situation for a particular system and supply. Synchronization does not harm the performance of the regulator. Interestingly, at least one manufacturer of pulse-width-modulation IC modules recognizes the possible benefits that might be forthcoming from synchronization. The Signetics NE5560N IC features a special terminal (9) for the application of a TTL-level synchronizing signal. Other manufacturers of similar modules will undoubtedly adopt this scheme as the benefits of synchronization become more widely appreciated. In the meantime, a little experimentation is usually sufficient to attain synchronization where a special terminal is not provided in the control module.

In a regulator-control IC, such as the Motorola 3420 or 3520, the ramp-in pin (terminal 5) is a good circuit juncture to inject the synchronizing signal. When this IC is so deployed, the usual connection between ramp-out (terminal 8) and terminal 5 is not made. Synchronization most readily occurs when the free-running rate of the internal ramp generator is made slightly lower than the frequency of the synchronizing signal (external resistance and capacitance set the ramp rate).

Some notes on shielding

The basic idea behind shielding is to cause an interfering electromagnetic wave to impinge upon a conducting medium with an effective impedance much lower than that of the wave itself. In such a situation, part of the energy in the wave will be reflected, part will be absorbed; only a small part, hopefully, will pass through the shield. Perfect shielding is impossible to obtain. Invariably, the best results are with mutual shielding, where both the source and the sensitive circuits are shielded. Also, the effectiveness of shielding techniques depends considerably upon the aforementioned impedances.

Electromagnetic waves in free space all have an effective impedance of approximately 377 Ω, provided that monitoring, detection, or measurement is made far from the source of such waves. "Far" in practice is 10 or more wavelengths distant. If, however, the interference is from electromagnetic waves that are a fraction of a wavelength from the source, a different situation prevails. Whereas the "far-field" impedance of 377 Ω pertains to the "radiation field," the "near-field" ("induction field") impedance depends on how the energy is generated. If generated at high voltage and low current, the impedance is high and the EMI is said to owe its existence to an *electric field*. If the converse is true, the EMI is then primarily derived from a *magnetic field*. Although these two terms are only relative, the effectiveness of shielding for the EMI that they designate depends somewhat on different approaches.

The so-called electric field is readily reflected by relatively thin metallic walls or partitions. Even sprayed or painted partially conductive surfaces can effectively reflect such waves. Although reliance is primarily based on reflection, absorption increases with both frequency and thickness of such "electric" shielding. A too-thin conductive coating on plastic might not be adequate when the frequency of a powerful EMI electric field is low.

Next, consider the induction field of an EMI source in which the energy is produced by the action of low voltage, but high current. Shielding against this magnetic field requires a different approach; for now shielding reliance must be primarily premised upon absorption, rather than reflection. This calls for thick conductive walls or partitions, because such magnetic or low-impedance EMI penetrates, rather than reflects from conducting surfaces. In practice, it is not always feasible or economical to provide sufficiently thick shielding. In such cases, additional attenuation can be obtained from magnetically diverting the flux in magnetically permeable sheets or enclosures. Unfortunately, highly permeable materials saturate much more easily than less permeable materials. Saturation renders the magnetic bypassing ineffective, for then the permeability of magnetic materials is reduced to a very low value—theoretically that of air (unity).

From the above considerations, it generally is true that if an enclosure effectively shields against the lowest frequencies present when EMI emanates from a low-impedance magnetic source, such an enclosure will likely provide effective attenuation of all higher frequencies, regardless of the nature of their sources. An exception might occur, however, when the enclosure has holes or openings in it.

In switching power supplies, most of the electromagnetic energy that surrounds the supply is of the "magnetic" mode and is at the switching frequency (20 kHz or higher). However, the EMI that disturbs other equipment is probably at some higher harmonic of the switching frequency and is commonly in the RF spectrum of one megahertz to hundreds of megahertz. This energy responds well to shielding, but it often "leaks" from the supply through cooling vents, control openings, windows, mating surfaces, meters, doors, panels, and other cabinet apertures. Conductive gaskets, conductive plastics, netted wire mesh, and numerous coatings and caulking compounds are widely advertised as remedies.

Attenuation of noise spikes with ferrite beads

At initial thought, it might seem that the large electrolytic capacitors used in the output filter system of switching regulators would effectively bypass high-frequency transients and noise spikes. Real-world capacitors cannot do this because of their internal resistance and inductance (ESR and ESI). The ESI, especially, permits noise spikes to appear in the output, no matter what the capacitance rating is. Often, a small ceramic or mylar capacitor that is connected across the filter capacitor provides some degree of bypass action for high-frequency noise. An even more effective technique is shown in Fig. 3-20. Here, a ferrite-bead "inductor," in conjunction with a ceramic capacitor, form an LCR network for attenuating high-frequency energy. The network resembles both a low-pass filter and a wave trap. It operates in a somewhat different manner from either the ordinary low-pass filter or the conventional wave trap.

The "R" of the LCR network is actually the core loss in the ferrite bead. Although core losses in selected ferrite materials are low in the 20- to 100-kHz region, these losses can mount up in the megahertz frequencies where much of noise energy resides. Besides, manufacturers have processed special ferrite materials that are purposely made lossy at higher frequencies. Not only that, but the permeability of such material goes down with frequency. These two effects prevent any sharp resonant effects within the usually offending noise spectrum. To high-frequency noise, the network actually appears as an RC low-pass filter, but without the dc voltage drop that necessarily would stem from use of a physical RC filter. This noise-suppression network can be called an *absorptive low-pass filter*. Such nomenclature pays homage to the circuit configuration and to the dissipative function of the network.

Figure 3-20B is an approximate equivalent circuit of the ferrite-bead network. Unlike the ordinary low-pass filter, where every effort is made to keep dissipation in the inductor at a negligible level, the contrary is true with the ferrite-bead network; the noise energy is absorbed in the equivalent resistance of the ferrite material, itself.

Figure 3-21 shows typical lossy ferrite beads that are intended for use as absorptive filters in suppressing parasitic oscillations, EMI, and noise transients. Where extended low-frequency attenuation is needed, the multiple hole beads can be used. Another technique consists of stringing two or more single-hole beads on a single conductor. Sometimes the volume resistivity of the ferrite material used in

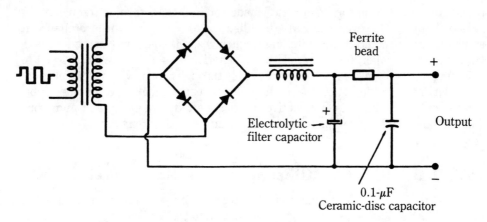

(A) High-frequency energy is dissipated in an absorptive low-pass filter comprised of a ceramic capacitor and a lossy ferrite-bead "inductor."

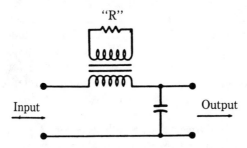

(B) Approximate equivalent circuit of the ferrite-bead network. The absorptive feature is clearly evident here.

3-20 An effective method of attenuating the energy in noise spikes.

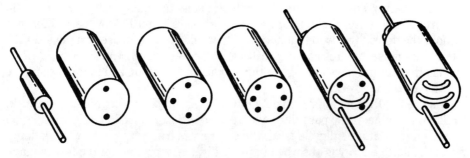

3-21 Examples of lossy ferrite beads and wire thread-through techniques.

these beads is relatively low. Such beads should be insulated or spaced so that they do not simultaneously touch two or more wires. Table 3-7 lists the magnetic and other characteristics of a family of lossy ferrite beads. The comparative attenuation characteristics of these beads are depicted in the chart of Fig. 3-22 (the imped-

Attenuation of noise spikes with ferrite beads

Table 3-7. Properties of lossy ferrite beads.

Property		64 material	43 material	73 material	75 material	77 material
Initial perm		250	850	2500	5000	1800
Maximum perm		375	3000	4000	8000	6000
Saturation flux density	gauss	2200	2750	4000	3900	4600
Residual flux density	gauss	1100	1200	1000	1250	1150
Curie temp	°C	210	130	160	160	200
Vol resistivity	ohms/cm	1×10^8	1×10^5	1×10^2	5×10^2	1×10^3
Specific gravity	1	4.7	4.7	4.7	4.7	4.7
Loss factor	$\dfrac{1}{u_o \times O}$	100×10^{-6} 2.5 MHz	120×10^{-6} 1.0 MHz	7×10^{-6} 0.1 MHz	5×10^{-6} 0.1 MHz	4.5×10^{-6} 0.1 MHz
Coercive force	oersteds	1.40	0.30	0.16	0.18	0.22
Temp coeff of initial perm	% °C	0.15	0.10	0.80	0.90	0.60
Freq range	MHz	above 200	50 MHz to 200 MHz	below 50 MHz	5 MHz to 15 MHz	below 50 MHz

Courtesy Amidon Associates, Inc.

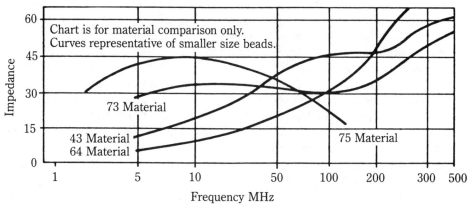

3-22 Relative attenuation of various ferrite-bead materials.

ance shown for the ordinates of these curves is the equivalent series impedance when the beads are used as absorptive filters. For practical purposes, this impedance can be construed as "relative attenuation" in this chart).

Sometimes these bead filters are operated in a different mode, particularly in the 1- to 10-MHz region where their dissipative losses are low. The ceramic capacitor is then selected to produce series resonance in the region of the offending noise spike, or to one of its lower harmonics, where considerable energy might be vested. When so operated, the bead filter is more suggestive of a wave trap. It might still function well as an absorptive filter for higher frequency noise. Experimentation can be worthwhile for individual EMI problems.

Feedthrough and connector-pin filters

A number of integrated EMI filters have appeared on the market during the past few years. These essentially consist of a conductor surrounded by sleeves of magnetically permeable and dielectric materials. The permeable material is generally a ferrite ceramic and is often lossy at high frequencies so that noise energy is absorbed as thermal dissipation. In practice, no discernible temperature rise results from this phenomenon because of the relatively low energy that is usually contained in noise and transients (in principle, however, heat removal might be required in a very large supply that was plagued with a continuum of high-frequency noise). These filters are made in two basic hardware formats. One type is in the form of a feedthrough device. Installation is by means of a hole in a metallic partition or wall. It is secured in place by a locking nut. Electrically, the feedthrough filter can resemble L, pi, or sometimes, T low-pass filters. Considerable shunt-arm capacitance is obtained from barium titanate high-K dielectric material. Physically, these filters resemble the popular feedthrough capacitors, which sometimes provide acceptable EMI attenuation by bypass action alone.

Although these filters can be described in terms of familiar discrete-component filter configurations, their operational mode differs somewhat from what you might expect from the use of discrete inductors and capacitors. The difference comes about from the aforementioned high-frequency dissipation in the lossy ferrite material (sometimes, the dielectric losses contribute appreciably to this noise-absorption process). Additionally, the physical construction of the device makes it behave more like a section of coaxial line (i.e., its parameters are distributed rather than "lumped"). Regardless of the equivalent circuitry, however, the frequency-discrimination action is always that of a low-pass filter. As such, the manufacturers offer various "cut-off" frequencies, attenuation rates, and ultimate attenuation levels. These filters display "clean" attenuation curves in the sense that internal resonances are virtually absent. High attenuation generally exists well into the microwave region. Because of the negligible dc voltage-drop, these filters are often used directly at the output terminal of switching supplies.

Physically more compact versions of the feedthrough filter are made for installation in multiple connector plugs. These connector-pin filters are very effective in cleaning up noise and transients on signal and control lines. They are fitted into the plug in place of the conventional pins or immediately behind them. Both hardware versions are shown in Fig. 3-23. Figure 3-24 shows the attenuation characteristics of a family of connector-pin filters. You should incorporate these connector pin filters in operating cables of both switching supplies and the powered equipment. The filters do not necessarily make cable shielding unnecessary, but they often relax the burden on shielding. These absorptive filters have distributed parameters and behave somewhat like a section of lossy coaxial transmission line. Best of all, they fit directly into the connectors.

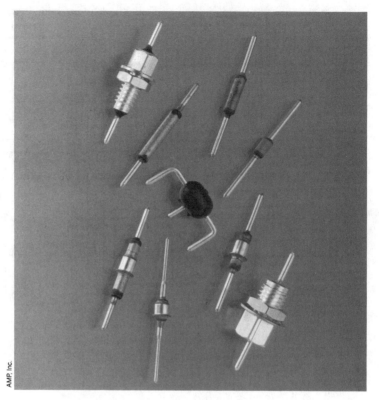

3-23 A few of the "quiet-line" filters made by Amp.

Connector-pin filters are not only used in high-density Amphenol and Cannon connectors, but also on pc board connectors, flexible wire harnesses, and on the multiple conductor connecting links that are used to interconnect pc boards.

Electrical noise in high-voltage supplies

High-voltage power supplies become sources of additional conductive and radiated electrical noise. Commencing at low levels at several kilovolts, such voltage-dependent noise is more evident at several tens of kilovolts. One of these noise sources is "spray radiation" from high-voltage stick rectifiers. This noise consists primarily of radiated RF energy and is derived from the rapid turn-on and turn-off of the numerous rectifying diodes that compose the stick. It is an occasional source of interference in TV sets—being particularly bothersome on weak stations in the low VHF band. Figure 3-25 shows RF spray radiation superimposed on the horizontal sync pulses in a TV receiver, in which interference from this type of noise was quite noticeable. As might be surmised, the received picture was subject to horizontal instability and could not be reliably locked in position on weak VHF stations (at UHF frequencies, the amplitude of the radiated RF is relatively low because the fundamental frequency of the offending harmonics is 15735 Hz, the horizontal repetition rate).

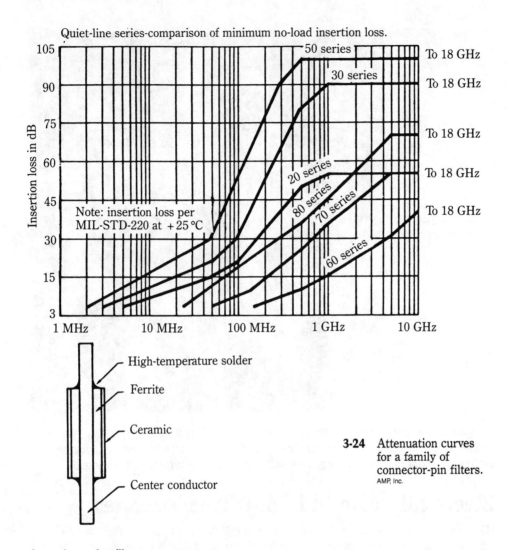

3-24 Attenuation curves for a family of connector-pin filters.
AMP, Inc.

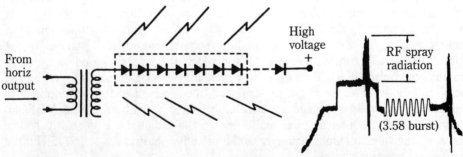

3-25 Rectifier spray radiation on the horizontal synch pulse of a TV receiver.

The solution to this problem in TV receivers is practically achieved via several approaches. First, the rectifying diodes can be processed with "softer" reverse-recovery characteristics, which would thereby minimize spray radiation at its source. Shielding, bypassing, and parts-placement techniques play a large role in reducing vulnerability. The use of RF filters and wave traps helps to attenuate the received RF. The antenna installation and the performance of the RF amplifier are important, because a strong input signal will confer immunity to such interference, other things being equal. Paradoxically, this type of interference doesn't so much plague the modern TV set, itself, as much as other electronic equipment in the immediate vicinity. What has been said pertaining to TV receivers is also true with other cathode-ray tube equipment, such as data displays and computer terminals. However, if high-voltage rectification is achieved at 60 Hz, less trouble will arise from this phenomenon than when the stick rectifier is driven by higher frequencies.

A second source of RFI and EMI, as well as conducted electrical noise, that is peculiar to high-voltage supplies is *corona*. Corona is gaseous ionization that involves the air or other gases, such as those that emanate from various insulating substances. Corona discharges require energy and tend to be destructive, although generally not immediately. Rather, the discharges exert a continuing electrochemical erosion that eventually weakens the insulation until a disruptive and destructive discharge path is produced. This process is more injurious inside than outside of an encapsulated package and inevitably gets its start within tiny voids in the encapsulating material. *Ozone*, one of the constituents of electrified air, is a very active gaseous agent and greatly accelerates the decomposition of many insulating materials. Corona is a function of both the voltage gradient and the rate of voltage change. Consequently, corona troubles are most likely to appear in tightly packaged high-voltage supplies that also involve high frequencies.

Corona discharge excites various resonances in stray circuit parameters and often generates a background of hash. The interferences might not be consistent, being altered by changes in temperature, atmospheric conditions, and by the manner in which the supply is used. Generally, the interference becomes worse with age, because the once-benign effect steadily consumes more electrical energy as it approaches catastrophic disruption. Corona can be alleviated considerably by avoiding small-diameter wires and sharp edges, where high voltage exists between conducting surfaces. *Corona balls* lower field stress because their relatively large surfaces decrease the density of the electric field. These often take the form of small copper or brass spheres and they might be as small as $1/4$ inch in diameter. Other things being equal, the greater the diameter of the corona ball, the more effective it will be in reducing or eliminating corona. The best remedy, where cost and specifications permit, is adequate spacing. Where reliance bears heavily on the insulating properties of an epoxy, resin, or silicone rubber compound, it is very important that no voids exist in the material. This is usually attained by performing the encapsulation in an evacuated chamber.

Although the mild corona outside of a high-voltage package might be benign, it is generally wise to keep it at minimal levels. The prolonged effect of the accompanying ozone, and even ultraviolet light, all too frequently decomposes the insulating material in unexpected places in the equipment.

A third source of electrical noise that is found in high-voltage power supplies is caused by the electrification of plastic cabinets and other insulating materials that are directly or indirectly exposed to the high-voltage electric field. These substances become charged to a high potential, but subsequently "bleed" much of their charge to ground planes and to adjacent objects. This kind of interference is more erratic in nature than disturbances that emanate from well-established corona sites. Many conductive sprays on the market effectively prevent such charge accumulations. These spray-on products derive their conductivity from graphite or from a dissolved metal. By so rendering the surface of insulating material conductive, a useful amount of shielding is also attained. Although the conductivity is not high in terms of ordinary current-carrying ability, it is more than sufficient for the purpose.

In some high-voltage supplies, particularly those that use vacuum tubes as rectifiers, as high-voltage switching devices, or as series or shunt regulating elements, a significant amount of x-ray radiation might be present. The x-ray radiation might not, itself, cause electrical noise (except where high-intensity x-ray energy directly ionizes the air), but it will accelerate insulation decomposition. Thus, corona might be worsened, together with its attendant electrical noise. In a completely solid-state high-voltage supply, x-ray radiation might still be present from cathode-ray tubes. Such x-ray radiation is relatively "soft" and can be readily attenuated by sheet-metal partitions. At voltages that are higher than, say, 30 kV, lead shielding will be more effective for a given thickness.

The combined effects of corona, ozone, ultraviolet, and x-ray radiation might directly or indirectly contribute to the electrical noise that is produced by a high-voltage power supply. The disturbances thereby created in other electrical equipment might be far more serious as an indication of progressively deteriorating insulation. Eliminating the received disturbance is usually only half the required task—it is also important to deal effectively with the corona and, in some cases, with the x-ray radiation.

It is necessary to also consider the thermal conductivity of high-voltage encapsulating compounds. The ill effects of any internal corona become more agitated at higher temperatures. Ordinary RTV silicone, cast polyurethane, and nitrite rubber have relatively low thermal conductivities. Special "loaded" silicone rubber materials are available with about three or four times better thermal conductivity. Such insulating materials will be subject to considerably less temperature rise if heat-removal from the walls of the package is provided for—this situation is usually natural. Cooler insulation also discourages *surface-leakage*, which is sometimes a source of electrical noise. At high voltages, surface leakage currents do not merely depend on the cleanliness of the surface, but they are a strong function of temperature. It is better to have the encapsulating material at near-ambient temperature than at an elevated temperature.

Synchronous rectifiers

Untold millions of switch-mode regulated power supplies have been designed and applied with PWM switching rates that are somewhere in the 20- to 40-kHz region. These supplies generally delivered 5 to 15 V of either or both polarities to loads that consumed anywhere from several watts to more than one kilowatt. Depending on required performance, the rectifiers have been general-purpose silicon pn diodes, fast-recovery diodes, or silicon Schottky diodes. For 5-V loads, the Schottky diode has contributed to high rectification efficiency for the double reason that these diodes exhibit lower forward-voltage drops than do pn diodes and the absence of reverse-conduction phenomena in Schottky diodes circumvents the reverse-recovery losses at high frequencies that are inherent in ordinary pn diodes (even fast-recovery types). Figure 3-26 compares forward voltage drops in similar current-rating Schottky and ordinary silicon pn diodes.

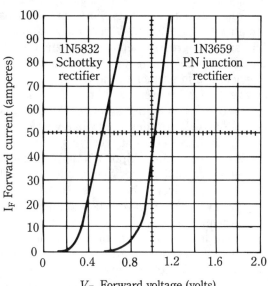

3-26 Comparison of forward-voltage drop in Schottky and pn-junction diodes.

The practical question is: where do you begin to encounter troubles with these hitherto-satisfactory rectifying diodes? With regard to switching losses from the reverse-recovery effect, the 50- to 100-kHz region already makes it desirable to consider the use of fast-recovery diodes or Schottky diodes. It is not always an easy choice for the designer, however. On the one hand, the fast-recovery characteristic trades lower switching losses for higher forward-conductive losses in pn silicon diodes. On the other hand, the Schottky diode is limited in its ability to withstand peak reverse voltage. This limitation might not manifest itself for 5- or 15-V loads, but after 40 or 50 V, the reverse-leakage current can impose serious losses. If it is desired to approach 75 V, Schottky rectifiers become costly and might require inordinately effective heat-removal hardware.

Certainly, by the time you get into the several-hundred kHz switching region, rectification efficiency can become a major headache—particularly if the needed output voltage is appreciably greater than 5 to 15 V; the situation is further aggravated if the output current exceeds several tens of amperes. Moreover, even for high-frequency 5-V supplies, where the Schottky diode offers both lower switching losses and lower conductive losses than pn diodes, the conductive losses can still be formidable in high-current supplies. With all its features, the Schottky diode still develops a higher forward voltage drop than the germanium diode. Although the germanium rectifying diode has good current capability and fair voltage ratings, it is a "slow-recovery" device and cannot compete with its silicon counterparts in frequency capability. Clearly, a solution to the dilemma is needed—is a rectifier more suitable for high-frequency operation than any of these mentioned?

From the framing of this question, the reader must anticipate an affirmative answer and, indeed, it will be shown that for certain applications, efficient rectification is forthcoming from a probably unsuspected source. Also, added to the already mentioned application area is yet another one—loads and systems operating from less than five volts. For example, it has been established that computer circuitry designed for 3.3-V operation can operate faster, with less power dissipation, and with greater memory and logic density than the long-prevailing 5-V format. However, looming up as an immediate obstacle is the burden on power supplies, and in particular, the rectifiers. It can be readily seen that the forward voltage drop in "conventional" rectifiers results in percentage-wise greater power loss for 3.3-V than for 5-V systems. Nonetheless, the rectification technique about to be discussed, provides a reasonably practical solution to this problem too.

The fast and efficient rectifying devices alluded to are not diodes, but are active devices such as bipolar transistors and power MOSFETs. Both of these devices are capable of providing *less* forward voltage drop than diodes. In order to cause active devices to behave as rectifiers, the base or gate signal must be synchronized with the ac wave to be rectified. This accounts for the name of the process—synchronized rectification; the circuitry for accomplishing such ac to dc conversion is appropriately called a synchronous rectifier. To a first approximation, a synchronous rectifying circuit is not unlike a class-B amplifier, at least in its half-wave form. Although a single device class-B amplifier would not be suitable for audio use, its rectification behavior is evident from the unidirectional pulses it would deliver to a load. Whereas the class-B amplifier operates in its linear region, we shall see that the synchronous rectifier works in its *saturated* region.

Although the concept of synchronous rectification is simple enough, it can be appreciated why the application had no merit during the reign of vacuum tubes. Tubes developed tens and hundreds of volts drop between their cathode and plate elements, and therefore could not confer benefits in rectification over their two-element rectifying tubes. Nor was there any great demand for synchronous rectification until switch-mode solid-state power supplies began to move upward from the long-dominant 20-kHz designs, until high-amperage, 5-V loads became more universal, and until dedicated low R_D power MOSFETs became available. Only *then* did it become economical to replace conventional rectifying diodes with synchronous rectifiers.

Refer now to the simple half-wave synchronous rectifier shown in Fig. 3-27. A germanium power transistor is used because of the low $V_{C(sat)}$ attainable in such bipolar transistor is generally less than the forward voltage drop of a PN diode made of the same material. Because of the phasing of the two secondary windings on the input transformer, it can be seen that proper collector polarity (negative for

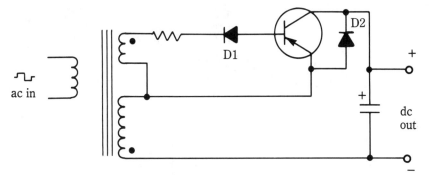

3-27 Half-wave synchronous rectifier that uses a germanium power transistor.

the PNP transistor) and forward base-emitter bias occur at the same time in the ac cycle. This is why the process half-cycle unipolar pulses at the output. Suitable filtering for many purposes can be provided by a single filter capacitor. For less output ripple, an inductor can be inserted ahead of the filter capacitor. However, the lowest inductance and capacitance values that are consistent with satisfactory output should be used. The idea is to protect the transistor from penetrating its SOA.

In order to further protect the transistor, silicon diode, D2, is included. When the rectifier is initially turned on, the filter capacitor appears as a short circuit and the transistor cannot operate in its saturated region. Therefore, until the filter capacitor is almost fully charged, penetration of the SOA could occur with attendant damage or destruction of the transistor. However, during the capacitor charging time, D2 is forward-biased and it absorbs much of the power dissipation that would otherwise be the burden of the transistor. After completion of the charge process, diode D2 becomes inactive; it needs 600 mV to conduct, but is then bypassed by the collector-emitter section of the transistor, which only develops 100 to 300 mV, as previously mentioned.

Diode D1 also protects the transistor. In this case, this diode prevents the avalanche or zener phenomenon from occurring in the base-emitter section of the transistor during application of reverse bias.

A wide variety of germanium devices continues to be marketed by Germanium Power Devices Corp. These still do yeoman service in industrial applications and often in electric vehicles. Effective heat removal is a must for germanium devices, but the main disadvantage for switch-mode power supplies is the poor frequency capability. Trouble from excessive switching losses can be anticipated in the several-kHz region or less.

A full-wave version of the simple synchronous rectifier is shown in Fig. 3-28. In this case, two silicon npn power transistors are used. Accordingly, this circuit is practical for switch-mode supplies in the 20- to 75-kHz range, or somewhat higher. It is true that $V_{C(sat)}$ of silicon transistors is higher than that of the erstwhile germanium transistor, but this is offset by the much greater frequency capability and the superior temperature characteristics of silicon devices. The circuit of Fig. 3-28 could be considered as replacement for a similar circuit that used fast-recovery

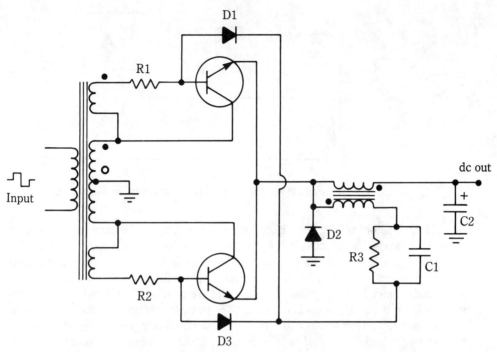

3-28 Full-wave synchronous rectifier that uses a silicon npn power transistor.

diodes on the basis of better rectification efficiency. This could be forthcoming because $V_{C(sat)}$ of properly chosen power transistors can be lower than the forward voltage drop of diodes (this is especially true with regard to fast-recovery diodes, where higher forward voltage drop is a trade-off for lower switching losses).

The circuit of Fig. 3-28 has some "bells and whistles." R1 and R2 are current-limiting resistors, which also permit speed-up pulses to be applied to the bases of the transistors to promote faster turn-off. The speed-up pulses are derived from the secondary winding that is placed on the filter inductor; a one-to-one turn ratio is usually sufficient. Diodes D2 and D3 steer the negative speed-up pulses to the transistor bases. C1 provides speed-up of the fed-back pulses around current-limiting resistor, R3. Free-wheeling diode, D2, delivers remnant energy that was stored in the filter choke to the load, thereby contributing to overall rectification efficiency. Although conductive losses in this circuit will not be quite as low as with Schottky diodes, the voltage-handling capability is much better. Thus, it also could be considered for supplies that have an output higher than several tens of volts.

To obtain optimal performance from this circuit, manufacturer's data should be scrutinized for silicon power transistors with inordinately low $V_{C(sat)}$ values. Some firms, such as Unitrode Corp., market transistors that are specifically intended for synchronous-rectifier applications. A bit of experimentation is usually needed to maximize the performance of synchronous rectifiers. The main goal is to ascertain that sufficient base drive is available over all operating conditions. At the same time, overdrive can be counterproductive because it slows down the turn-off time of the transistors. Although square waves are easiest to work with, this circuit has been successfully used in PWM switch-mode supplies.

Consider the full-wave synchronous rectifier circuit that is shown in Fig. 3-29. Superficially, it closely resembles the circuit of Fig. 3-28. The look-alike aspect is even closer if the filter-choke secondary circuit of Fig. 3-28 were omitted, because it is not needed for basic operation. Then, it would appear that the only real difference between the two synchronous rectifiers is that bipolar transistors are used in one and power MOSFETs are used in the other. If this was true, you could envision a compromise transformer design that would allow plugging in either type of active device. There is more than meets the eye, however.

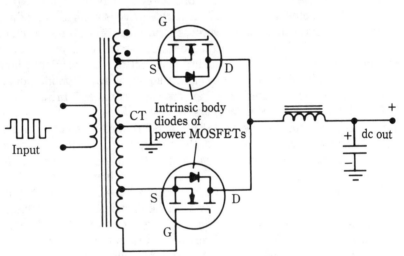

3-29 Full-wave synchronous rectifier that uses power MOSFETs. Unlike most MOSFET circuits, the N-channel MOSFETs in this application operate in their third-quadrant region (i.e., conduction occurs with negative drain voltage!).

A closer study of the two circuits reveals a paradox. Both circuits use N-type devices—npn bipolar transistors on the one hand, and N-channel power MOSFETs on the other. Notice, however, that the source-drain portion of the bipolar transistor circuit is connected opposite to the emitter-collector portion of the bipolar transistor circuit. This difference, in itself, would not necessarily prevent operation of the MOSFET circuit; it would, you might suppose, cause the dc output to be negative. Because both circuits are depicted with positive output polarity, you might suspect a mistake on the part of the designer or perhaps the draftsman. After learning that no such mistake has been made, the situation poses a dilemma; can both circuits actually operate as depicted? The paradox is heightened by the fact that most amplifiers and switching circuits can operate in quite similar fashion by exchanging appropriate bipolar transistors and power MOSFETs. Minor changes in drive or bias circuitry can perform acceptably from either type of active device in many applications.

The answer to the inconsistency lies in a little-known behavior of power MOSFETs. Although not always evident from data sheets, MOSFETs can be operated in the third, as well as more commonly, the first quadrant of their characteristics. Thus, an N-channel MOSFET can operate in certain applications with positive source and negative drain. Surprisingly, conduction still depends on the gate being positive, with respect to the source. After accepting this mode of operation as fact, the internal diode is still so polarized as to short the active device itself. What, indeed, goes on here?

The solution to the internal-diode problem is to keep it electrically inactive; this is accomplished by ensuring that the voltage drop across the power MOSFET is less than the turn-on bias of the internal diode (approximately 0.6 V). Thus, proper operation of the MOSFET synchronous rectifier requires that MOSFETs with a low R_D are selected, and that sufficient gate drive is applied to produce deep saturation of drain current. In other words, the product of R_D times drain current must be less than 0.6 V. The fact that this voltage drop can be kept well below this 600-mV limit is why the MOSFET synchronous rectifier has displaced both fast-recovery and Schottky diodes in many high-frequency switch-mode supplies. A nice thing about this new technique is that some semiconductor firms now market power MOSFETs as *synchronous rectifiers*.

An interesting feature of the MOSFET synchronous-rectifier circuit of Fig. 3-29 is that if the gate drive was removed, the circuit would continue to operate as a full-wave rectifier and deliver positive voltage to the load. However, the internal diodes of the MOSFETs would now be involved as ordinary diode rectifiers and the MOSFETs proper would be inactive. Such operation would be accompanied by relatively high conductive losses, and for high frequencies, high switching losses as well. Despite lowered rectification efficiency, the circuit would still be workable. Some nice features of MOSFET synchronous rectifiers are that they don't short, reverse polarity, or make other drastic operation changes.

Because of this feature, it can be quickly determined whether the circuit is performing as intended. If the voltage drop across the drain-source terminals exceeds 600 mV, the circuit is not operating as a synchronous rectifier. On the other hand, if this voltage drop is in the several-hundred-millivolt region, synchronous rectification is definitely occurring. Even so, further optimization can often be realized by experimenting with gate drive, free-wheeling diodes, and filter parameters. Other things being equal, the voltage rating of the MOSFETs should be no higher than necessary because best R_D values are found in the lower-voltage MOSFETs. Also, good heatsinking can be important because R_D increases with temperature.

Once the principle of operation of the full-wave MOSFET synchronous rectifier of Fig. 3-29 is understood, it is possible to deduce the working mechanism of the bridge circuit shown in Fig. 3-30. It too uses MOSFETs and performs as a full-wave rectifier. Notice, however, the complementary-symmetry aspect of this circuit—both N-channel and P-channel power MOSFETs are used.

Remember that all of the MOSFETs in the bridge configuration operate in their third quadrant. Thus, the N-channel MOSFETs will conduct when their drains are negative with respect to their sources and when their gates are positive with respect to their sources. No conduction can occur with conventional (first quadrant) drain-source polarities because their gates will, at that time, be negative. Similarly, the P-channel MOSFETs will conduct when their drains are positive with respect to their sources and when their gates are negative with respect to their sources. Again, no conduction can occur with conventional (first quadrant) drain-source polarities because their gates will, at that time, be positive. Fortunately, the internal diodes are reverse biased when the MOSFETs are polarized for first-quadrant operation!

Like the center-tapped full-wave circuit of Fig. 3-29, the internal diodes of the MOSFETs will be inactive as long as the voltage drop across the MOSFETs can be held to less than 600 mV. This, again, is accomplished by selecting low R_D MOSFETs and by adequately driving their gate circuits. As a practical matter, the R_D values tend to be a bit higher for P-channel MOSFETs than for N-channel MOSFETs, but this factor is not troublesome, as long as all forward-voltage drops can be maintained well below 600 mV.

As shown in Fig. 3-30, no output filter is used. When this circuit is used with 50% duty-cycle square waves, as are derived from some inverters, it might, indeed, be feasible to dispense with filtering. For most applications, some filtering is required, but such a rectifier generally produces less ripple than is encountered with pn diodes. Even Schottky-diode rectifiers often produce high ripple because of their reverse-current leakage at high voltages and/or high temperature.

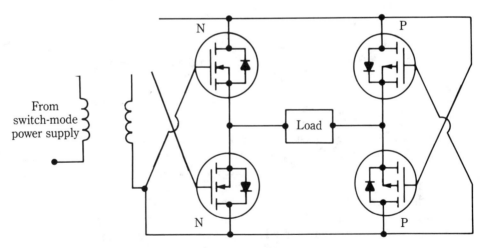

3-30 Synchronous bridge rectifier that uses complementary-symmetry MOSFETs. Notice the arrow directions in the symbols of the N-channel and P-channel power MOSFETs.

In the event of insufficient gate drive, this circuit will continue to behave as a bridge rectifier, but rectification will then be accomplished by the internal diodes; rectification losses will then be relatively high—especially at high switching rates. Because two, rather than one, forward voltage drops are involved in this bridge circuit, the full-wave circuit of Fig. 3-29 might be more desirable for 3.3-V supplies.

The circuit of Fig. 3-31 operates as a half-wave synchronous rectifier, although such operation is achieved somewhat differently than in the previously discussed synchronous rectifiers. In common with the other circuits, the power MOSFET in Fig. 3-31 provides third-quadrant conduction and requires a MOSFET with very low R_D in order to keep the internal diode from becoming forward-biased into its conductive region. The positive-going turn-on signal for the gate is derived in a unique way, however.

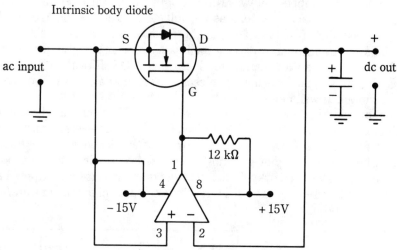

3-31 Synchronous rectifier that uses an electronic-timing technique. The power MOSFET operates in the third quadrant with conduction occurring when the drain is negative, with respect to the source.

A comparator is connected so that it senses the source-drain voltage of the N-channel MOSFET. The requisite positive turn-on signal is applied to the gate only when the source of the MOSFET is positive with respect to its drain. Thus, third-quadrant conduction is achieved. No such gate signal is delivered by the comparator for "conventional" or first-quadrant operation. This is the way rectification is produced by the power MOSFET. Notice that the internal diode is reverse-biased during the blocking portion of the cycle, so it is then out of the way. An auxiliary dual-polarity 15-V low-current supply is needed for the voltage comparator.

This circuit is probably easier to coax good results from a sine-wave input source than from more conventional synchronous rectifier circuits, because a high-level rapid-rise and -fall gate signal is precisely developed by the comparator. For the same reason, this circuit is worthy of consideration for PWM waves when a considerable duty-cycle range must be accommodated.

As shown, no galvanic isolation is between input and output portions of the circuit. A possible remedy is to use a transformer at the input. Such a transformer could be physically small at high frequencies. In order to ascertain proper circuit operations, monitor the peak voltage that appears between the source and drain when the output is fully loaded. This voltage should be well below 600 mV to ensure that the rectification is not being performed by the internal diode of the MOSFET. If this condition cannot be met under worst conditions (full-load and maximum operating temperatures), more effective heat removal might be necessary (R_D of power MOSFETs has a positive coefficient of resistance with respect to temperature). Stronger drive, as can be obtained from a ±20-V auxiliary supply, might be necessary. Finally, you might need to select a power MOSFET with lower voltage drop under operating conditions.

A subtle pitfall with regard to regulation specifications

Switching power supplies commonly provide more than one output voltage. For example, a popular combination of output voltages composes 5 and ±12 V. In such a situation, the 5-V output is likely to be the actually regulated voltage; that is, the electronic regulation involves direct sensing of the 5-V circuitry. The 12-V circuits "come along for the ride" via transformer coupling to the 5-V regulated circuitry. Although advertising literature commonly depicts all of the outputs as being regulated, it would probably be more accurate to describe the 12-V outputs as being *semiregulated*. In any event, the 12-V outputs will not exhibit the tightest regulation available from the 5-V output. Moreover, the 12-V outputs will provide regulation, not only as a function of their loading, but it will be considerably dependent on the loading of the 5-V output as well. This sometimes leads to unexpected malfunctioning of the power system. This situation is again alluded to in chapter 4, but here it is discussed in detail.

In order to better understand the nature of such power supplies, recall the basic regulating mechanism of the sensed output; in this example, it is the 5-V output. Increased loading of the 5-V output reduces the voltage that is actually delivered to the load. However, the pulse width of the PWM-switched waveform within the supply broadens. This maintains the output voltage at its preset level, for example, at exactly 5 V. For decreased loading, the pulse width narrows, and thereby prevents the rise in output voltage that would otherwise occur. Again, the output voltage is maintained at (ideally) 5 V.

Consider, now, that one of the 12-V outputs is delivering current to a fixed load. As alluded to, this voltage will be semiregulated in the sense that nominally, 12 V will be delivered if the fixed load is made variable. The regulation will not be as good as provided by the 5-V output because the 5-V level is actually sensed by the electronic circuitry. However, the situation is actually worse than this; the 12-V output depends not only on its own nonsensed voltage drops, but will also be considerably affected by changes that occur in the loading of the 5-V output. If, for example, the load in the 5-V system becomes lighter, the power supply will produce a narrower pulse width in response. This is fine from the standpoint of maintaining regulation in the 5-V load, but it is obvious that it will cause an "uncalled for" reduction in voltage to the 12-V load! This interdependency is illustrated in Fig. 3-32.

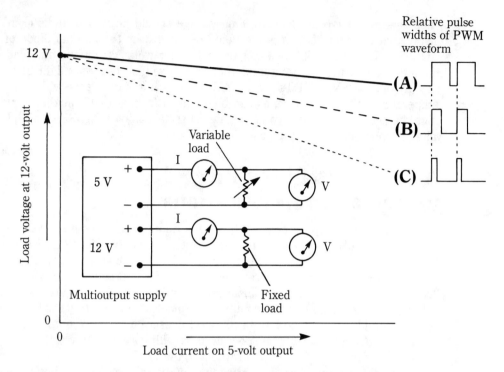

(A) Regulation curve of 12-volt output with 5-volt output fully loaded.

(B) Regulation curve of 12-volt output with 5-volt output operating into one-half of its fully rated load.

(C) Regulation curve of 12-volt output under worst conditions of load variation at 5-volt output.

3-32 Effect of 5-V load variation on 12-V load in multioutput supply. In a typical multioutput supply, the 5-V output is directly regulated. The 12-V output is indirectly regulated through the output transformer.

The existence of such interdependency is not always easy to extract from technical specifications and it is even less obvious from advertising literature. For additional cost, supplies with individually regulated outputs are available.

An unexpected pitfall that results from the semiregulation of auxiliary outputs on a multioutput switching supply is illustrated from the following scenario. A systems designer finds from a catalog that two models, A and B, appear to fit voltage, current, and power requirements. Power supply A has a 5-V regulated output that is rated at 10 A, just right for the load under consideration. Additionally, it provides 10 V at 0.5 A, which is ample for the linear circuits of the system. Power supply B is similar, except that the 5-V output is rated for 15 A. Which is the best choice?

Here, common sense might suggest that supply B would merit selection, providing that higher cost and, perhaps, that the greater bulk can be accommodated. It would appear that reliability would be enhanced because the larger supply would operate well below its rated current and power capability. Consider, however, an important difference in electrical performance that might ensue from the two supplies.

In both supplies, the PWM regulation technique narrows the switching wave if the 5-V output rises. In this regard, notice that supply B operates at relatively light loading of its 5-V output, compared to supply A for the same current drain. Thus, the switching pulse in supply B will be narrower than in supply A. Because of this, the 10-V output of supply B will be narrower than in supply A, other things being equal ("other things" implies similar 10-V output circuitry and the same 10-V loading). Thus, the choice of supply B could result in unacceptably low "10-V" from the auxiliary (nonregulated) output. This situation will be aggravated for heavy loading of the 10-V output. Summing up, light loading of the primary 5-V output, together with heavy loading of the 10-V auxiliary output could cause an inordinately low "10-V" from the auxiliary output. Interestingly, this situation was engendered in this particular case by selecting a more amply-rated supply. Even without the added change of heavier loading for the 10-V output, the choice of supply B possibly borders on trouble.

A factor that saves the day in the above-described situations is that in many systems, the 5 V applied to digital and logic circuits is more in need of close regulation than the less-critical 10, 12, or 15 V that are applied to analog circuitry. Even so, you should be careful, because supplies are marketed so that the 5-V output "comes along for the ride"; that is, the higher output voltage is the one that is directly sensed and regulated. This is done to attain galvanic isolation of the 5-V output without using an optoisolator in the feedback path of the 5-V circuitry.

4
Implementation of regulation techniques

THIS CHAPTER COVERS THE VARIOUS SITUATIONS, CIRCUITS, AND PRACTICES THAT are relevant to regulated power supplies. These subjects are discussed to provide penetrating insights and to broaden your viewpoints.

Dissipation control

Because the various aspects of series-type regulators were covered earlier, this section commences with a few remarks about the practical implementation of such regulator design. It has been assumed that the inherent operation of such regulators enables them to cope easily (via manual adjustment or electrical programming) by increasing line voltage or reducing dc output voltage. Actually, the series-pass element only needs to accommodate a higher voltage drop across its terminals. Except for small power supplies, this task is easier said than done. In a transistor regulator, it is necessary to maintain only a couple of volts across the collector-emitter terminals of the series-pass transistor. If the load current is one ampere, the series-pass transistor must dissipate a mere two watts (assuming that a 2-V drop is across transistor). Now, suppose that the output voltage is adjusted down from, say, 35 to 20 V to supply a new load condition that now draws one ampere at the reduced voltage. The pass transistor must now drop 17 V and must be capable of dissipating 17 W. A simultaneous rise in ac line voltage can increase this requirement to 20 W, for example. Obviously, you now have a different "ball game" insofar as concerns the type of transistor used and its heatsink requirements. If you provide a larger and more costly transistor and a massive heatsink, you must remain haunted by the fact that the larger package and reduced efficiency are really not necessary for a load with small power dissipation.

Dissipation limiting techniques

Obviously, a means for limiting power dissipation in the series-pass transistor is worthwhile in all except the very small power supplies. Even when the output voltage is fixed, it is desirable to spare the series-pass transistor the burden of dissipating extra power as the consequence of high line voltage (also, the reduction of unnecessary voltage drop across the series-pass transistor helps make this vulnerable element less susceptible to failure). One of the simplest approaches to the dissipation problem is the use of a tapped (variable) transformer, which is mechanically ganged to the potentiometer that controls the dc output voltage of the regulator. Figure 4-1 illustrates such a scheme, which uses a variable autotransformer connected between the power transformer and the rectifier bridge. By choosing an appropriate taper for the potentiometer, it is not difficult to obtain the requisite tracking between the ganged elements. Additional tracking manipulation is caused by inserting a small resistance in series with a potentiometer terminal or by connecting a large resistance from its wiper to one or the other of its terminals. Usually the objective is not to produce a constant voltage across the series-pass transistor, but rather to limit its maximum value.

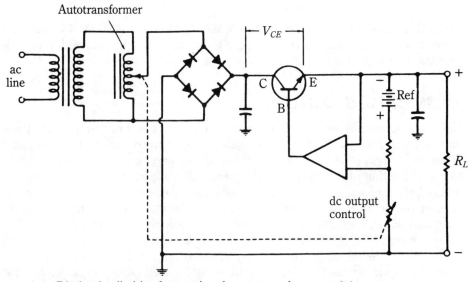

4-1 Dissipation limiting by ganging the autotransformer and dc output control.

In Figs. 4-2, 4-3, and 4-4, the same objective is attained by electronically controlling the ac voltage that is applied to the rectifier. The control transistor in Fig. 4-2 is properly polarized for both half-cycles via the commutating action of the diodes in the bridge rectifier. This transistor acts as a variable resistor and dissipates much of the heat that the series-pass transistor would otherwise be forced to handle. Although the overall efficiency of the supply is not appreciably changed by this method, it is still beneficial to operate the series-pass transistor under relaxed thermal and electrical conditions.

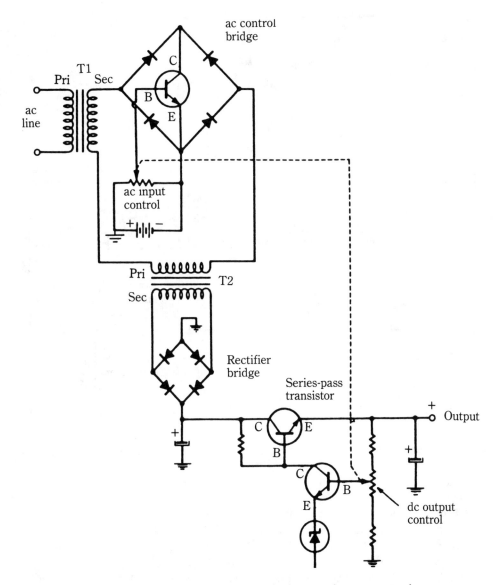

4-2 Dissipation limiting by ganging the input and output controls.

The scheme shown in Fig. 4-3 uses a silicon-controlled rectifier (SCR) to produce similar results to that described for the circuit of Fig. 4-2. However, the SCR is phase-controlled and thereby operates much more efficiently than the transistor in the circuit of Fig. 4-2. The SCR is at all times either fully on or fully off; it blocks an adjustable portion of the sine wave to control the rms value of the wave that is applied to the full-wave rectifier bridge. Because of its on or off conduction state, the I^2R loss that is developed in the SCR is relatively low compared to that which the transistor of Fig. 4-2 must dissipate. Therefore, the efficiency of such phase

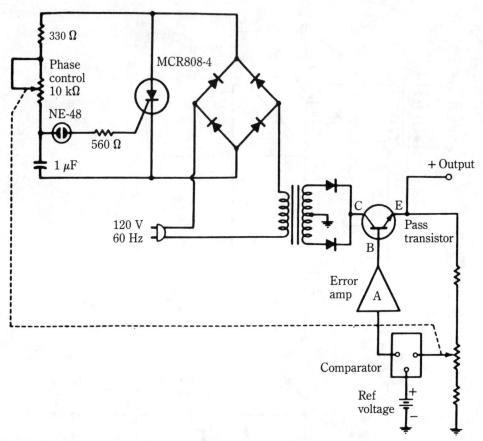

4-3 Dissipation limiting by ganging the phase-controlled ac input and dc output control.

control is high and the method is well-suited to control very large amounts of power.

The operation of the circuit shown in Fig. 4-4 is basically similar. Here, the use of a triac eliminates the need of a diode commutating bridge (or of two SCRs). This circuit and that of Fig. 4-3 deliver a distorted wave to the rectifier/filter circuit. This action can add noise components in the output of the regulator. Strong harmonics throughout the audio spectrum are generated in increasing intensity as the conduction angle of the SCR or triac is reduced. Even more troublesome is the RFI that is generated from the abrupt turn on of these devices. One remedial measure is to incorporate a low-pass filter. Such an RFI filter is shown in the circuit of Fig. 4-4.

Dissipation control is often provided by a magnetic amplifier (Fig. 4-5A). This arrangement consists of a "coarse" regulator followed by a "fine" regulator that uses a series-pass transistor. As with the preceding circuits, the mechanical coupling between the output controls limits the dissipation that is allowed for the series-pass transistor. In this system, the overall regulation (for low frequencies at least) is enhanced by the preregulation that is provided by the magnetic amplifier.

Dissipation control **169**

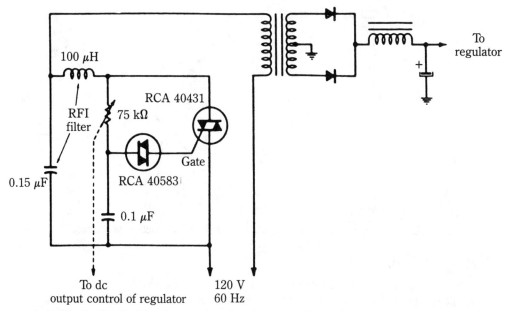

4-4 Dissipation limiting by ganging the triac-controlled ac input and dc output control.

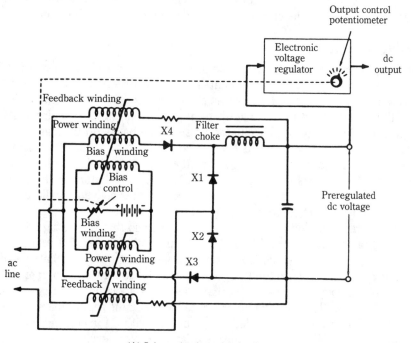

(A) Schematic of control circuit.

4-5 Dissipation control by a magnetic amplifier.

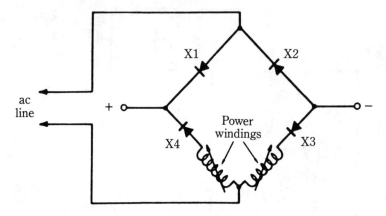

(B) Part of circuit (A) rearranged to illustrate bridge rectifier function of diodes.

4-5 Continued

In Fig. 4-5B, the actual connections of the diodes are retained, but the layout of the circuit is changed to make the principle of operation more evident. Here, the diodes are in a full-wave bridge-rectifier configuration with variable inductances (the power windings of the magnetic amplifier) in each half-wave section. The inductance of the power windings is varied automatically by the feedback (control) windings and manually by adjusting the current that flows through the bias windings.

Double regulation

Double regulation is a particularly useful means of controlling dissipation in the series-pass element. The arrangement in Fig. 4-6 reveals the basic idea that is involved in this technique. Here, the ac power that is delivered to the bridge rectifier is the amplified output of an audio oscillator. In the simplified circuit shown, a series-tuned Colpitts (Clapp) oscillator is used. The base-emitter bias of oscillator transistor Q1 is partially derived from an amplifier that senses the voltage drop across the series-pass transistor, Q4. Feedback is such that a higher dc voltage drop across Q4 reduces the output amplitude of the oscillator. At the same time, the regulator circuit that is associated with Q4 operates conventionally to maintain a constant output voltage from the supply.

The net result is that an equilibrium is reached so that the voltage drop across Q4 is maintained at a low value and the output voltage remains regulated. The time constants that are associated with the variable-amplitude oscillator result in a relatively slow correction rate. This, together with the lack of a voltage reference in the feedback loop, endows the oscillator and its associated circuitry with coarse regulating action. This is satisfactory because the tight regulation demanded at the output of the supply is provided by Q5 and X1 with Q4 in conventional voltage-regulating action. With this scheme, regulated voltage levels of several kilovolts can be obtained with the use of low-voltage pass transistors.

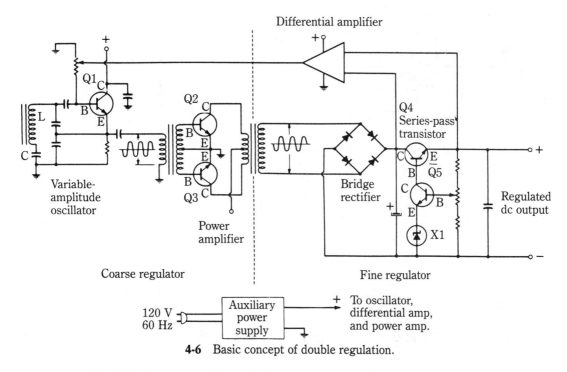

4-6 Basic concept of double regulation.

Dissipation control with a preregulating transistor

The scheme shown in Fig. 4-7 is a special case of double regulation; it is considerably simpler than the generalized arrangement of Fig. 4-6. Whereas the coarse regulator of Fig. 4-6 operates via an ac (usually higher than line frequency) portion of the circuit, coarse regulation in Fig. 4-7 is achieved at dc. Here, Q2 is the conventional series-pass transistor. Transistor Q1 is actually an emitter-follower voltage regulator that "sees" the collector-emitter terminals of Q2 as its load. Transistor Q1 operates to maintain approximately M_2 volts across Q2. This being the case, Q1 must dissipate all of the power it absorbs from Q2. This situation would not be bad, but the addition of resistance R makes the overall situation even better; now Q1 is relieved of considerable dissipative burden (it is less costly and more reliable to absorb power in a resistance than in a transistor). The net result of this arrangement is that the maximum power that needs to be dissipated in Q1 or Q2 is about 25% of the power that would have to be dissipated in a regulator using Q2 alone. Of course, resistance R degrades the regulating ability of Q1. This loss is not of much importance, however. As long as Q1 retains the ability to maintain the voltage across Q2 approximately at the low value (often two or three volts) set by M_2, the basic objective is achieved.

In actual circuits, M_2 is replaced by a zener diode or by a series string of ordinary diodes that are operated in the forward-conduction mode. Again, the soft-knee and high dynamic impedance that such reference sources might have are of

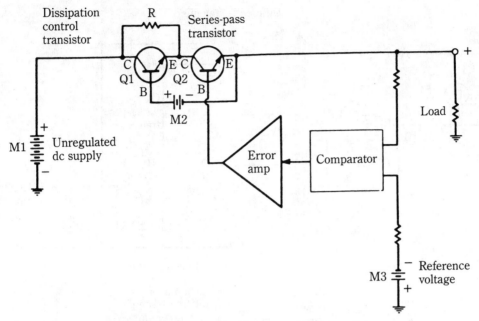

4-7 Dissipation control with preregulating transistor.

little consequence to the desired result. In this scheme, even though the dc behavior of Q2 is modified, its action, which provides a low dynamic output impedance, is not appreciably affected. Moreover, Q1 also improves performance by smoothing out the ripple and line disturbances that are impressed upon series-pass transistor Q2.

The ferroresonant constant-voltage transformer

Dissipation control (in the sense that the series-pass or shunt-regulating device is protected from assuming the burden of high line voltage) can be provided by special regulating transformers. Typical of these is the *ferroresonant transformer*, known also as the *static-magnetic regulating transformer* and the *constant-voltage transformer*. In this transformer, the flux in the primary magnetic circuit is accommodated by unsaturated core material, as in a conventional transformer. The secondary flux, however, encounters saturated iron; therefore, a change in primary flux produces far less than a proportionate change in secondary flux. Thus, the secondary induced voltage remains relatively independent of the voltage that is impressed upon the primary winding.

This action can be seen in Fig. 4-8. A *magnetic shunt* between the two windings enables much of the secondary flux to be decoupled from the primary winding. Moreover, capacitor C causes a large reactive current in the secondary

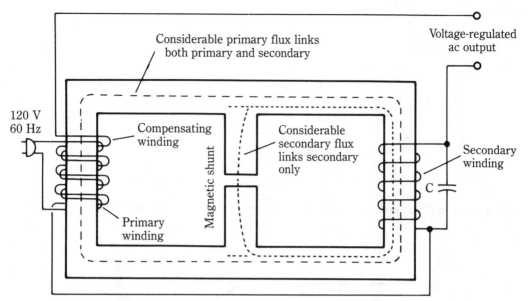

4-8 The magnetic circuit of the ferroresonant transformer.

winding, which thereby saturates the secondary magnetic circuit. Such saturation increases the magnetic isolation between the two windings because more primary flux then takes the path through the magnetic shunt. Of course, complete decoupling would isolate the secondary from its source of energy, the primary circuit. For this reason, and also because the transition from the unsaturated to saturated core region is rather gradual, the regulating action is less than perfect. An additional compensating winding considerably improves the circuit. This winding carries ac load current and opposes the primary winding flux. This, in essence, constitutes regulation; the higher secondary voltage increases the ac load current, but this current produces further decoupling action, which reduces the voltage induced in the secondary.

The use of such a transformer also limits dissipation from over-current or short-circuit load. Excessive current demand greatly diminishes the mutual flux between the primary and secondary windings, and the secondary voltage rapidly falls to zero (Fig. 4-9B).

The secondary capacitor does not resonate the inductance of the secondary winding. Rather, the secondary voltage is higher than would be indicated by the secondary-primary turns ratio because of the increased secondary flux that is caused by the capacitor current. Thus, the term *ferroresonant* is used.

Typical ferroresonant transformers provide a 10-to-1 improvement over the variations in raw power-line voltage. They also produce surprisingly good load regulation, ±2% being typical. On the other hand, the ferroresonant transformer is considerably larger and heavier than a conventional transformer of the same power capacity. Their response time is slow (2 to 5 cycles of ac frequency). They also tend to be sensitive to supply-line frequency, but for most applications this sensitivity is

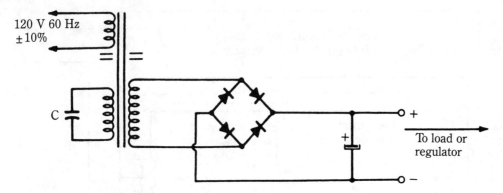

(A) Substitution of ferroresonant transformer in conventional bridge rectifier circuit.

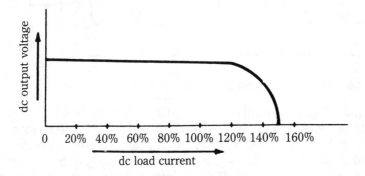

(B) Voltage collapse from excessive load current.

4-9 Dissipation control and preregulation with a ferroresonant transformer.

probably not of consequence. They deliver a waveform that approaches a square wave, rather than a square wave. Actually, the harmonics of such a waveform yield readily to filtering. At the same time, rectifying elements receive less stress from peak voltages than when connected to a sine-wave source. A more recent development is the *Paraformer transformer*, which is discussed later in this chapter.

The bipolar voltage-regulated power supply

The *bipolar power supply* can deliver a continuous range of output voltages of either polarity, with respect to a common terminal. Such a supply provides greater functional flexibility when used as a laboratory source of regulated voltage. When made electrically programmable, the bipolar supply becomes an excellent power amplifier, with frequency capability down to zero frequency (dc). Some loads generate reverse currents for which a flow path must be provided in order for the load to operate properly. An example of such a load is the grid-cathode circuit of a class-C amplifier, which is biased from a voltage source. Unless the voltage source used for this purpose can provide the required path for reverse current flow, the desired grid bias cannot be properly developed. Inductive and switching circuits often have transients that tend to reverse the current through the voltage source. In such cases, the situation can endanger the series-pass transistor if no reverse path is available. The bipolar supply handles such situations nicely because it can accommodate current in either direction.

The simplified diagram of Fig. 4-10 illustrates the basic concept of the bipolar voltage-regulated power supply. Notice the use of both npn and pnp series-pass transistors. Control potentiometer R allows the selective adjustment of output voltage level of either polarity, with respect to the common terminal (ground). The range of voltage excursion that is enabled by potentiometer R can be controlled by the setting of R_f. Except when providing a return path for load-generated reverse currents, one series-pass transistor is always in its off state. The output capacitor in a bipolar supply finds a discharge path through the "idle" series-pass transistor when the supply is operated as a power amplifier or during the programming of its input. This operation leads to greater bandwidth and faster programming speed.

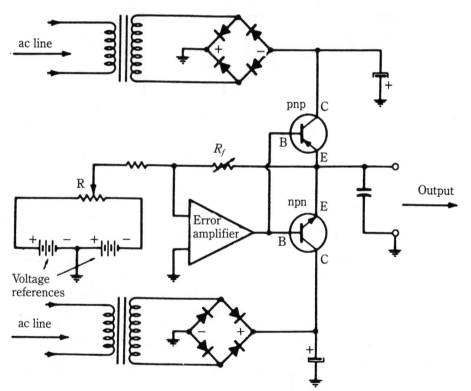

4-10 Basic concept of a bipolar voltage-regulated power supply.

Dual output tracking

A different sort of dual regulated supply is shown in Fig. 4-11. This arrangement involves separate output terminals for supplying positive and negative regulated

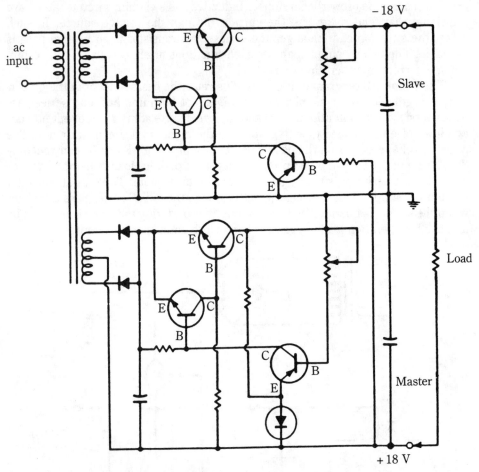

4-11 Dual output-tracking arrangement.

voltages, with respect to a common terminal (ground). However, the arrangement is not simply two regulated supplies that are packaged together. The positive supply is conventional in that it contains the usual voltage-reference diode. The negative supply, however, is devoid of a reference diode. Nonetheless, it operates as a voltage-regulated supply, the required reference voltage is obtained from the output of the positive supply. Because of this arrangement, both supplies increase or decrease their output voltage simultaneously. This simultaneous change preserves the balance, with respect to common (ground), both statically and dynamically. Such performance is beneficial to certain loads, such as the operational amplifier. Although the voltage changes from a regulated supply might be relatively small, an operational amplifier is often required to impart precise high gains to small input signals. Under such conditions, the balance and common-mode rejection of the operational amplifier input stage can be adversely affected by nonsymmetrical operation of the positive and negative power supplies.

The shunt voltage regulator

The *shunt voltage regulator* is derived by substituting a fixed resistance for the series-pass element and connecting the control element directly across the output of the supply (of course, other appropriate changes must also be made in the associated circuitry). The basic concepts of sensing, comparing, and forming a negative-feedback system remain valid. However, the output voltage is now varied by deliberately loading the output and thereby dropping a controlled amount of voltage across the fixed series resistor. Figure 4-12 illustrates two shunt-type voltage regulators. This method of producing regulated voltage can be just as good as with the series type. Several points of shunt voltage regulators are:

- The operating efficiency is very low at zero and at light loads because the shunt element then dissipates maximum or near-maximum power. From this consideration, the shunt regulator should be designed into systems that consume more than half of the rated current of the power supply. Indeed, the closer the load demand is to the rated value, the more efficient the operation of the shunt regulator will be.

- The shunt regulator is inherently protected from overload or short circuit of the output; no auxiliary circuitry needs to be incorporated for such protection. In the event of a short, the shunt transistor does not overload; it is simply deprived of operating voltage.

- The shunt regulator will provide a path for reverse current (i.e., it will absorb power from the load, as well as deliver power to it). This is excellent for operation of dc motors, class-C amplifiers, and certain switching and inductive circuits.

- The shunt regulator is particularly adapted to high-voltage vacuum-tube regulator circuits because of the wide availability of tubes that are capable of withstanding high voltages under sustained operation.

- The shunt regulator is more effective than the series regulator in preventing load-current transients from being fed back to the unregulated dc source, and, therefore, to the ac power line. This is particularly true of the shunt regulator in Fig. 4-12B.

In Fig. 4-12A, the single transistor combines the functions of the comparator, the error-signal amplifier, and the shunt-control element. Figure 4-12B is an expanded version in which gain is imparted to the error signal by a separate amplifier, as is common with the more familiar series-type regulators. There is, however, a qualitative, as well as a quantitative, difference in the performance of these two shunt regulators. In the simple configuration of Fig. 4-12A, it is necessary to associate the reference-diode circuit with the unregulated supply. If R1 was connected to the output side of R_D, insufficient voltage would be available to operate the reference zener diode. Thus, the current through the reference diode is completely susceptible to changes in the voltage that is delivered from the unregulated source. Thus, the zener diode of regulator A is not aided by the shunt transistor insofar as

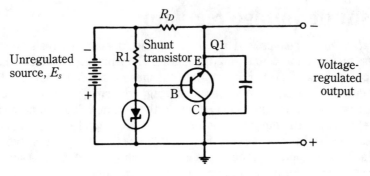

(A) Single transistor version.

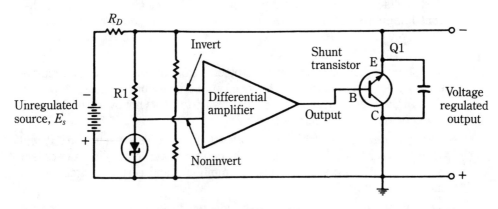

(B) Expanded version.

4-12 Basic concept of shunt voltage regulator.

the line-voltage variation is concerned. The output impedance is approximately the dynamic impedance of the reference diode divided by the beta (β), of Q1.

Because an amplifier is interposed between the base of Q1 and the reference diode in Fig. 4-12B, the regulated output voltage can be higher than the reference voltage. This voltage allows R1 to be connected to the regulated-output-voltage side of R_D. This not only improves the load regulation and drift, but it greatly enhances line-voltage regulation over that which is attainable with the simple regulator of Fig. 4-12A. In Fig. 4-12B, both Q1 and the amplifier that precedes it are instrumental in stabilizing the output voltage against line-voltage variations.

Figure 4-13 illustrates a simple shunt regulator that uses tubes. Actually, the simplicity of this circuit masks its unique capability. Klystrons, traveling-wave tubes, backward-wave oscillators, and voltage-tunable magnetrons often require about 1000 V from a regulated source that has a several-milliampere capability. This range is awkward; solid-state circuits might not be at their best at such high voltages—good protection must be incorporated and reliability suffers at higher temperatures.

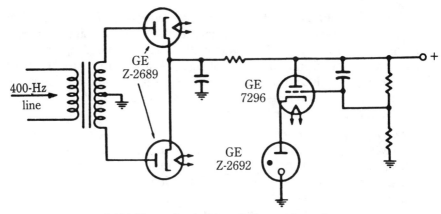

4-13 Example of shunt regulator using tubes.

On the other hand, conventional tubes are bulky, fragile, and generally require at least an additional amplifier stage over that shown in Fig. 4-13. The tubes that are designated in Fig. 4-13 are ceramic types. They are physically small, extremely rugged, and capable of operating at temperatures far exceeding that of transistors. Moreover, the triode types have transconductances up to 50,000 $\mu\mho$, which results in more-than-adequate regulation and output impedances for these applications. Although both the shunt regulator and the voltage-reference tube shunt the output, the circuit action is substantially the same as in the transistor shunt regulators that were previously shown with only the shunting element across the output. Because of the high-frequency capability of the ceramic triode, no output capacitor is needed. The small capacitor (from plate to grid) provides a low-attenuation feedback path for greater ripple suppression. When powered from a 400-Hz line, such a regulator has provided very satisfactory performance in airborne and space vehicles.

Junction field-effect transistors in dissipative regulators

The *junction field-effect transistor* received its major development after the bipolar transistor had already become a commercial success; the latter quickly became "traditional" in certain circuit techniques. For this reason, the relatively late-evolving FET encounters a psychological inertia when designers consider its use. With regard to power supplies, some of the characteristics of the FET are indeed intriguing to the technically alert.

FET characteristics important in regulated supplies

- *Thermal stability.* The FET possesses a negative gain/temperature coefficient, which prevents "thermal runaway." Therefore, when used as a series-pass element or as a shunt element, a better use of the dissipation capability can be obtained.

An interesting aspect of this thermal behavior is the FET's capability at depressed temperatures. At low temperatures where the bipolar transistor is virtually dead, the FET is superior to that at room temperature. For example, in the vicinity of liquid-nitrogen temperature, $-160\,°C$, the ON resistance of the FET would be about 30% of its minimum value at room temperature; moreover, its transconductance would approach a threefold improvement at the lower temperature!

- *Freedom from second breakdown.* Because the output circuit of the FET does not involve a pn junction, the second breakdown phenomenon that plagues bipolar transistors at high-power dissipation is entirely absent in the FET. Consequently, maximum allowable voltage does not need to be derated as the power is increased.
- *Ease of paralleling.* When similar bipolar transistors are paralleled to obtain greater power-handling capability, more drive power is required. This, more often than not, invokes both technical and economic headaches. The FET, on the other hand, is like a vacuum tube in that negligible drive power is required. The probability is that, whatever the nature of the preceding stage, more than enough drive will be available for a number of paralleled FETs.
- *Excellent constant-current characteristics.* Although the junction transistor is often compared to a constant-current generator, particularly in the common-base configuration, the FET has unique constant-current behavior that is more readily exploited for power-supply applications. For example, the FET source-drain (output) current is primarily established by the gate-source (input) voltage and it does not run away with increasing load demand. The practical consequence is that an FET series-pass element automatically provides its own protection under the short-circuit load condition. Because it limits its maximum output current and because of its negative gain/temperature coefficient, the FET's operating temperature soon attains a safe equilibrium following the short-circuit application. For the bipolar series-pass element, auxiliary circuitry must be incorporated for such protection. Unfortunately, such protective techniques are sometimes too slow in response or are otherwise inadequate to prevent catastrophic damage to the transistor.

Using FETs in the regulator circuit

When used expressly as a constant-current source, the FET outperforms the bipolar transistor for most practical purposes because it is much easier to achieve a high order of temperature immunity. The bipolar transistor by itself is not often used as a source of constant direct current. Rather, it is associated with a zener diode, and sometimes with a forward-conducting diode for temperature compensation. Conversely, the FET can provide constant current without depending on a voltage reference. When so used and when the gate source voltage, V_{GS}, is properly adjusted, an essentially zero-temperature coefficient of drift can be achieved.

This value of V_{GS} is readily obtained by adjusting resistance R_S in Fig. 4-14B. The zero-temperature-coefficient mode of operation will result when the difference between the manufacturer's specified pinch-off voltage (V_P) and the voltage that is developed across R_S (V_{GS}) is between 0.6 and 1.0 V. This zero-temperature coefficient pertains to output current as a function of temperature. The negative gain/temperature coefficient that immunizes the FET against thermal runaway remains in effect.

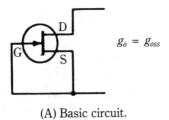

(A) Basic circuit.

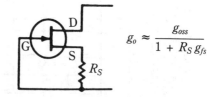

(B) Circuit with source resistor added.

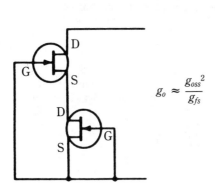

(C) Circuit using two transistors.

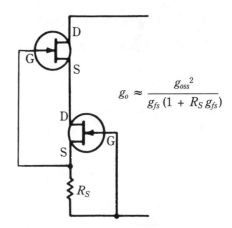

(D) Circuit using two transistors and source resistor.

4-14 Constant-current circuits that use the field-effect transistor.

In Fig. 4-15, the near-horizontal portions of the drain-current curves reveal the constant-current behavior of the FET. Experimental verification of this intrinsic current regulation can be made with the circuit of Fig. 4-16. Here, it is readily established that the output current, I_D, is substantially constant, despite wide variations in V_{DS} and/or R_L. All FET equations for constant-current operation are valid only when I_D is less than I_{DSS} and V_{DS} is greater than V_P.

Practical implementation usually assumes the form of one of the circuits shown in Fig. 4-14. Notice that these are all two-terminal arrangements. Circuit B is commonly used. With presently available high-transconductance (g_{FS}) FETs,

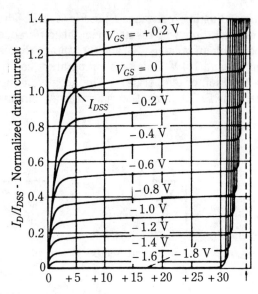

$$I_{DS} = I_{DSS}\left(1 - \frac{V_{GS}}{V_p}\right)^2$$

$$V_p = \frac{2 I_{DSS}}{gm @ I_{DSS}}$$

$$gm = \frac{2\sqrt{I_{DSS} I_D}}{V_p}$$

$$V_{GS} = V_p\left(1 - \sqrt{\frac{I_D}{I_{DSS}}}\right)$$

$$gm @ I_{DSS} = \frac{-2 I_{DSS}}{V_p}$$

I_{DSS} = Drain current when gate is connected to source.

4-15 Typical constant-current curves of the junction field-effect transistor.

Basic circuit equation: $V_{GS} = V_p\left(1 - \sqrt{\frac{I_D}{I_{DSS}}}\right)$ Where V_p and I_{DSS} are obtained from specification sheet

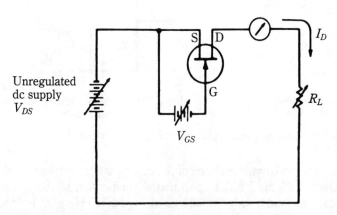

Change in output current as function of change in supply voltage:

$$\Delta I_D = \Delta V_{DS}\left(\frac{I_D}{I_{DSS}}\right)(G_{OSS})$$ where I_{DSS} and G_{OSS} are obtained from specification sheet

4-16 FET constant-current circuit.

considerable current-adjustment range can be had by means of R_s, consistent with a high-dynamic output impedance (i.e., low g_o). For many applications, the zero-temperature-coefficient mode of operation that corresponds to a discrete value of R_s is not necessary. In the formulas given in Fig. 4-14: g_o is the output conductance of the circuit; g_{oss} is the small-signal common-source short-circuit output conductance; g_{fs} is the small-signal common-source short-circuit forward-transfer conductance (the same as g_m in vacuum-tube circuits).

An interesting and useful application of the constant-current property of the FET is the simulated zener diode of Fig. 4-17. Here, R_S can operate at, or in the vicinity of, the zero-temperature-coefficient of drift. Then, R_L is a precision resistance of such magnitude as to yield the desired voltage reference. Notice that the burden of temperature stability of this arrangement falls on the resistances, R_S and R_L. One approach is to secure resistances with a very small temperature coefficient. Another scheme is to compensate the relatively small temperature coefficient of the resistances by finding an adjustment of R_s that endows the FET with a temperature-coefficient of the opposite sign. Although several trials might be necessary to bring both the desired reference voltage and the requisite stability into coincidence, the result can be worthwhile. This is particularly true for references that are below five volts, where both zener diodes and forward-biased diodes leave something to be desired.

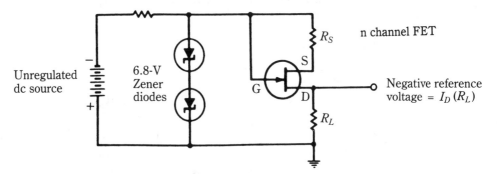

4-17   FET voltage-reference circuit.

In general, to use the potential stability of the circuit in Fig. 4-17, the input impedance of the stage that samples the reference voltage should be at least 100 times greater than R_L. This impedance is readily achieved by FET or vacuum-tube stages, bipolar stages with bootstrap circuits, or with feedback or Darlington techniques that raise the input impedance. Also, an operational amplifier can provide a high-impedance input because of the high values that are often permissible for the feedback resistances and because the "virtual ground" at the summing point does not consume current. The circuit of Fig. 4-17 will furnish a positive reference if the polarities of the supply and zener diodes are reversed and a p-channel FET is substituted.

The techniques in the voltage-regulated power supplies shown in Figs. 4-18 and 4-19 utilize these FET characteristics. In Fig. 4-18, bipolar transistor Q1 is the

184 *Implementation of regulation techniques*

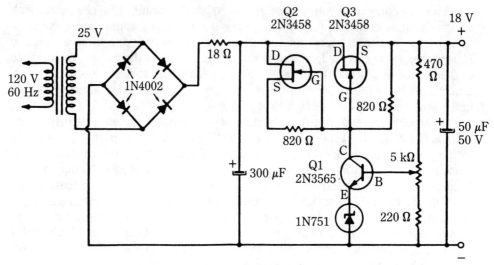

4-18 FET as a series-pass element in a voltage-regulator circuit.

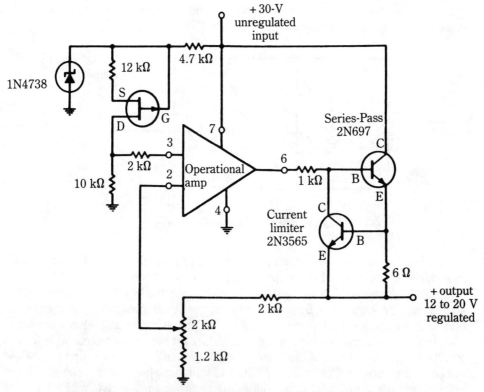

4-19 Series-pass voltage regulator with P-channel FET reference source.

comparator and error-signal amplifier. The high transconductance of the bipolar transistor as well as the fact that high input impedance is not required in this circuit position account for its use here. Q2 is an n-channel FET that is operated as a constant-current source for the collector of Q1. This source enables Q1 to develop high-voltage gain, which is readily implemented because of the inherently high input impedance of FET Q3. In the event of a short circuit across the output terminals, Q3 will deliver maximum current of approximately I_{DSS}, which is the "nature of the beast." Thus, no auxiliary current-limiting circuitry is needed. Insofar as Q3 is concerned, short-circuit operation can endure indefinitely. As previously pointed out, this circuit has no tendency toward thermal runaway.

The circuit of Fig. 4-18 does not need to be limited to the 75 mA or so that are available from a "small-signal" FET, such as the 2N3458. Power MOSFETs with current ratings of 12 A and higher are now available. These devices (initially introduced by Siliconix Inc.) are now manufactured by a number of firms. Although most of the developmental effort was initially expended on the n-channel MOSFET, an increasing number of p-channel types have been appearing on the market. The salient feature of the junction FET, negligible drive power, is even more dramatic in the power MOSFET. Moreover, some of these new devices carry voltage ratings of 500 V and greater (drive voltage is not negligible—in the 10- to 12-V range). MOSFETs are particularly well-suited to switching regulators and are discussed in chapter 5.

The voltage-regulated power supply of Fig. 4-19 uses a voltage-reference source that is constructed around a p-channel FET. Although satisfactory operation is obtained with little attention directed toward temperature stability, enhanced performance for rigorous thermal environments could readily be caused by appropriately selecting the source and drain resistances to cause the FET to operate near its zero-temperature coefficient of drift.

Switching-type regulators

The rheostat-like action of the series- or shunt-type regulators is not the only way that output current or voltage can be controlled or stabilized. The substitution of a switching device for the series-pass element is another way to vary the dc output level. In using the switch for this purpose, its duty-cycle or repetition rate can be the variable parameter whereby control is exerted over the dc output level when the input to the switch is a dc source. When the input to the switch is an ac source, the average value of ac output can be controlled by varying the time of switch closure; rectification and filtering occur after the switching process. The salient feature of all switching-type regulators is their improvement in efficiency over the continuous-duty types. In the zero-loss ideal switch, no I^2R is lost when it is closed, because it makes perfect resistanceless contact (i.e., R is zero). Conversely, when the ideal switch is open, it allows zero-current passage, which again results in zero power dissipation.

General characteristics and use of switching-type regulators

The following list gives some of the main characteristics of switching-type regulators and contrasts their performance with that of the dissipative series and shunt types.

- Because of their greater efficiency, switching-type regulators excel over dissipative types in packaging considerations. Considerable savings in volume and weight can be attained—especially at higher output power levels. Often, thermal considerations are of relatively little consequence.

- Relative to dissipative regulators, the switching-type regulator provides a coarsely regulated output. The switching technique inherently precludes high performance in response time, output impedance, and bandwidth (this can be largely overcome, however, by the use of high switching rates). Also, certain practical matters render the attainment of close regulation difficult. For example, one difficulty arises because of the filter that follows a dc switching device.

- The switching-type regulator can serve the unique function of a dc-to-dc transformer.

- Unless deliberate precautions are taken, the switching-type regulator can be the source of serious electrical noise and electromagnetic radiation. This regulator often requires adequate shielding and the insertion of line filters.

Combined use of dissipative- and switching-type regulators

A unique deployment of the switching-type regulator is as a preregulator, in conjunction with a dissipative regulator. Often, the association between the two regulator types is the relationship of double regulation, rather than a mere tandem arrangement. As previously shown, double regulation automatically enables the coarse regulator to reduce dissipation greatly in the series or shunt type. Then, all of the dynamic advantages of the dissipative regulator are obtained with minimal dissipation. The overall result can be a considerably smaller and lighter package than would be feasible with the dissipative regulator alone. Although components are added, the overwhelming factor is generally the reduction of operating temperature in heatsinks, blowers, and other devices.

Phase-controlled switching regulators

The principle that is embodied in the phase-controlled switching regulator is that the average value of an ac wave can be changed by varying the time allotted for passage of the wave through a switch that has been synchronized to the wave itself. This principle can be visualized as a form of duty-cycle modulation. It is easy to accomplish with such switching devices as thyratrons, (and particularly) SCRs, and triacs. A typical voltage regulator that uses SCRs is shown in Fig. 4-20. The salient feature here is the substitution of SCRs (X1 and X2, for their diode counterparts) to complete the bridge-rectifier circuit. The SCRs block passage of current

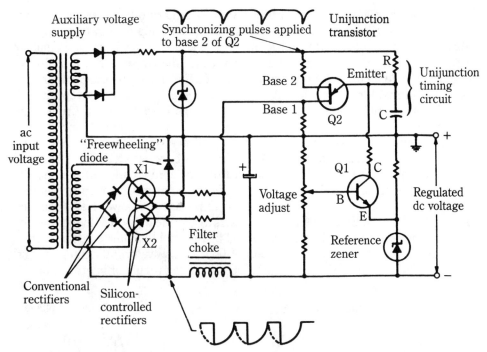

4-20 Voltage-regulated supply that uses phase-controlled switching technique.

until their gates are triggered. If their gates are triggered at the beginning of each half-cycle, the bridge rectifies in the same manner as a conventional four-diode bridge. If, however, the triggering signals occur later (i.e., after the sine wave has crossed the zero axis), only incomplete portions of each half sine wave will appear at the output of the bridge. The unidirectional pulses then delivered to the filter will emerge as a dc output level of smaller average amplitude than if a portion of each pulse had not been eliminated by the delayed conduction of the SCRs. Thus, dc output is controlled by varying the conduction time of the SCRs.

In Fig. 4-20, the timing of the SCR gate signals is established by a unijunction transistor that, in turn, generates the gate signal for the SCRs. Transistor Q1 functions as a comparator by varying its emitter-collector resistance, in response to the difference between a sample of dc output voltage and the voltage provided—by the reference zener. More specifically, a rise in output terminal voltage causes increased conduction in Q1, which thereby slows the rate that C can charge through R. Thus, the unijunction transistor fires later in the ac cycle and the attendant late conduction of the SCRs results in a restorative decrease in dc output voltage. The converse sequence of events occurs if the output voltage attempts to decrease.

To ensure that the SCRs are triggered at definite, rather than random, times, with respect to the power-line frequency, the gate pulses delivered by the unijunction transistor must be synchronized to the line frequency. This is brought about in a simple way. No filter capacitors are used in the auxiliary voltage supply that provides operating voltage for the unijunction stage. The ripple from this supply is great enough to appreciably decrease the voltage that is applied across B1 and B2 of the unijunction transistor 120 times per second (the consequence of full-wave rectification and a 60-Hz line). When its base voltage is low, the unijunction transistor will fire with relatively low voltage developed across charging capacitor C. Such firing will not, however, cause a large enough output pulse to trigger the gates of the SCRs. Rather, the voltage across C will start to charge from the same near-zero level, 120 times per second. This constitutes synchronization to the line; with Q2 in a given conduction state, the SCRs will be alternately fired, 60 times per second and each firing will be delayed from the zero crossing of the line frequency by the same phase angle. The significant point of this operation is that charging capacitor C repeats its cycle each time that the power-line frequency crosses its zero axis. Each time the cycle starts again, the delay that is required for a firing of one or the other SCRs is governed by the effective time constant (established by R, C, and the conductive state of Q1). The waveform diagrams of Fig. 4-21 show the operation of a phase-controlled switching regulator in greater detail.

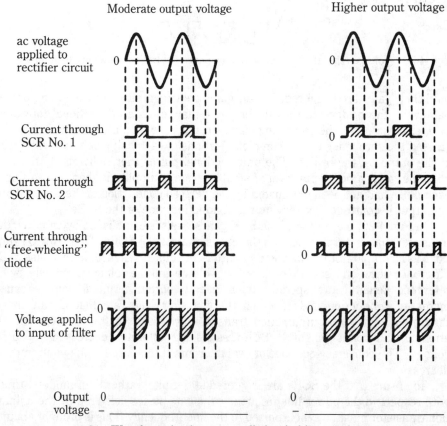

4-21 Waveforms in phase-controlled switching regulator.

The dc switching regulator

The phase-controlled switching regulator operates at relatively high efficiency because the control element(s) absorb a negligible portion of the controlled power. As has been explained, this is because these control elements are either in their fully ON or OFF states. Because in-between conductive times are negligible, the I^2R switching losses are very low. However, the need to provide subsequent filtering for the multiples of line frequency also means that the transient response of such a regulator must necessarily be slow. Thus, you are naturally led to contemplate a switching regulator, in which the ON-OFF repetition rate is much higher than line frequency.

A practical embodiment of such philosophy is the dc switching regulator. This regulator provides switching after rectification and therefore the necessity to synchronize it to the line frequency is eliminated. In the most general case, the dc load current is interrupted at an audio or ultrasonic rate and is controlled by duty-cycle variation. The switching voltage waveform is a rectangular wave in which the ON or OFF portions are duration-modulated by the error signal. A subsequent low-pass filter is required to average the dc level and remove the switching frequency and its harmonics. However, this filter, which is designed for a high-frequency fundamental, allows much better transient response than the similar filter in the phase-controlled regulator. The block diagram of Fig. 4-22 illustrates the basic concept of the dc switching regulator. Figure 4-23 shows how this operating principle can be implemented by appropriate circuitry.

Ideally, the dc switching regulator should excel other regulator types in all performance parameters. All that should be done, you might suppose, is to make the switching rate very high. However, when this is attempted beyond 20 or 25 kHz, the area of returns diminishes—turn-on and turn-off times then become an appreciable fraction of ON time. Thus, switching losses are no longer negligible and the prime virtue of this type of regulator is lost.

This problem has several causes. First of all, a power transistor other than a very expensive RF type, delays the rise and fall time. Secondly, the presence of the filter inductor renders the approach to step-function chopping exceedingly difficult. Nor is this problem circumvented by reducing the size of the inductor and simultaneously increasing the size of the output filter capacitor. This technique fails because the inductor must have a certain minimum inductance, the *critical inductance*, in order to produce a continuous current (this principle is similar to that which pertains to the design of the filter inductor in a simple brute-force rectifier system). Moreover, a capacitor input filter is not the answer either. Such a filter might allow faster switching transitions, but these are defeated by high-peak current, which might otherwise be conferred by the reduction in rise and fall time. A third difficulty with very high switching rates is the general difficulty in handling the high-order harmonics in circuitry that must operate efficiently at dc, audio, and radio frequencies. This is by no means beyond the state of the art. However, cost factors often limit switching rates. Although cost, efficiency, and physical size are often satisfactory for audio-frequency switching rates, such regulators can be acoustically noisy if appreciable power is involved. Switching rates in the vicinity

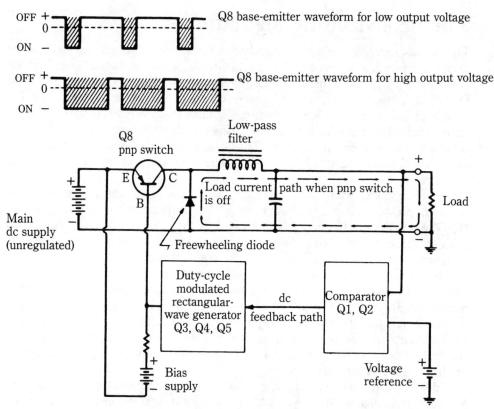

4-22 Block diagram of dc switching regulator (*Q* numbers refer to Fig. 4-23).

of 20 kHz have a good balance of all important performance and evaluation factors. The trend now is for higher switching rates. These rates are more readily accomplished with power MOSFETs, rather than bipolar transistors.

Some comments on inverter circuits and power switches

The chopping circuits of switching power supplies are variously referred to as *power switches*, *inverters*, or *converters*. These terms are sometimes used interchangeably. However, the term *converter* actually implies that a power switch or inverter is in conjunction with an output rectifier circuit so that both input and output voltages are dc. Inverters and power switches are similar in that they both "chop" dc, which thereby "inverts" to ac. In order to regulate, the ac waveform must then be pulse-width or duty-cycle modulated by an error signal that is sampled from the dc output of the supply. It is not always easy to distinguish between the power switch and the inverter. Generally, a single transistor that is used to interrupt the dc in a controlled manner is called a *power switch*. Pairs of transistors, especially when used in push-

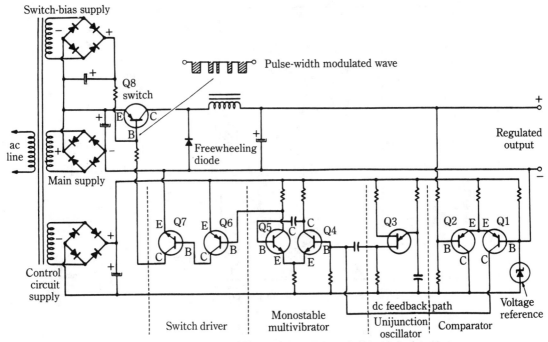

4-23 Generalized schematic of dc switching supply.

pull or half-bridge configurations, appear more deserving of the term *inverter*. This term applies whether such paired transistors are self-oscillatory or driven from a clock or waveform generator. The same is true for a full-bridge configuration that is composed of four transistors or active "choppers."

In modern regulated supplies, an important function of power switches and inverters, other than providing the means of current control, is to enable the switching frequency to be much higher than the 60-Hz frequency of the power line. Switching rates in the vicinity of 20 kHz have long been popular and the trend is toward much higher frequencies. The rewards of using high switching rates are drastically reduced sizes of transformers (and other magnetic-core devices) and of filter capacitors. Shielding and filtering against electromagnetic interference is generally easier. Heat removal improves, if for no other reason than because more "real estate" is available for heatsinks and other thermal hardware. Finally, with many manufacturers competing to provide the needed semiconductors, core components, and capacitors, the modern high-frequency "switcher" has become a very cost-effective means to regulate dc voltage and current. From a circuit approach, the inverter or power switch is the heart of such equipment. A brief look at some of these circuits will pave the way for the "guided tours" for switching power supplies in chapter 7.

A basic inverter circuit is shown in Fig. 4-24. This circuit is *self-oscillatory* and its operation depends on the magnetic-saturation characteristics of the transformer. Because the oscillation frequency is not primarily a function of an inductance-resistance time constant or a capacitive-resistance time constant, it is not a

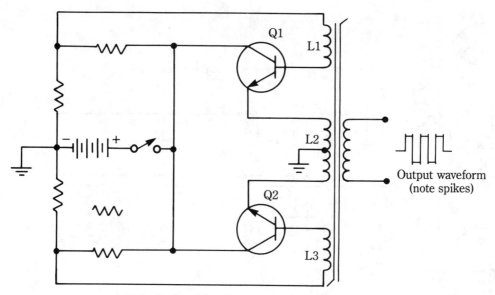

4-24 Basic self-oscillating saturable-core inverter.

multivibrator (it resembles a multivibrator only because it produces a square-wave output). Nor is it a member of the conventional feedback-oscillator family; no LC resonance is involved in setting its oscillation frequency. Feedback windings are used, but the oscillation mode is quite different than in LC oscillators. Finally, although it schematically resembles the magnetostriction oscillator, it is not of this variety either.

Suppose that the circuit is initially powered from its dc supply. A perfect balance of transistor characteristics would not likely exist and one transistor could be expected to draw collector current at a faster rate than the other. Assume the current hog to be transistor Q1. As the Q1 collector current builds in the upper portion of primary winding L2, it induces base-feedback voltages in windings L1 and L3. The polarities of these feedback voltages are such that forward-bias is applied to the base of Q1, but reverse-bias is applied to the base of Q2. These polarities reinforce the situation where Q1 is accelerated into hard conduction, but Q2 is definitely turned off. However, hard conduction in Q1 is also accompanied by magnetic-core transformer saturation. Thus, this feedback cannot be maintained indefinitely because transformer action ceases with the onset of core saturation.

Magnetic saturation collapses the magnetic field and this, in turn, induces voltages in all of the windings of opposite polarity to those that prevail while Q1 was ON and Q2 was OFF. Now, Q1 and Q2 exchange conduction states and the core becomes saturated in the opposite direction. The magnetic field again collapses and it can be seen that the process is both regenerative and repetitive. Because of regeneration, each switching event reinforces its own action so that the alternation of transistor switching states occurs very abruptly. The relationship and sequence of these events is illustrated in the diagram of Fig. 4-25.

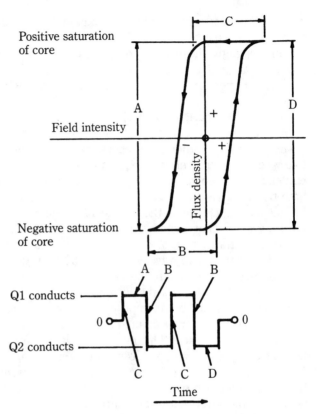

4-25 Relationship of core hysteresis loop to generated waveforms.

Steep sides of the generated wave correspond to saturation states of the core; horizontal portions of the wave correspond to hysteresis transition states. One complete cycle of the generated wave corresponds to one complete traverse around the hysteresis loop.

The oscillation frequency of this *saturable-core inverter* is:

$$f = \frac{E \times 10^8}{4 \times N \times B_{max} \times A}$$

where:

f is the frequency in hertz,
E is the dc voltage applied to the transformer,
N is the number of turns in one-half of the transformer primary winding,
B_{max} is the saturation flux density of the core material,
A is the cross-sectional area of the core.

This equation is derived from the classical relationship between transformer parameters. In practice, the dc voltage of the supply and E will differ slightly because the transistors, being imperfect switches, will incur a voltage drop, V_{CEsat}. However, once designed, the frequency of oscillation will be directly proportional to the applied voltage. In using this equation, the ordinary rules of dimensional consistency must be followed. For example, if B_{max} is expressed as lines per square inch, then A must be expressed in square inches.

Interestingly, the design of ordinary transformers always entails precautions to avoid operation in, or even too close to, the magnetic-saturation region. Conversely, for inverter operation, penetration into the saturation region of the BH hysteresis curve is a must. This penetration, however, is accompanied by the drawback that the transformer designer apprehends—degradation of efficiency. Thus, the self-oscillating saturable-core inverter necessarily sacrifices efficiency (in the form of hysteresis loss) in order to be operational. This, among other reasons, excludes this type of inverter from the larger power-supply circuits. Other disadvantages are tendencies for "spike" generation and a vulnerability to simultaneously conduct both transistors—often destroying them. The voltage spikes are generally caused by leakage inductance (in the transformer), which can penetrate the SOA rating of the transistors. Simultaneous conduction is very much a function of the turn-off speed of the transistors. Various circuit techniques are available to help produce a relatively clean square wave with nearly 50% duty cycle. The trick, however, is to prevent much further operating inefficiency.

You might ponder how the duty-cycle of the self-oscillating inverter can be controlled as a regulated power supply. The answer is that such a process is not feasible and is rarely, if ever, attempted. Nonetheless, this simple circuit is quite nice for output-level modulation. This modulation is accomplished in a "brute-force" way by controlling its dc voltage. To be sure, this voltage will also vary its frequency. However, the frequency change is generally inconsequential in the overall operation of the regulated power supply where this scheme is used. The voltage that is applied to the inverter can be controlled by a single transistor "power switch," which is duty-cycle or pulse-width modulated.

In order to circumvent the sacrifice in efficiency, which the saturating output transformer entails, and to escape spike generation, an inverter circuit (Fig. 4-26) can be used. Its operation is essentially similar to that of Fig. 4-25, but the saturating core element is now in the low-power-level base circuit. The output transformer now no longer saturates, but merely delivers the square-wave power to the load—in true transformer fashion (no saturation). This inverter arrangement is generally capable of better all-around performance than is readily attainable from the simpler one-transformer prototype.

The driven inverter

The majority of inverters that are used in switching-type regulated power supplies are not self-oscillating; they instead depend on a duty-cycle or pulse-width modulated waveform that is obtained from logic circuitry and driver stages. If the driving-power requirement is not too great, a single dedicated IC module is sufficient. In essence, the inverter then functions as a power amplifier and its configuration and operation both resemble that of a conventional class-B audio amplifier. This resemblance is especially true when deploying a nonsaturating output transformer. There is, however, no need for the linearity that is so important in class-B

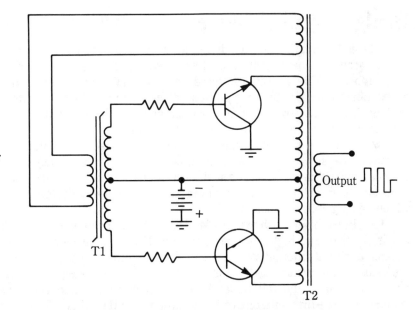

4-26 The two-transistor saturable-core inverter.

stages and every effort is made to ensure the transistors are saturated. This saturation further increases the overall efficiency of such an inverter. Under these operating conditions, the inverter operates as a class-D, rather than a class-B, amplifier. Figure 4-27 depicts a simplified example of such a driven inverter.

In addition to high operating efficiency and relative immunity to transistor-destroying spikes, an important feature of this type of inverter is its freedom from simultaneous conduction at maximum duty cycle. This comes about via the stepped waveform shown. This wave incorporates a "dead-zone" of sufficient duration to encompass the worst possible lag of turn-off time in the switching transistors.

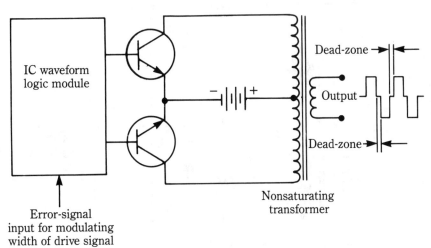

4-27 Example of a driven inverter.

Bridge-circuit inverters

Both the self-oscillating and the driven inverters that have been described are configured around the push-pull connection of two transistors. Such a circuit arrangement appears to be a straightforward and essentially simple method for energy conversion. However, the push-pull circuit has some built-in disadvantages. One problem is that each transistor must withstand twice the dc supply voltage plus the voltage spikes. Another drawback is that the utilization factor of the transformer is only 50%; each half of the primary winding is only used one-half of the time. In driven inverters, where core saturation of the output transformer is undesirable, the push-pull circuit often allows such saturation to occur because of inequalities in transistor parameters. Such action in turn produces collector-current spikes, which endangers the transistors.

The driven inverter shown in Fig. 4-28 not only overcomes these difficulties, but it is capable of higher speed operation. Known as a *half-bridge circuit*, this scheme operates differently than the push-pull circuit, yet it produces essentially the same overall result. Each transistor "sees" approximately the dc supply voltage, rather than double this value. Also the collector winding of the transformer is in constant use—first with one transistor, then with the other. The effects of leakage inductance generally do not appear as voltage spikes across the transistors, but tend to be dissipated in the output capacitor of the power supply (or in such a capacitor as C2 in the diagram). No net dc core flux is in the transformer and therefore it has no collector current spikes.

Implementation of this circuit might admittedly be more difficult than the push-pull circuit. The isolated base drive of the transistors is necessary. Also, capacitor C1 must be considered, particularly with regard to its current-carrying capability. Finally, the commutation diodes must have sufficiently rapid recovery characteristics to enable the circuit to be driven at the designated switching speed.

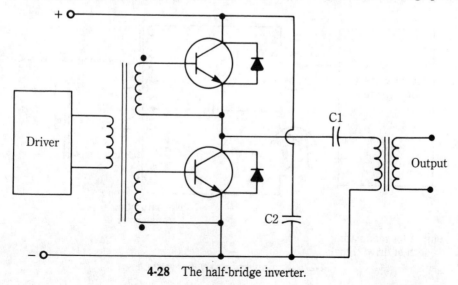

4-28 The half-bridge inverter.

The full-wave bridge inverter shown in Fig. 4-29 operates similarly to the half-wave arrangement, as long as the proper base-drive logic is available. Twice the output power will be developed, other things being equal. As is true in the half-wave bridge, the peak-to-peak transformer primary voltage is twice that of the dc supply. Yet, each transistor "sees" only approximately the supply voltage. Thus, in high-power regulated supplies, this inverter is often selected. The timing and isolation complications of the drive circuitry are readily resolved by the special IC waveform modules that have become available for just such purposes.

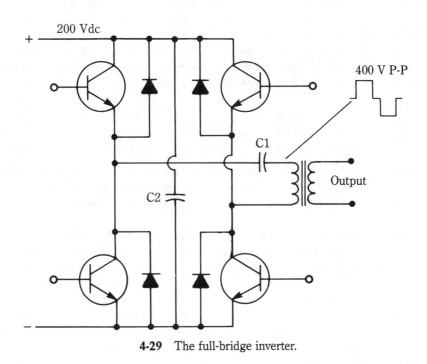

4-29 The full-bridge inverter.

Although the bridge inverters can be made to self-oscillate by the addition of appropriate feedback windings, this practice is not commonly encountered in regulated power supplies. The reason is that the same IC waveform module that is used for the drive logic conveniently provides the duty-cycle or pulse-width modulation that underlies most regulation techniques. A self-oscillating bridge inverter would require a "brute-force" approach to accomplish regulation—usually a power-switch circuit to modulate the dc voltage that is applied to the inverter. Such a scheme has already been alluded to in the section on the simple push-pull saturable core inverter. It appears to be an unsophisticated approach, but such "amplitude modulation" of any self-oscillating inverter produces a fixed harmonic spectrum. Conversely, the harmonics from duty-cycle or pulse-width modulation is "all over the place." This technique sometimes leads to knotty interference problems in the powered equipment.

Symmetry correction circuit

Although simultaneous conduction in push-pull power stages and inverters can be effectively eliminated by imposing dead time in the waveform, another malperformance often plagues regulators. Inequality of the volt-second product in the two halves of the output-transformer primary winding manifests itself as a net dc magnetizing current. This, in turn, causes *core saturation* with its attendant collector-current spikes. As with simultaneous conduction, asymmetrical operation can easily penetrate the SOA limits of the transistors and either slowly damage them or, as is more likely, cause quick catastrophic destruction.

Bridge and half-bridge power stages circumvent this kind of malperformance because such circuits use a capacitor to block any net dc current. If, however, you want to use the push-pull configuration, the following techniques can be implemented to minimize the effects of asymmetrical operation:

1. The transformer can be air gapped to forestall saturation.
2. The push-pull power stage can be fed from a current-limiting dc source.
3. The power transistors can be matched for saturation voltage and for speed.

Any or all of these methods can be used with varying degrees of success. When working with high power levels, however, these approaches tend to produce undesirable designs or operational side effects, which involve impracticalities and cost disadvantages.

A much better approach is to devise control circuitry for actually equalizing the volt-second products over a complete cycle. This idea is not new, but it has been unwieldy in discrete-circuit designs. The advent of the MC3420 IC (see chapter 5) control module has made symmetry correction relatively straightforward. It is readily implemented in any regulated supply that already uses the MC3420. Indeed, from a practical design standpoint, it is a good idea to get the supply operational without symmetry correction first (it can be evaluated under relaxed line and load conditions). Then, symmetry correction can be added.

With regard to the scheme to be described, the objective is to equalize the volt-second products over a complete operational cycle, but not necessarily to produce equal conduction times of the two power-output transistors. Thus, an oscilloscope inspection of the output waveform might not reveal wave symmetry that would please the artistic eye. A closer evaluation of the waveform would, however, reveal that the areas of the positive and negative portions are equal. This information is tantamount to volt-second symmetry and is the condition that must be attained to prevent any net core magnetization. The concept of symmetry correction is illustrated in Fig. 4-30, where the characteristics of the two push-pull output transistors are dissimilar.

In the circuit of Fig. 4-31, only the relevant portion of the MC3420 control IC is shown. If no symmetry correction had been implemented, pins 4 and 16 would have been tied together. Freeing pin 16 allows the pulse width of output transistor Q2 to be independently controlled (the pulse width of output transistor Q1 remains under the control of the supply's regulatory feedback loop).

Symmetry correction circuit

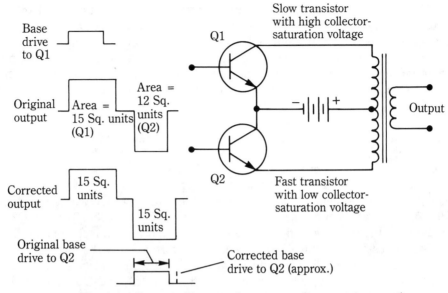

4-30 Waveform diagrams illustrate the concept of symmetry correction.

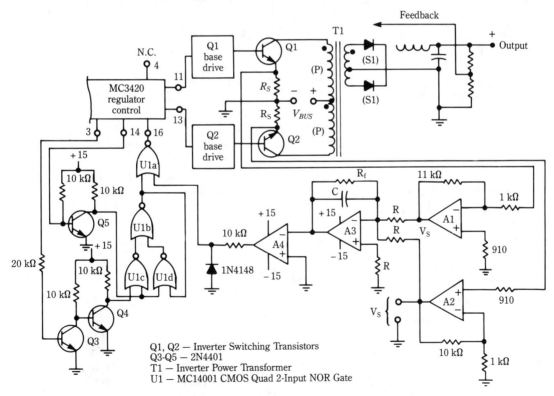

Q1, Q2 — Inverter Switching Transistors
Q3-Q5 — 2N4401
T1 — Inverter Power Transformer
U1 — MC14001 CMOS Quad 2-Input NOR Gate

4-31 Symmetry correction circuit for use with MC3420 control IC.

The primary currents of output transformer T1 are sensed as small voltage drops across the matched sensing resistances, R_S. These voltages are amplified in op amps A1 and A2. These amplified voltages are impressed at the input summing node of integrator A3. Because A1 produces polarity inversion and A2 does not, integrator capacitor C is charged in one direction when transistor Q1 conducts, but in the opposite direction during the conduction of transistor Q2. The charge accumulated in C during conduction of Q1 represents the volt-second product in the upper half of the primary winding of T1. The basic idea now is to use this stored information to cause the volt-second product from conduction of Q2 to have the same value. Then, no net magnetizing current will saturate the core of T1. To accomplish this, the conduction time of Q2 will have to be "metered."

When integrator capacitor C charges during the conduction of Q1, it causes the output of voltage comparator A4 to be low. This inhibits NOR gate U1 so that a "turn on" voltage (positive) cannot be applied to pin 16 of MC3420. Thus, Q2 is, for the time being, prevented from being driven into its conductive state. This enables a preset "dead time" to elapse. The information that represents dead time, as well as the basic operating frequency, is derived from pins 3 and 14 of MC3420. The circuitry (composed of transistors Q3, Q4, and Q5, and the four NOR gates) makes the lapse of turn-on action of Q2 for the dead time interval possible. It also synchronizes Q2 to the switching rate of the system, once the dead time is over.

After the lapse of dead time, the output of U1a will be high and will thereby cause power transistor Q2 to turn on. How long Q2 will be allowed to remain in its conductive state will now be determined by the opposing charging action of integrator capacitor C. As soon as C's "memorized" charge is neutralized, A4 will revert to its low output, and thereby cause the output of U1a to go low and turn off Q2. Because the original charge on C (from conduction of Q1) represented the volt-second product as a result of Q1's current in its half of the transformer winding, the discharge of capacitor C during Q2's conduction causes Q2 to be turned off at a time such that the volt-second product in the two halves of the primary winding of T1 is the same. With no net volt-second product over a complete operational cycle, the transformer core is not subject to any dc magnetizing effect.

Resistance R_f is high enough that this action suffers little detrimental effect. At the same time, R_f ensures that at power supply start up C will have no "memorized" volt-second product, which helps guard against one form of turn-on transient.

The three basic power switches

The power switch is often thought of as an on-off device that controls the average direct current passing through it with duty-cycle or pulse-width modulation. This is essentially true and references are made to just such a power switch in Fig. 4-22. Actually, this power switch is but one member (Fig. 4-32A) of the family of three power switching circuits (Fig. 4-32). A different arrangement of similar circuit components produces the flyback configuration (Fig. 4-32B). Whereas circuit Fig. 4-32A always develops less output voltage than dc input voltage, the flyback circuit generates a higher output voltage than the applied dc input voltage. A third

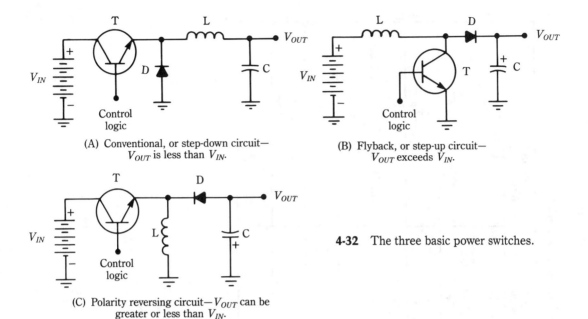

4-32 The three basic power switches.

permutation of circuit components results in the polarity-reversing power switch of Fig. 4-32C. Here, the numerical value of the output voltage can be either less than or greater than the dc input voltage.

Table 4-1 provides useful design formulas for these three power switches. These relationships are generally used for "ballpark" performance. Then, the empiricist takes over and experimentally finds the desired performance. Such procedure is used because users rarely care to know the exact characteristics of the respective components. Even if considerable effort was expended in measurements and calculations, various compromises were probably mandatory when the parts catalog was consulted. Also, one or more of the design parameters are usually initially assumed. The assumptions are generally made on the basis of practical experience. For example, you might assume a V_{ripple} value of 30 mV peak-to-peak because a lower ripple voltage would likely produce problems with ESR values of readily available filter capacitors and could also play havoc with feedback-loop stability. Moreover, the decision to design for a 30-mV ripple voltage would usually appear compatible with respect to the needs and tolerances of digital loads.

In using the design formulas listed in Table 4-1:

- V_{OUT} is in volts. Where V_{OUT} appears in the various design formulas for the polarity-inverting circuit, the simple numeric value should be used. If, for example, the output voltage is -10 V, the number 10 should be "plugged into" the formulas wherever V_{OUT} is indicated.
- C is in microfarads,
- L is in henries,
- I_{pk} is the peak inductor current in amperes.

Table 4-1. Design equations for the three power switches.

Design Parameter	Conventional circuit (voltage step-down)	Flyback circuit (voltage step-up)	Polarity reversing circuit (can be step-up or step-down)
V_{OUT}	$\dfrac{(L)(I_{PK})}{t_{OFF}} - V_D$	$\dfrac{(L)(I_{PK})}{t_{OFF}} - V_D + V_{IN}$	$\dfrac{(L)(I_{PK})}{t_{OFF}} - V_D$
I_{PK}	$2(I_{OUT})\dfrac{(V_{OUT}+V_D-V_T)}{V_{IN}-V_T}$	$2(I_{OUT})\dfrac{(V_{OUT}+V_D-V_T)}{V_{IN}-V_T}$	$2(I_{OUT})\dfrac{(V_{IN}+V_{OUT}+V_D-V_T)}{V_{IN}-V_T}$
t_{OFF}	$\dfrac{(L)(I_{PK})}{V_{OUT}+V_D}$	$\dfrac{(L)(I_{PK})}{V_{OUT}-V_{IN}+V_D}$	$\dfrac{(L)(I_{PK})}{V_{OUT}+V_D}$
$\dfrac{t_{ON}}{t_{OFF}}$	$\dfrac{V_{OUT}+V_D}{V_{IN}-V_T-V_{OUT}}$	$\dfrac{V_{OUT}+V_D-V_{IN}}{V_{IN}-V_T}$	$\dfrac{V_{OUT}+V_D}{V_{IN}-V_T}$
f	$\dfrac{1}{t_{OFF}+t_{ON}}$	$\dfrac{1}{t_{OFF}+t_{ON}}$	$\dfrac{1}{t_{OFF}+t_{ON}}$
L	$\dfrac{V_{OUT}+V_D}{I_{PK}}(t_{OFF})$	$\dfrac{V_{OUT}-V_{IN}+V_D}{I_{PK}}(t_{OFF})$	$\dfrac{V_{OUT}+V_D}{I_{PK}}(t_{OFF})$
C	$\dfrac{(I_{PK})(1/f)}{(8)(V_{RIPPLE})}$	$\dfrac{(I_{PK}-I_{OUT})^2(t_{OFF})}{(2)(I_{PK})(V_{RIPPLE})}$	$\dfrac{(I_{PK}-I_{OUT})^2(t_{OFF})}{(2)(I_{PK})(V_{RIPPLE})}$
Efficiency	$\dfrac{V_{IN}+V_D-V_T}{V_{IN}} \times \dfrac{V_{OUT}}{V_{OUT}+V_D}$	$\dfrac{V_{IN}-V_T}{V_{IN}} \times \dfrac{V_{OUT}}{V_{OUT}+V_D-V_T}$	$\dfrac{V_{IN}-V_T}{V_{IN}} \times \dfrac{V_{OUT}}{V_{OUT}+V_D}$

Regulators based on parametric power conversion

The Wanlass Electric Company, patent assignee for the Paraformer transformer,[1] provided information for this section. By virtue of its *mutual inductance*, the transformer is able to transfer electrical energy from one winding to another. This, of course, is "old hat"—so much so that it is remarkable that so few drastic innovations of the transformer have emerged throughout the years. In any event, the Paraformer is certainly one of these. This device has a laminated core structure as well as a primary and secondary winding. However, this superficial resemblance to a conventional transformer is misleading; for the operation of the two devices is derived from different effects. As will be shown, the Paraformer transformer possesses inherent properties that considerably improve several operating characteristics of regulators.

To best analyze the Paraformer transformer, the operational concepts that underly the old standby, the conventional transformer, should first be reviewed. Figure 4-33 depicts a simple transformer that is composed of a toroidal core of ferromagnetic material (powdered iron, for example) and two windings, which for the sake of simplicity, can be assumed to have the same number of turns. The following definitions apply to terms used in the formulas and graphs that appear in Fig. 4-33. E is the battery voltage; e_R is the instantaneous voltage that is developed across resistance R; e_L is the instantaneous voltage that is developed across the primary winding and is also equal to e_p; e_s is the instantaneous voltage across the secondary; N_s is the number of turns in the secondary winding; Φ is the flux-linking primary and secondary windings.

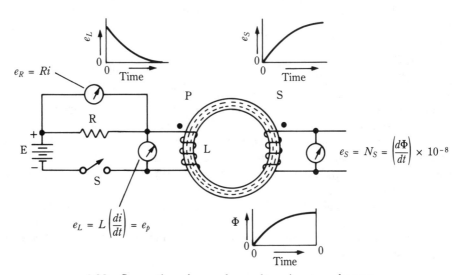

4-33 Generation of secondary voltage in a transformer.

---

[1]This device is presently being manufactured by Altech Power, Inc., Anaheim, California.

After closing switch S, current will build up in the primary winding and finally attain a steady value of $\frac{E}{R}$, where, again (for the sake of simplicity), it is assumed that the dc resistance of the primary is negligibly small. The rate of current growth will be governed by the time constant, $\frac{L}{R}$, where L is the inductance of the primary. Because the secondary is open-circuited, it does not enter into this cause-effect relationship. The current (i) at any time (t) is given by:

$$i = \left(\frac{E}{R}\right)(1 - e^{-Rt/L})$$

This is the exponential law of growth of current in a series LR circuit. If t is large, the term $e^{-Rt/L}$ becomes small, and i approaches its Ohm's law value, $\frac{E}{R}$. The interest here is the phenomena that occur within several time constants after closing the switch. For practical purposes, the current in the primary windings has attained its steady-state value after 9 or 10 time constants. In transformer action, only the changing primary current is of interest in this case.

The equation that defines circuit action after switch closure is $E = Ri + L\left(\frac{di}{dt}\right)$. The changing current designated by $L\left(\frac{di}{dt}\right)$ produces a proportionately changing flux, Φ, which links both the primary and secondary windings. As a consequence of this flux linkage, an induced voltage appears across the terminals of the secondary winding. This induced voltage has an instantaneous value of $N\left(\frac{d\Phi}{dt}\right)10^{-8}$ volts. Because of its unique involvement in this cause-effect relationship, $L\left(\frac{di}{dt}\right)$ is identified as the flux-coupling term. From what has been described so far, it would appear that a mutual and varying flux threading through the turns of both primary and secondary windings is the only way that electrical energy can be transported from a primary to a secondary winding. However, this energy can be transported in at least three ways. Two of these mechanisms for energy transfer are *electromagnetic radiation* and *capacitive coupling*. These two methods would, indeed, be feasible at radio frequencies. But, how else can they be coupled at lower frequencies?

The third additional way in which energy can be transferred from a primary to secondary winding that lacks a mutual flux linkage is by parametric coupling. The possibility of such coupling can be inferred by a restatement of the voltage equation following switch closure. Mathematically, its complete form is

$$E = Ri + L\left(\frac{di}{dt}\right) + i\left(\frac{dL}{dt}\right)$$

The added term, $i\left(\frac{dL}{dt}\right)$ is generally omitted because all but the comprehensive treatises on circuitry only cover the cases where it is assumed that the inductance L remains constant, thereby making $i\left(\frac{dL}{dt}\right)$ equal to zero. In the Paraformer transformer, primary current modulates the secondary inductance. Moreover, the secondary winding has a capacitor connected across it. Thus, the secondary is a resonant-tank circuit. The energy needed to oscillate this resonant tank and to supply useful power to a load is derived from the primary winding via the parametric coupling term, $i\left(\frac{dL}{dt}\right)$. The inductance that is involved in establishing resonance is the unmodulated inductance value of the secondary winding. In essence, this device is more closely related to an oscillator than to a transformer. Like more conventional oscillators, the energy at the resonant frequency of the tank circuit builds up initially from a small level.

Figure 4-34 shows the Paraformer is suggestive of a transformer. However, the configuration of this device is contrary to the philosophy of transformer design. The flux paths of the two windings have independent existences, rather than the all-important mutual linkage of transformers! Nonetheless, primary flux does pass through a segment of the core, which is associated with the secondary winding. Because of this and because the permeability of iron is a function of its flux density in a magnetic circuit, the desired variation of secondary inductance is caused. At the same time, there is negligible induced voltage developed in the secondary winding from the familiar transformer action.

The Paraformer must not be confused with the ferroresonant transformer or other so-called constant-voltage transformers. These modified transformers transfer energy in conventional transformer fashion—via a flux path mutual to the pri-

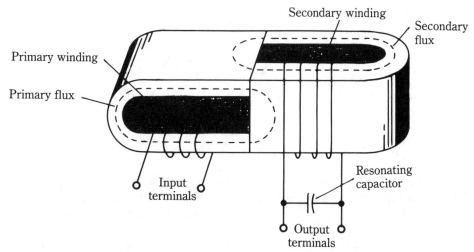

4-34   The construction of the Paraformer-type transformer.

mary and secondary windings. Also, they regulate by means of core saturation. This method results in a highly distorted output waveshape and in an inordinately high ratio of weight-per-watt.

Unusual characteristics of the Paraformer transformer

The Paraformer transformer displays characteristics that are ideally suited for use in regulated power supplies. Unlike a conventional transformer, its secondary reproduction of noise and transients that accompany the primary voltage source is practically nil. Thus, the Paraformer transformer dispenses with the need for RFI filters in the power line. Moreover, this suppression operates in both directions: a switching transient that emanates from the load will not be propagated back into the primary circuit. Figure 4-35 shows that the secondary circuit "sees" only the sine-wave component of the primary voltage. As will be discussed later, this inherent bandpass filtering action still operates effectively when the primary voltage is a square wave (the secondary voltage wave remains sinusoidal).

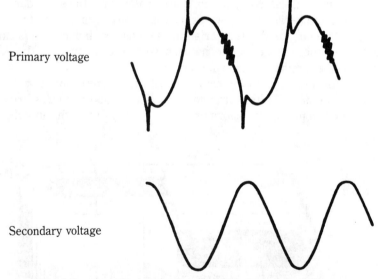

4-35 Inherent noise immunity of the Paraformer transformer.

The Paraformer transformer is an excellent line-voltage regulator. Figure 4-36 depicts the inordinately wide regulatory range that is attainable. This property alone would make it useful as a means for dissipation control in conventional series-pass voltage regulators. When used in this manner, the Paraformer transformer would spare the series-pass transistor from the burden of high power-line

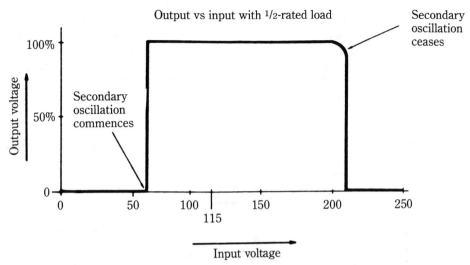

4-36 Typical line-voltage regulation of a Paraformer transformer.

voltage. The line-voltage regulation of a Paraformer transformer operates into half its rated current load and can typically be in the vicinity of ±0.1% over a line voltage range of 100 to 130 V.

At the same time, the Paraformer transformer provides good load regulation, as can be seen from Fig. 4-37. This regulation can be made arbitrarily smaller by increasing the wire size of the secondary.

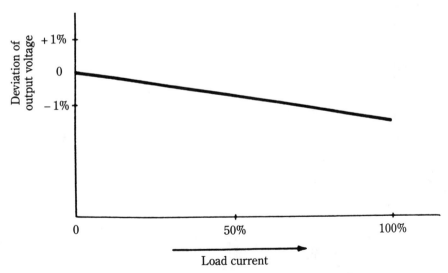

4-37 Typical load regulation of the Paraformer transformer.

208 *Implementation of regulation techniques*

As shown in Fig. 4-38, the Paraformer transformer is well-adapted to dissipation control and preregulation when used in the place of a conventional transformer. When so used, it is possible to dispense with the auxiliary circuitry that is often incorporated to protect the series-pass element from overcurrent or short circuits. As can be seen from Fig. 4-39, excessive current demand simply causes the output voltage to collapse to zero.

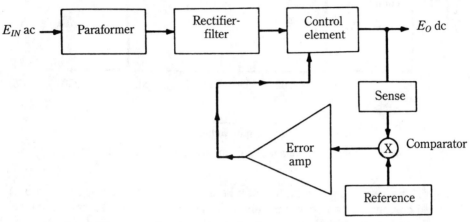

4-38 Dissipation control and preregulation by use of a Paraformer transformer.

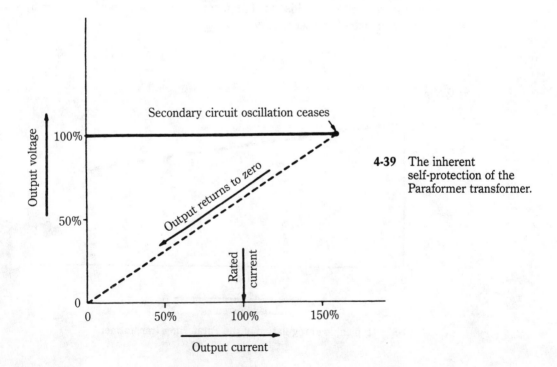

4-39 The inherent self-protection of the Paraformer transformer.

A particularly valuable feature of the Paraformer transformer that is relevant to its use in dc regulated power supplies is its capability for highly efficient energy transfer when the primary is excited by a square wave with a repetition rate many times that of the power-line frequency. Thus, a unit that is designed for operation from a 10-kHz source permits a dramatic reduction in physical size and weight compared to 60-Hz equipment. Moreover, the square-wave generator is also compact and efficient—its switching elements, being either ON or OFF, contribute very little to dissipative losses.

As if all of these characteristics are not sufficient for you to consider the Paraformer, this device seems to be intended for a regulatory feedback loop because of its control characteristics. Although the secondary voltage exhibits remarkable immunity to variations in primary voltage and to load changes, it is very much a function of the frequency of the primary excitation. This suggests a closed-loop technique where the dc output voltage is sensed and made to vary the repetition rate of a square-wave generator (inverter), which is connected to the primary of the Paraformer. Figure 4-40 is a block diagram of a dc voltage-regulated supply that uses such a technique.

A disadvantage of the Paraformer is that its primary draws appreciable current, even at no load. However, the overall efficiency improves at higher ratings. Another factor is that the starting current tends to be high.

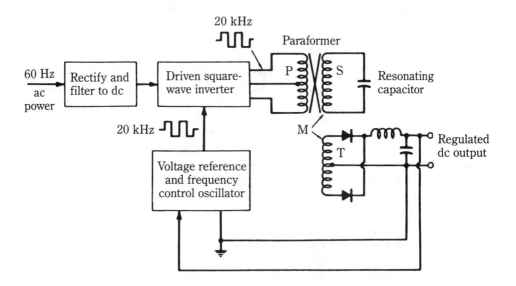

P = Primary or input winding
S = Secondary or resonating winding
T = Tertiary or output winding

4-40 Switchmode supply with Paraformer transformer as control unit.

The RF power supply

Embracing both vacuum-tube and semiconductor technologies, the RF power supply has waxed and waned with the passing years. Because it possesses some inherent advantages, the RF supply has never lapsed entirely into obsolescence. As new devices and upgraded components become available, the shortcomings of this type of power supply appear less formidable. The salient features of the RF supply are:

- It is normally compact with lightweight construction.
- It is a relatively safe high-voltage source because small filter and output capacitors are sufficient for filtering.
- Its cost compares at least favorably, and probably advantageously, with brute-force transformer step-up of the 60-Hz power-line frequency.
- The feedback loop is discontinuous at dc; which permits considerable design and construction flexibility and allows greater freedom from high-voltage insulation problems. Some of the knotty problems that are encountered in the adaptation of conventional series-pass regulators to high-voltage control are neatly circumvented.

Surely then, the RF supply must have some severe disadvantages to account for its relatively modest use. Indeed, it does; however, some of the alleged shortcomings are controversial or are shared to a considerable extent by other types of supplies. The one major obstacle that is associated with the RF supply is its tendency to radiate or otherwise inject high-frequency interference into powered circuitry or other equipment. By the time suitable shielding and decoupling techniques have been invoked to minimize such interference, the cost factor might no longer retain its initial appeal. Although these problems might deter its use in modern television receivers, other equipment (such as Geiger-Mueller counters) are less susceptible to radio frequency interference, and might thrive very well in association with the RF-type regulated supply. Moreover, it is not necessarily true that other types of supplies (notoriously, the phase-controlled SCR regulator and often the dc-switching regulator) do not generate equally severe interference.

The RF-type regulated supply (Fig. 4-41) is basically an old design approach. However, the elimination of high-voltage rectifier tubes and the substitution of a zener-diode voltage reference for the much used VR tubes (and even neon bulbs) considerably improve the performance and relax the packaging requirements. In older designs, the rectifier tubes derived their filament power from a several-turn link that was coupled to the tank circuit. This means was not suitable for securing long life from the tubes, nor was the interaction with the RF-energy supply desirable. Also, the filament power directly reduced the available output current.

The supply of Fig. 4-41 uses a Meissner oscillator circuit that is built around V3. This circuit is particularly suited to the purpose because oscillation occurs at, or close to, the frequency at which high-voltage winding L2 resonates with its own distributed capacitance (generally in the 100- to 300-kHz range). Meissner oscillator-coil assemblies are commercially available with various step-up ratios. The

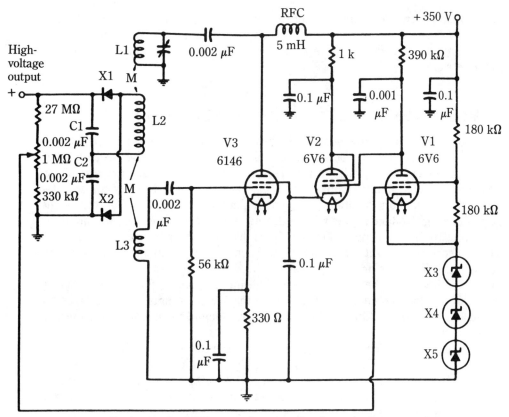

4-41 Basic circuit of RF-type voltage-regulated supply that uses tubes.

induced RF voltage in L2 is rectified by a full-wave voltage-doubler circuit with silicon diodes X1 and X2 and capacitors C1 and C2. A sampled portion of the high-voltage dc output is returned to the control grid of comparator V1. Zener diodes X3, X4, and X5 provide a voltage reference of about 150 V. The amplified error signal that is developed in the plate circuit of V1 is impressed at the control grid of cathode-follower V2 and finally it exerts control over the amplitude of the RF oscillations that are generated by V3. This action is accomplished by varying the screen-grid current of V3.

Many different tube lineups can be successfully used, in addition to that shown in the circuit diagram. For example, V3 can be a TV horizontal output tube and V1 and V2 can be almost any of the miniature pentode types that are commonly used in tube TV sets (the suppressor grid is then connected to the cathode). Also, V2 can be a triode. The use of a beam-power tetrode for V3 is desirable because of the relatively loose coupling between L2 and L1—a tube with too low power capability can limit output current capability and restrict the experimental optimization of overall performance. As shown, the circuit is suitable for dc outputs on the order of 5 kV. Higher voltages (to 30 kV, for example) can be produced with an appropriate step-up ratio between L1 and L2. For the higher voltage levels, the series string of 50-V zeners should be increased so that an inordinate loss in loop gain does not result through the necessity of an excessively high divide-down factor in the sampling network.

The solid-state RF high-voltage supply

Solid-state regulated high-voltage supplies have been implemented by "brute-force" techniques that incorporated two design approaches. On the one hand, they have used 60-Hz step-up transformers. However, much can be gained in size reduction, reduced cost, and personal safety. The last item is extremely important—60-Hz filter capacitors inevitably store enough energy to be lethal. Conversely, the output of a high-voltage supply with high-frequency step-up methods will not be as dangerous, unless the supply also has been designed for high-current capability. This is because a tiny filter capacitor is sufficient at high frequencies. Other things being equal, it is only necessary to store $1/1000$ of the energy in a 60-kHz output filter capacitor as for the same degree of filtering in a supply that uses 60-Hz step up (even if a low-frequency supply can be made with a small output capacitor because of "electronic filtering," lethal energy is bound to be stored elsewhere in the circuit).

The other brute-force technique has been the use of series-pass regulating transistors directly in the high-voltage line. This technique can be used because the transistor only "sees" the several tens or several hundreds of volts that it drops. Of course, the trick here is to devise reliable protective circuitry that will protect the series-pass transistor in the event of a high-voltage short circuit.

A much better approach involves the basic idea used in the simplified circuit of a solid-state regulator (Fig. 4-42). The overall action is similar to that of the tube circuit of Fig. 4-41, even though it does not have a one-to-one correspondence. All transistors operate at low voltages and, because of the high frequency (nominally 100 kHz), the energy storage is relatively low. Nonetheless, use extreme caution when working with high voltages. Much gray area divides the irritating shock from the lethal shock. For example, not too many deaths result from contact with automobile ignition voltages or with TV high-voltage supplies, but both of these sources are only marginally "safe." Both have histories of lethal consequences.

In the circuit of Fig. 4-42, regulation occurs as a result of controlling oscillation amplitude with a series-pass transistor. Notice that the oscillator and power amplifier stages are not actually functionally distinct stages. Rather, the entire RF circuit is involved in the oscillatory process by virtue of the inductive coupling between the oscillator tank coil, N, and the primary winding, N1, of the RF step-up transformer. This technique dispenses with the need for separate tuning of the oscillator and amplifier tank circuits. In practice, capacitor C is selected so that the oscillation frequency coincides with the self-resonant frequency of secondary winding N2. This can be ascertained by severing the feedback path of the regulator and then selecting C for the greatest high-voltage dc output. An inherent feature of this type of supply is that a short circuit or a heavy overload that is applied

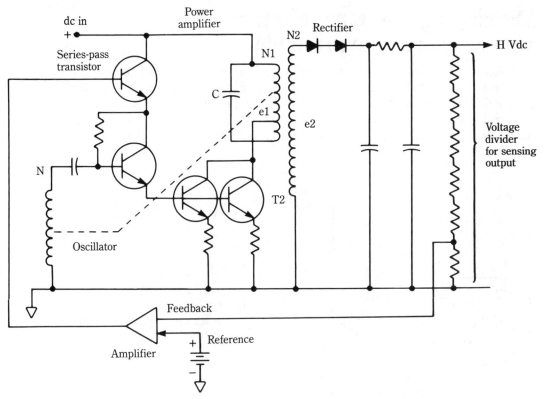

4-42 Simplified solid-state high-voltage regulated supply circuit.

to the output will not cause any damage. Generally, the oscillator will only provide 15 or 20% more current and the output voltage will drop so that the supply will not be subject to excessive power dissipation.

A very important component in this type of supply is the output voltage divider. For best results, it should be a specially made high-voltage type with a low temperature coefficient. Once this requirement is met and a stable voltage reference is used, the regulating ability of the supply can readily rival that of its low-voltage counterparts.

Switching-type regulated supplies have not yet made deep inroads in high-voltage technology. It appears that RF techniques readily yield such features as low cost, good regulating ability, low energy storage and reliability, and compact packaging. However, with switching rates now invading the RF spectrum, it is likely that pulse-width modulation will find applications in high-voltage supplies.

A useful circuit building block for the high-voltage designer is the voltage multiplier (such as is commonly used in TV receivers). These multipliers usually provide step-up ratios of three or five and relax various difficulties that often accompany the production of high voltage. Although intended for 15-kHz operation, these multipliers are also suitable for use at the frequencies that are ordinarily used in RF power supplies.

Developing a voltage reference via the energy-gap principle

An interesting circuit function that is readily achieved with monolithic technology is represented by the simple arrangement shown in Fig. 4-43. The basic circuit can be associated with other circuitry to provide either constant current or stabilized voltage. Its configuration appears deceptively primitive in view of the exceedingly useful performance it can provide. Although qualitatively similar operation is

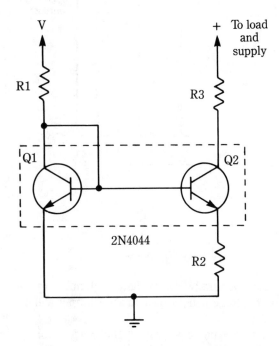

4-43 Monolithic transistor pair as a current source or voltage reference.

obtained from discrete transistors that are similarly connected, the monolithic version is incomparably superior in predictability, stability, and temperature coefficient. The reason is that the transistors of the monolithic pair are nearly perfectly matched in transconductance and other parameters. Also, the thermal coupling between them is much better than can be obtained practically with discrete transistors. This monolithic technique should not merely be considered as an alternative to a zener diode or a constant-current source, because it has the following unique characteristics:

- It can provide constant current in the low-microampere range—even though its own internal resistances do not exceed a few thousand ohms. This is particularly useful in monolithic modules because the fabrication of high resistances is difficult. The output current is a linear function of the absolute temperature. Good linearity prevails over a wide temperature range—typically −55 °C to over +100 °C.

- When used as a current source for the emitters of a differential amplifier, a nearly logarithmic change of current is produced, with respect to dc supply voltage. This maintains the collector current and voltage gain of the differential amplifier practically constant despite wide variation of supply voltage.
- When used in conjunction with a resistance to provide a voltage reference (R3 in Fig. 4-43), a positive temperature coefficient is obtainable that can compensate the negative temperature coefficient that is generally displayed by a transistor regulator circuit. Therefore, the entire regulator can be made to have a near-zero temperature coefficient.
- When functioning as a voltage reference, much lower stabilized voltage values are available than can be obtained from a zener diode.
- This circuit produces only a small fraction of the noise that is generated by zener and avalanche phenomena.

Referring to Fig. 4-43, diode-connected transistor Q1 operates at a much higher collector current than transistor Q2. The collector current to Q1 is determined by V and R_1. The base-emitter differential voltage, ΔV_{BE}, of the two transistors is developed across R2. This differential voltage enables the two identical transistors to operate at different collector currents. The two important equations that govern the operation of this circuit are:

$$\Delta V_{BE} = \frac{kT}{Q} ln \frac{I_{C1}}{I_{C2}} \quad (1)$$

where:

ΔV_{BE} is the base-emitter differential voltage of transistors Q1 and Q2,
k is Boltzmann's constant $= 1.38 \times 10^{-23}$ (watt-sec per °K),
Q is the charge of the electron $= 1.6 \times 10^{-19}$ (coulomb),
T is the absolute temperature in degrees Kelvin (°K), (25 °C is approximately equal to 300 °K),
I_{C1} is the collector current of transistor Q1 in amperes,
I_{C2} is the collector current of transistor Q2 in amperes.

$$R_2 = \frac{V_{BE}}{I_{C2}} \quad (2)$$

R1 is selected to produce a current in Q1 that is perhaps 50 times greater than the current, I_{C2}, that Q2 will be required to stabilize. Next, V_{BE} is calculated from equation 1. Then, equation 2 is used to derive the value of R2. Transistor Q1 can be thought of as the biasing transistor, and Q2 as the current-source transistor. When Q2 acts as a current source for another circuit, R3 is generally zero. Resistor R3 is used when it is desired to utilize the potential drop across it as a voltage-reference source. Whether Q2 is used as a current source or, in conjunction with R3, as a voltage source, the collector of Q2 must be positively biased. This occurs naturally when Q2 is the current source for the emitters of an npn differential amplifier.

If R_2 is zero, ΔV_{BE} vanishes and Q2 receives all of the base-emitter bias that is developed by Q1. Under this condition, $I_{C1} = I_{C2}$ and the temperature coefficient of the current in the collector circuit of Q2 is zero over a wide temperature range. Indeed, it would be very difficult and impractical in production to approach such electrical and thermal tracking with discrete transistors.

The off-line switching regulator

Among the most significant advances in regulated power supplies have been novel circuit implementations of switching regulators. Indeed, much of the reductions in size and weight that are attributed to the "switcher" actually stems from such unique circuit arrangements. Even though the simplest switching regulator that uses a series "chopper" already surpasses a linear-type supply in the realm of compact packaging, it, too, suffers a common shortcoming along with the linear supply. The physical parameters of both types are very much determined by the 60-Hz input transformer and the relatively large filtering components. The new schemes that have come into prominence dispense with 60-Hz magnetics altogether. Reductions in size, weight, and cost have been dramatic.

A typical way to accomplish this is shown in Fig. 4-44. Although a "power transformer" remains in the system, it is of minimal size and weight because it

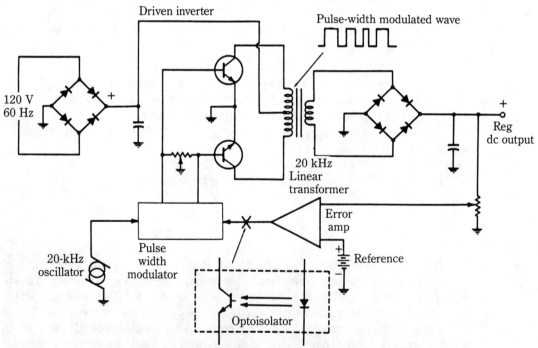

4-44 Simplified circuit for one type of line-operated switching regulator.

operates at a much higher frequency than 60 Hz (generally near 20 kHz). At first, you might assume that the direct operation of such a regulator from the ac power line is a "good-economy/bad-engineering" technique, such as those used in radios and TV sets. However, the conductive isolation of the load circuit from the power line in such a switching regulator can be just as good as is attainable from any regulated supply that uses a 60-Hz input transformer. To accomplish this, just insert an inexpensive optoisolator in the feedback path (Fig. 4-44).

In addition to the elimination of the 60-Hz transformer, the regulated supply in Fig. 4-44 has other attractive features. Because the input rectifying system operates at a relatively high voltage, the filter capacitor is often physically smaller than in supplies in which the line voltage is stepped down. Notice also that, regardless of polarity, operation can also be obtained from a dc power line. The transistors in the driven inverter work in conjunction with a nonsaturating output transformer, which implies negligible spike production. Also, the linear operation of this transformer results in lower core losses than prevails in saturating transformers. Conspicuous by its absence is the free-wheeling diode that is found in series-chopper switching power supplies. This basic configuration provides the highest operating efficiency of all regulated power supplies. The output rectifying circuit in units with 5-V output is, in the interest of efficiency, often the full-wave center-tap arrangement, rather that the bridge type. Schottky rectifying diodes are becoming popular for this purpose because of their lower forward voltage drop, compared to silicon pn-junction diodes.

With the advent of numerous types of optoisolators, this device will probably become a normal inclusion of switching regulators. Even where power-line isolation is not of prime importance, the use of an optoisolator frees the power-supply designer from problems that relate to dc levels in the feedback loop. The optoisolator is less likely to disturb the phase conditions in the feedback loop than other methods of complying with the often incompatible ac and dc requisites of the feedback path. Practically, this method has less trouble from self-oscillation and other instabilities.

Another implementation of the line-operated switching regulator is shown in Fig. 4-45. Similar overall operation is obtained here, but somewhat at the expense of efficiency. As a trade-off, however, some advantages might be meaningful in some applications. Instead of a driven inverter, a self-oscillating type is used in this scheme and the output waveform is always a symmetrical square wave. Also, the square-wave repetition rate is fairly constant. Regulation is obtained by varying the squarewave amplitude. As a result, the frequencies and amplitudes of harmonics are not as "wild," as with pulse-width modulation. In practice, this often means that the regulated dc output will be easier to filter. Although the power switch does use pulse-width modulation, its effect upon the rectifier output filter is much less than in switching circuits, where the output transformer handles the pulse-width-modulated wave directly. The series chopper also generally operates at a high frequency; experience has shown that the operating frequency can be in the vicinity of 20 kHz.

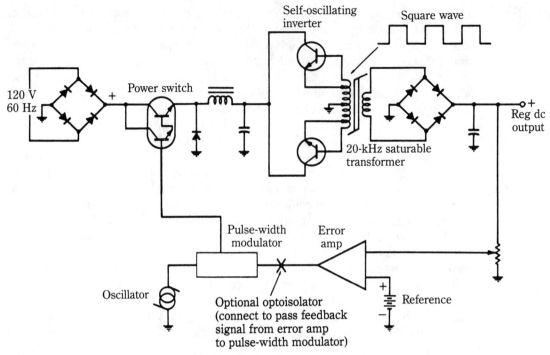

4-45 Simplified circuit for another line-operated switching regulator.

Pulse-width modulation

It is evident that the *pulse-width modulator* is an important building block in switching-type regulators. Probably the best way of implementing this function is to use one of the relatively new ICs that are specifically designed for this purpose. Two of these, the Motorola MC3420 and the Silicon-General SG1524, are described in chapter 5. However, pulse-width modulation schemes have been successful before the advent of these dedicated ICs. Several of the methods that use op amps are depicted in Fig. 4-46. The block diagram in Fig. 4-46A represents a straightforward method of generating a waveform in which the duty-cycle is varied in response to the changing dc level that is obtained from the error amplifier of the switcher. With this method, the summing amplifier must be kept from being driven too hard—it must operate within its linear range at all times. Figures 4-46B, 4-46C, and 4-46D show relevant op-amp circuits useful that are for such a pulse-width modulator. The LM3900 IC is very convenient to use for this purpose because it is composed of four independent op amps. Moreover, these op amps are intended for single-supply operation. The arrow on the op-amp symbol indicates the deployment of the *current-mirror principle* to achieve the noninverting input function. For practical purposes, the behavior of these op amps is similar to those that use the more conventional differential-input stage.

Pulse-width modulation 219

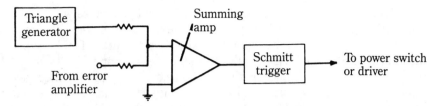

(A) Functional block diagram of a pulse-width modulator.

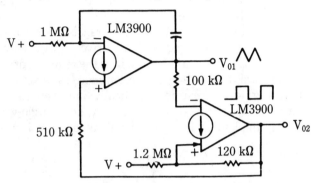

(B) Triangle/square generator using the LM3900 op amp.

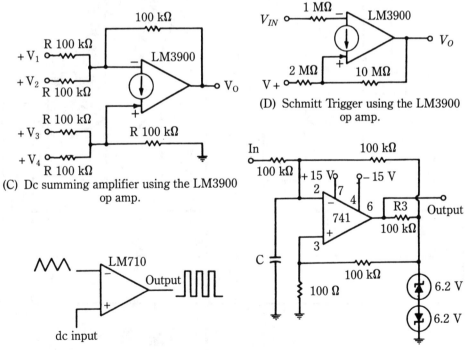

(C) Dc summing amplifier using the LM3900 op amp.

(D) Schmitt Trigger using the LM3900 op amp.

(E) Simple modulator obtained by driving the amplifier into saturation.

(F) Self-oscillating modulator does not require triangle generator.

4-46 Methods for implementing pulse-width modulation.

By driving the summing amplifier hard, a simpler pulse-width modulator is obtained (Fig. 4-46E). Finally, by making the amplifier self-oscillating, it is even possible to dispense with the separate triangle-wave source. Thus, the pulse-width modulator of Fig. 4-46F is actually the ultimate in simplicity. As might be suspected, however, the simpler methods might not always readily meet such operational requirements as range or circuit flexibility. Here, as elsewhere in this book, the venerable 741 "workhorse" op amp can be replaced by the newer BiMOS types. These use MOS input, and bipolar output transistors. Offset current and voltage are very low in these devices.

Driven inverters that use a pair of push-pull transistors (such as is used in the regulating system of Fig. 4-44) require two drive signals of opposite phase. Although various methods can be used to produce the required drives from the single-ended circuits shown in Fig. 4-46, it is often more satisfactory to use a pulse-width modulator (Fig. 4-47) that is specifically intended for push-pull drive applications. Because this circuit is self-oscillating, no triangle-wave generator is needed. To properly implement this pulse-width modulator, it must be operated

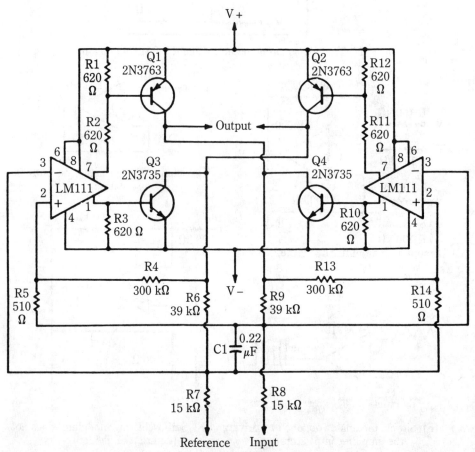

4-47 Pulse-width modulator for push-pull inverters.

from a "floating" dc source—one electrically isolated from both the ac line and the dc output circuitry of the regulator (a single 10- to 15-V unregulated supply is sufficient). The reference lead connects to the same ground point that is shown connected to the emitters of the inverter transistors in Fig. 4-44. Finally, the output of the error amplifier connects to the input lead of the pulse-width modulator.

Compliance with these procedures prevents part of the output circuit of the pulse-width modulator from shorting. Such special handling stems from the use of a bridge output circuit, rather than the more commonly encountered push-pull configuration. The self-oscillation frequency of the modulator can be adjusted via timing capacitor C1.

Line-operated power supplies using both linear and switching techniques

Other arrangements can eliminate the need for a massive 60-Hz transformer. These arrangements often use various combinations of linear- and switching-type supplies. Two examples of this approach are shown in Fig. 4-48. The scheme in

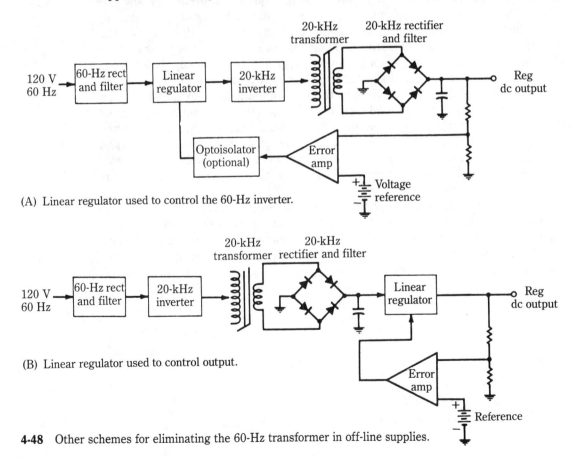

4-48 Other schemes for eliminating the 60-Hz transformer in off-line supplies.

Fig. 4-48A is similar to the system shown in Fig. 4-45. However, the functional block that controls the dc voltage that is applied to the 20-kHz inverter is now a linear, rather than a switching-type regulator. Although the overall efficiency is necessarily lower than the setup of Fig. 4-45, this method is preferable for installations where minimal switching transients can be injected into the ac power line.

A second configuration that uses a linear regulator is in Fig. 4-48B. Here, the linear regulator is placed at the end of the system, and the remaining functional blocks compose the unregulated dc supply. Although such an arrangement is not the most efficient, it is the best of the line-operated supplies for obtaining close regulation, low dynamic output impedance, and low ripple. Notice in Fig. 4-48B that the isolation between the dc output and the ac line is provided by the 20-kHz transformer—no optoisolator is required for this configuration.

Digitally controlled power supplies

You would hardly expect digital techniques to have overlooked regulated power supplies. In any event, it is often worthwhile to be able to control the output voltage or current digitally. You can then adjust the output quickly, conveniently, and precisely. More sophisticated behavior can also be exploited. For example, a digitally controlled power supply is ideal for automatic testing or for obtaining rapid measurements. The operation of motors and machine tools via computer programming is relatively straightforward when a digitally programmed power supply is used as a basic function block. A digitally controlled or programmed power supply regulates similarly to ordinary supplies with fixed outputs or with potentiometer-adjustable outputs. The essential difference is that the conventional adjustment potentiometer is replaced by some method of making discrete adjustments in pre-calibrated steps. These adjustments can be accomplished mechanically by using appropriate tap switches, relays, etc. On the other hand, more sophisticated approaches use IC logic, shift registers, memory, and other techniques that are derived from computer technology.

The prime reason that regulated power supplies respond favorably to digital control is that their outputs are predictable in terms of the resistances used in the feedback networks or in terms of the reference voltage. In this respect, the regulated supply performs exactly as an operational amplifier. In the simplest case, the basic op-amp equation is used:

$$E_{out} = \frac{E_{IN} R_f}{R_{IN}}$$

where:

E_{out} is the output voltage,
E_{IN} is the reference voltage,
R_f is the feedback resistance,
R_{IN} is the input resistance.

By appropriately selecting the value of any one (or more) of the three right-hand members of this equation, any desired sequencing of E_{out} value can be achieved. For example, you can simulate a sine wave or produce a stream of predetermined output levels. Whatever the programming format, such a controlled power supply behaves essentially as a *digital-to-analog converter* with high output capability. A simplified power-supply circuit that is suitable for digital control or load voltage programming is shown in Fig. 4-49.

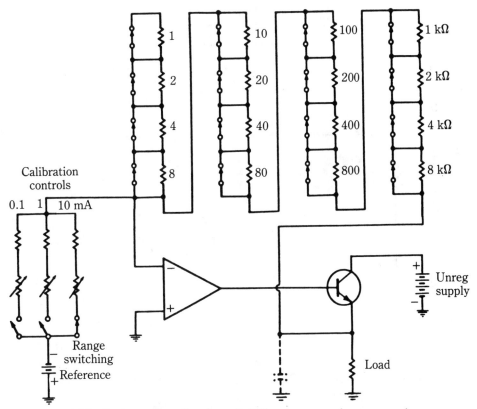

4-49 Basic configuration for a digitally programmed power supply.

Frequently, considerable speed is demanded from a digitally controlled power supply. In such instances, the sequencing of digital data by means of mechanical devices might not be feasible. Instead, the solid-state elements that are common in computer technology can then be used to control the power supply. Accordingly, such devices as bipolar and MOS transistors, IC logic, and optocouplers are in place of relays, switches, and various motor-driven mechanisms. The programs might originate from punched cards or from data buses in computer systems. Thus, the power-supply designer might find it difficult to implement a regulated supply that can follow the available bit rate. For example, the conventional output capacitor in regulator circuits limits the rate of change on the output voltage level (slew rate). If high speed is a prime objective, this capacitor must be of minimal size—if it can be eliminated, so much the better.

The simplest approach to an electronically programmed power supply in digital format is via a three-terminal regulator IC. Such an arrangement is shown in Fig. 4-50. The LM117 IC is well-adapted for this application because of the high impedance of its control (ADJ) terminal. The bipolar transistors perform the same function as the mechanical switch contacts that are depicted in the digital control scheme of Fig. 4-49.

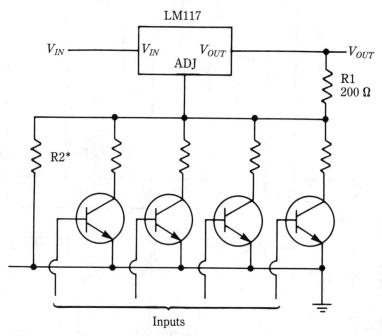

*Sets maximum V_{OUT}

4-50 Simple circuit for digitally selecting the output voltage of a power source. National Semiconductor Corp.

A somewhat more sophisticated circuit is shown in Fig. 4-51. Instead of the bipolar transistors, a single JFET IC module is used to select the binary-weighted resistances in the feedback network of the power op amp (LH0021). Notice that in this arrangement, the controlled resistances are in the input circuit, rather than in the feedback network. Accordingly, the lowest resistance selected corresponds to the highest output voltage from the supply. Do not be confused by the so-called "analog switch," which is used to perform digital switching operations.

A unique feature of this digitally programmable supply, one that lacks the simpler arrangement of Fig. 4-50, is the provision for an output with either polarity. This is made possible via another JFET analog switch, the AH0161, which has spdt "contacts" and is actuated by a differential amplifier. As can be seen in Fig. 4-51, the AH0161 enables either a positive or a negative reference voltage to be

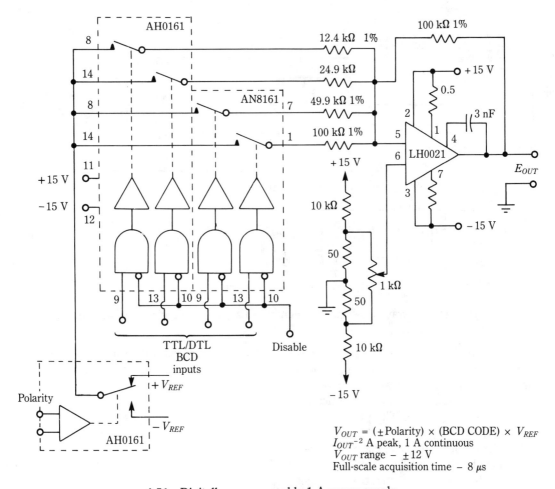

4-51 Digitally programmable 1-A power supply.

$V_{OUT} = (\pm \text{Polarity}) \times (\text{BCD CODE}) \times V_{REF}$
I_{OUT} – 2 A peak, 1 A continuous
V_{OUT} range – ± 12 V
Full-scale acquisition time – 8 μs

applied to the inverting input of the LH0021 power op amp. This technique can provide an output voltage with either polarity because the power op amp is energized from dual-polarity sources; the output voltage is thereby free to assume either positive or negative levels.

The block diagram of another digitally controlled power supply is illustrated in Fig. 4-52. In this novel arrangement, the use of optoelectronic couplers permits the DAC to float on the output voltage of the supply. Because of this, the same configuration could be adapted for a 500-V output by merely substituting an appropriate output transistor and an unregulated power supply. As can be seen, a 25-V ungrounded auxiliary voltage source is required. The MC1466L IC regulator is a floating type and is admirably suited for this application—especially because it has an internal current source and the MC1406L DAC provides an output current sink. In addition to this compatibility, the arrangement in Fig. 4-52 has no binary-weighted resistance network (ladder) because the apportioning is done within the

226 *Implementation of regulation techniques*

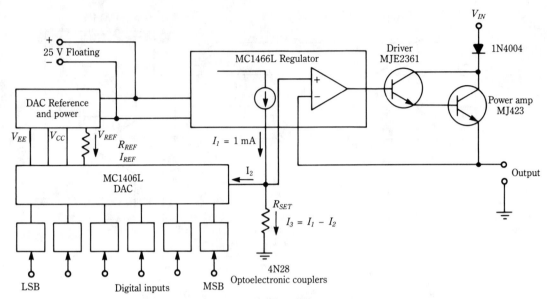

4-52 Block diagram of a digitally-programmed supply with optically isolated logic.

MC1406L itself. This dispenses with the headaches that are commonly experienced in obtaining precise-valued resistors with the required stability and temperature-tracking characteristics.

The schematic diagram of an optically isolated, digitally controlled power supply appears in Fig. 4-53. The functions of the three potentiometers are as follows: Resistor R1, in conjunction with the fixed 57.6-kΩ resistor, establishes the reference voltage for the MC1466L regulator as a result of the one milliampere of current that is provided by an internal source in the MC1466L. Thus, when R1 is adjusted to its approximate midrange setting, the total resistance in series with the internal current source will be 63 kΩ. This will produce a reference voltage of 63 V, which will also appear at the output terminal of the system, V_{OUT}. Nothing is special about the 63-V output level, and other values can be selected via R1. The circuit enables the output voltage to be adjusted in 64 one-volt increments to a maximum of 63 V via the digital "words" that are inputted to the optoelectric couplers. This harmonizes efficiently with the nominally 70-V unregulated supply, V_{IN}.

Resistor R2 adjusts the reference current that is required by the MC1406L DAC. This current is derived from the 1N827 temperature-compensated zener diode. When this reference current is properly adjusted, the digital programming of the optocouplers will enable the sink current, I_2, to vary between zero and one milliampere (Fig. 4-52). When I_2 is one milliampere, I_3 becomes zero, as does the reference voltage that is applied to the MC1466L regulator. The output voltage is also zero. Thus, R2 is used to set V_{OUT} to zero when all of the optocoupler switches are open. This setting corresponds to the digital word 111111.

Resistor R3 controls the current-limiting function in the MC1466L regulator. R3 enables selection of the maximum current output of from near zero to slightly

Digitally controlled power supplies 227

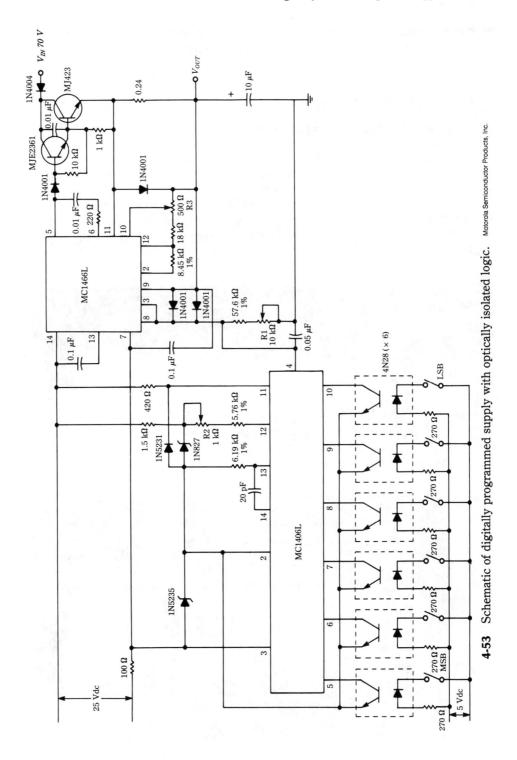

4-53 Schematic of digitally programmed supply with optically isolated logic. *Motorola Semiconductor Products, Inc.*

228 *Implementation of regulation techniques*

over one ampere. If the supply is operated for sustained periods at low output voltages at maximum current, adequate heatsinking must be provided for the MJ423 output transistor.

The systems-oriented digital control of dc power

A more elegant approach to digital control of power supplies uses a systems concept in which the controller or programmer is mated by design to the digitally controlled power supply(s). Kepco has been a pioneer in the development of such systems and markets a wide selection of dedicated subsystems that enable the user to both "talk to" and "listen to" the performance of remotely located power supplies. A digital controller and a triad of power-supply modules are shown respectively in Figs. 4-54 and 4-55.

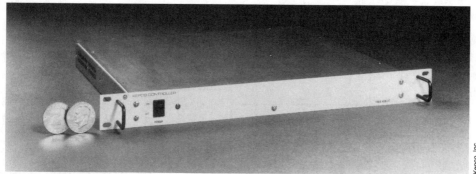

4-54 Kepco Model TMA 488-27 digital controller for a dc power system.

4-55 Kepco dc supplies that are mated to a digital controller.

Actually, up to 1000 ft of separation can exist between the controller and the controlled supplies, and up to 27 power-supply modules can be programmed, controlled, and monitored via the IEEE-488 (GPIB) bus. Over the RS232 communications link, Kepco also makes a plug-in board for personal computers and provides appropriate software to emulate the functions of the TMA 488-27 digital controller (Fig. 4-54). Such direct control dispenses with the need for the IEEE-488 bus.

In operation, the Kepco system provides 12-bit resolution and the controlled power supplies are polled every millisecond. The installation of such a mated system spares the user many headaches (such as ground conflicts, temperature tracking, crosstalk and noise problems, and line-voltage incompatibilities). A $3^1/_2$-digit output display is provided on the front panel of the power-supply modules, and it is switch-selectable between voltage and current. LEDs indicate the major operational modes, constant voltage, constant current, output polarity, and output enabled. The values at the power-supply outputs are read back to the IEEE-488 bus. All in all, there are more operational options, error-detection schemes, and protective techniques than can ordinarily be conveniently and economically implemented by pursuing a nonsystem approach to digitally controlled dc power.

Multiple output supplies

Regulated power supplies with more than one output voltage have become popular. In linear or series-pass regulators the format of dual voltages, center-tapped voltage, or tracking voltages is commonly used. Thus, you might have ±15 V available, with respect to a ground or a midpoint. A unique problem can arise in switching-regulator supplies that provide multiple voltage levels. In such supplies, it is common to provide, in addition to a ±15-V output, one or two other voltages (such as 5 V and 10 V). Often the 5-V output is considered to be the "main" output, because it has the greatest current capability. Accordingly, the feedback circuit of the supply operates from a sampled portion of the 5-V output. The other voltages are conductively isolated from the 5-V output, because it is derived from different transformer windings. Likewise, regulation is best for the 5-V output. This situation is illustrated in the simplified functional diagram of Fig. 4-56. Notice that only one output is sampled. The regulation of the sampled output is better than that of the other outputs and is likely to be accorded prominence in specifications sheets.

The specifications for these supplies are sometimes worded so that they can lead you to assume that the quoted line and load regulation applies to all of the output voltages. Actually, regulation of those that are not sampled, and thereby involved in the feedback operation, are not as good as is obtained for the 5-V (or whichever output voltage is sensed to produce the regulation of the supply). In most applications, especially in those that involve digital equipment, the lack of extremely tight regulation will not be of consequence—after all, the tendency is to provide supplies with tighter regulation than is actually needed. Here, however, the poorer regulation of the nonsampled output voltages implies a certain degree of interaction between the "isolated" outputs. In some cases, this interaction could produce adverse effects in the loads. It is as much as if the various loads were coupled together via a common impedance.

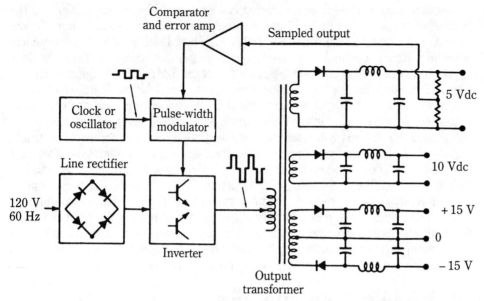

4-56 Functional diagram of a multiple-output switching regulator.

The user should be aware that some manufacturers have overcome this problem with a more complex (and costly) arrangement in which each output is separately regulated.

Hold-up time—uninterruptible operating power for awhile

Yet another important feature of off-line switching regulators exists. Because the filter capacitor(s) that are associated with the line rectifier operate at a relatively high voltage, their energy storage tends to be many times that which is ordinarily found in linear supplies. The significance of this is that such an off-line switching supply can maintain its output voltage for as much as 30 ms following loss of ac input power. This time is sufficient to enable volatile memories to preserve their data in the face of a large percentage of ac line disturbances. In essence, such a supply behaves somewhat as an *uninterruptible power source (UPS)* for many brownout and blackout conditions. In comparison, linear supplies ordinarily have a "carry-over" or "hold-up" of about 2 ms—an inadequate duration for typical utility-line malperformance.

The practical implementation of long hold-up time is based on two factors. First, energy storage in a capacitor is directly proportional to the square of the dc voltage. Thus, other things being equal, three times the dc voltage produces nine times the energy storage. Of course, other things are not all equal—a higher-voltage rating capacitor must be used and will necessarily be physically larger. However, a net improvement in capacitor dimensions is because the volume of electrolytic capacitors is approximately directly proportional to the voltage rating. Thus, the high-voltage capacitor-filter circuit still has a decided increase in the ratio of stored energy to capacitor size, when compared to the energy storage (obtained from the same volume of capacitance operating at lower voltage).

Important, too, is that duty-cycle-modulated switching regulators operate at about the same efficiency over a wide range of input voltages. In contrast, the efficiency of a linear regulator plummets drastically as input voltage is increased. This, of course, is another reason why it is practical to design the off-line switching supply to have a high hold-up time.

Regulation of ac

Regulation of ac voltage is generally accomplished with a ferroresonant transformer. However, other methods might be better suited to a particular application. The four simple zener-diode circuits shown in Fig. 4-57 stabilize the ac output voltage by clipping the peaks of the sine wave. Of course, these circuits do not use error-sensing and feedback. The clipped waveform varies in rms value, even though its peak-to-peak value might be maintained at a near-constant value. Nonetheless, the regulation is much better than it would be without zener diodes. All of the circuits provide line and load regulation, but the circuit in Fig. 4-57A is best for load regulation. Being shunt regulators, these circuits are also useful for removing transients from the power line.

The circuits in Figs. 4-57A and B are "brute force" in the sense that power must be deliberately thrown away in the series resistances. The circuit in Fig. 4-57C can be made to operate similarly to the circuit in Fig. 4-57A, but with much better efficiency because the series inductance wastes relatively little power. This method is good for stabilizing filament voltage in vacuum-tube final amplifiers in

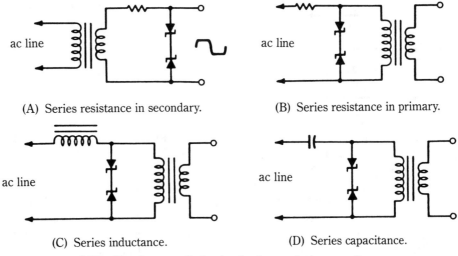

(A) Series resistance in secondary.

(B) Series resistance in primary.

(C) Series inductance.

(D) Series capacitance.

4-57 Simple zener-diode circuits for regulating ac voltage.

amateur transmitters. Such stabilization can result in prolonged tube life and consistent performance. In experimenting with such an arrangement, use a modern electronic rms voltmeter so that the measurement error does not accrue from the trapezoidal waveform. In order to obtain "on-the-button" voltage, you must have at hand a number of zener diodes to select from or be blessed with inordinate luck. A transformer with tapped windings is helpful in this regard.

The circuit in Fig. 4-57D operates similarly, but in practice is not as well adapted to high-current capability; it is more difficult to accommodate high current in readily available capacitors than in inductors. In using these circuits, the experimenter should be aware that zener diodes are available in power ratings from several hundred milliwatts to at least 50 W.

Figure 4-58 depicts another technique for accomplishing essentially the same result, but with greater precision and generally at a higher efficiency. Here, two three-terminal IC voltage regulators are operated from an ac source, rather than from the more-usual rectifier system. The near-trapezoidal output waveform results from series limiting, rather than from shunt clipping (as with the zener-diode circuits). The IC regulators make this approach simple, although the principle predates such modules.

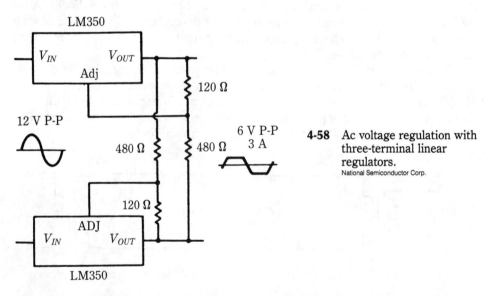

4-58 Ac voltage regulation with three-terminal linear regulators.
National Semiconductor Corp.

The more elaborate circuit of Fig. 4-59 uses SCRs and is, therefore, capable of operating at high power levels. The efficiency is high because the phase-controlled SCRs are two-state devices; they are either on or off, at which times their power dissipation is much less than the usual conductive state of series-pass transistors or zener diodes. This circuit is old, but it is a "classic." It is not critical and can accommodate many semiconductor devices of the same basic type of the replaced device. Its response to variations in line voltage or to load changes is relatively slow (on the order of 80 ms or so). Within this limitation, it provides excellent ac voltage stabilization. It is readily modified for operation at other voltages and

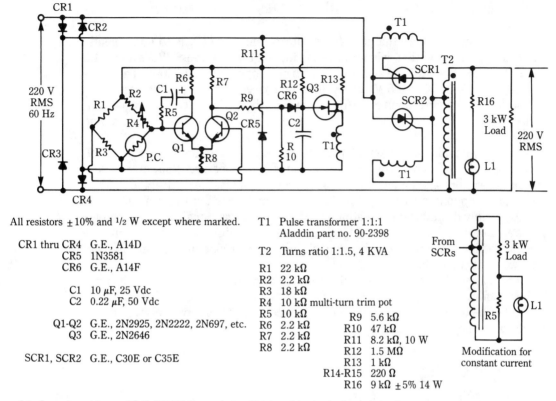

4-59 A representative RMS ac voltage-regulating circuit. General Electric Co.

power levels, and it can be made to regulate ac load current, rather than voltage. Also, a rudimentary low-pass filter in the output can restore the output waveform to a reasonable sine wave (however, too much reactance, either in a filter or in the load can interfere with proper commutation of the SCRs).

The error-voltage detector composes a hermetically sealed module that contains an incandescent lamp and a cadmium-sulfide photocell. The lamp filament operates at 2, rather than at its rated 5 V. Thus, its probable lifetime is at least comparable to the semiconductor devices in the circuit. As will be seen, the lamp samples ac output voltage and the photocell is connected to a differential amplifier via a bridge circuit.

The output from the differential amplifier controls the triggering time of the UJT relaxation oscillator (Q3), which, in turn, governs firing and conduction time of the two SCRs. This control technique is essentially similar to commonplace methods that are used in lamp dimmers and motor speed-control systems. In this particular application, the balancing and unbalancing of the bridge circuit (R1, R2, R3, and PC) always occur in such a manner as to counteract changes in the preset value of the ac load voltage. Resistor R4 in the bridge network sets the output voltage level.

The rationale behind the use of the autotransformer, T2, is to conveniently cause the circuit condition, wherein the conduction time for the SCRs is about 4.5 ms with nominal line voltage and with 220-V rms output with a full load. This, incidentally, corresponds to 165 V across the primary section of the autotransformer. These conditions correspond to the optimum sensitivity of the SCRs for phase control, considering that allowances have to be incorporated for both high and low line voltages.

Other companies manufacture incandescent-lamp cadmium-sulfide photocell modules, such as Clairex. An LED optoisolator cannot be used, however, unless circuit innovations are implemented to accommodate its fast response; the thermal lag of the lamp unit is part and parcel of the original design of this regulator.

A possibly desirable modification would have a conventional two-winding output transformer in place of the autotransformer. At a slight sacrifice in compactness and electrical efficiency, the valuable feature of isolation between line and load would then be obtained.

This basic scheme can be readily modified to produce constant load current, rather than voltage (Fig. 4-59). The sensing resistance must be rated to carry the desired load current and should have a resistance value so that two volts are developed across the lamp.

An ac voltage regulator with electronic sensing

The use of such devices as filamentary lamps, thermocouples, or thermistors to sense rms value in ac regulators is a straightforward way to deal with the nonsinusoidal waveform that is produced by phase-controlled thyristors. However, because of thermal lag, the response time of such systems is necessarily slow. Much faster response can be attained by sensing the thyristor waveform electronically. This task is readily accomplished with a sensing circuit that has a transfer function so that its averaged output is proportional to the square of its input voltage. A differential amplifier that has one transistor biased in the vicinity of its nonlinear cutoff region can be made to perform in this fashion. Moreover, if the two transistors are composed of a monolithic pair, the tight thermal-coupling between them will immunize the operation against temperature change. A subsequent RC filter averages the squared voltage values from the sensor.

It might be thought that the square root of the output signal is necessary, but it isn't for this purpose. No degradation in regulation follows from this apparent omission to comply rigorously to the mathematical root-mean-square concept. The mean-square value is sufficient for sensing and regulation purposes.

Figure 4-60 shows a complete closed-loop ac voltage regulator in which the rms voltage that is applied to the load is maintained constant. Transistors Q1A and Q1B constitute the nonlinear differential sensing amplifier. Here, resistor R11 holds the base of Q1A about 300 mV below that of linearly biased Q1B. The filter-

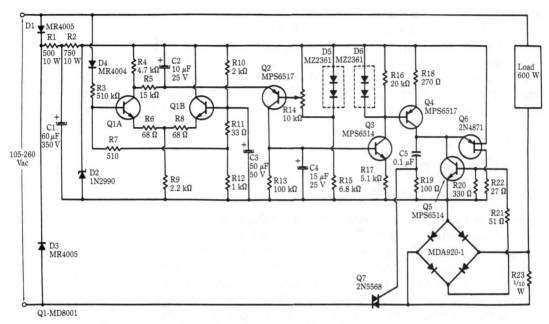

4-60 Ac voltage regulator with electronic RMS sensing. Motorola Semiconductor Products, Inc.

ing action of R4, R5, and C2 satisfies the mathematical requirement of having the average or mean output voltage proportional to input voltage.

The output signal from the differential amplifier is applied to Q2, which functions as a voltage comparator. This stage develops the error signal, which is then amplified by stage Q3. In turn, Q3 controls the charging current that Q4 delivers to timing capacitor C5. In this way, the triggering time of UJT Q6 is either retarded or advanced in compliance with the magnitude and polarity of the error signal.

The culmination of this chain of events is that the conduction time of the triac is varied by its UJT-derived gate signal and the rms load voltage is regulated. For this scheme to be workable, however, a synchronization circuit is also required so that the triac conduction time can be redetermined twice for every cycle of the ac voltage. The synchronizing circuit consists of R23, the diode bridge, R20, R21, and Q5. Transistor Q5 is not conductive when the triac is in its OFF state. Thus, timing capacitor C5 is free to charge to the triggering voltage of the unijunction transistor, Q6. This advent causes the triac to fire also. In turn, the voltage drop that is produced across sampling resistor R23 drives transistor Q5 into saturation, which thereby depletes the charge in C5. When the load voltage commences its decline again, Q5 comes out of conduction and C5 is allowed to charge for a new timing function.

Sampling resistor, R23, has two detrimental effects. First, the actual voltage regulation is not accomplished directly across the load, but rather across the series combination of the load and R23. Second, R23 must be selected for the particular load. A corollary of this is that a fairly constant load must be used; the regulator, although it provides excellent regulation against line voltage, does not do so well in regulation against load change. The experimenter probably can overcome these drawbacks by replacing R23 with an appropriate transformer and perhaps using zener diodes so that Q5 is properly controlled over a wide range of load currents.

A somewhat similar technique is already incorporated. Thus, the operating dc of the regulator is derived from diodes D1 and D3 (filtered by R1 and C1), but it is maintained at constant voltage by zener diode D2. This simple arrangement helps guard against disturbances in the UJT triggering time when variations are in the ac line voltage.

Regulating a permanent-magnet alternator's charging current

In marine and other nonautomotive battery-charging systems, the alternator often has a permanent-magnet field. This field is a problem in regulation, because the alternator output must be reduced when the battery attains its fully charged condition. On one hand, the battery voltage must be sensed; on the other, something must be done to the alternator to control its charging rate. It might seem that some kind of series interruptor could be devised for this purpose. Practically, the inductance of the alternator is high enough to produce some nasty problems with this approach.

A successful technique involves short circuiting the alternator when a high charging rate is no longer needed. Admittedly, your initial reaction to this approach is likely to harbor little enthusiasm. However, when implemented in the manner shown in Fig. 4-61, the charging current is automatically controlled with little dissipation and without detrimental effects to the alternator. Bipolar shorting of the

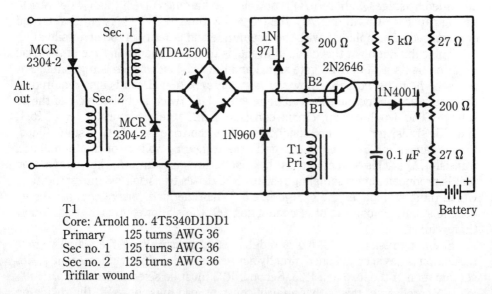

T1
Core: Arnold no. 4T5340D1DD1
Primary   125 turns AWG 36
Sec no. 1   125 turns AWG 36
Sec no. 2   125 turns AWG 36
Trifilar wound

4-61 Voltage regulation for alternator with permanent-magnet field. Motorola Semiconductor Products, Inc.

alternator's ac output is accomplished by the two SCRs. This is not damaging because the high-voltage condition of the battery (which is responsible for the SCRs being triggered into their conductive states) will not maintain itself; short circuiting of the alternator quickly removes the condition that is responsible for the high terminal voltage of the battery. Because SCRs that are impressed with ac are self commutating, the short circuit quickly extinguishes itself when the SCRs no longer receive gate signals.

Two interesting operational features of this regulator circuit might not be immediately obvious. One is that, because of the unilateral transfer characteristic of the bridge rectifier, the battery never experiences the short circuit that the SCRs place across the alternator. The other feature is the approximate temperature compensation between the 1N4001 diode and the emitter-base-1 section of the UJT. For practical purposes, only the temperature coefficient of the 1N960 zener diode is left to produce temperature dependency of circuit operation.

The 2N2646 unijunction transistor (UJT) has its base-2 voltage clamped at approximately 9 V by the 1N960 zener diode (the other zener diode has a 27-V breakdown rating and is not involved in ordinary circuit operation). Clamping of the base-2 voltage imparts a fairly precise voltage sensitivity to the UJT circuit; when battery voltage is low, the 1N4001 diode is forward biased and the UJT relaxation oscillator circuit is inactive. No gate pulses are available for the SCRs and the alternator is permitted to charge at maximum current. Such charging will ultimately cause the battery voltage to rise to meet the emitter triggering requirement of the UJT. When this happens, the 1N4001 diode becomes reverse biased and effectively drops out of the circuit. Relaxation oscillations are then generated by the UJT and audio-frequency gate pulses are delivered to the SCR gates through the pulse transformer, T1.

The shorting action of the gated SCRs halts the charging process and the battery voltage will gradually decline until the UJT oscillations are again inhibited by the forward conduction of the 1N4001 diode. Then, the alternator is again free to provide charging current. The regulation process is more of an off-on format than the continuous regulation that is generally encountered. The duty cycle of the charging periods is adjustable by the 200-Ω potentiometer.

In the event that corrosion or vibration disconnects the battery, the alternator voltage will increase. Its rise will be limited by the conduction of the 1N971 zener diode because this event will produce SCR gate pulses at the alternator ripple frequency. The rectifier, UJT, and associated components will thereby be protected from the higher voltage, which the unloaded alternator might otherwise develop.

Improving long-accepted inverter design

Because inverters often form the basic system component about which regulated power supplies are configured, it is profitable to investigate some of their properties somewhat further than has previously been done. Allusions will also be made to considerations that are relevant to powering electric motors. Such power supplies are often regulated and can be designed to deliver either ac or dc to appropriate motors. The previous one-transformer, two-transformer self-oscillating inverters, which used the saturating-core principle together with driven inverters, underlies a vast number, if not the preponderance of manufactured designs. Additionally, most of the self-oscillating circuits have been of the voltage-driven type. Here, *voltage-driven* implies that the circuit is self-oscillating, developing positive-feedback that is proportional to the voltage delivered to the load. This most "natural" type of circuit, paradoxically, has not been the best choice from a number of considerations. Representative voltage-driven inverters are depicted in Figs. 4-24 and 4-26.

As might be surmised, the alternate type of feedback circuit for inverters is the current-driven type. *Current-driven* implies that the circuit is self-oscillating and developing positive feedback that is proportional to the current that is delivered to the load. Both types use core saturation in the same manner to produce the switching sequence in the power transistors. Indeed, the schematic diagrams of the two circuitries are quite similar and it might well appear that it is splitting hairs to make any fuss over the two connections. It would seem that any way to bring about the switching interaction between the transistors and the saturable-core transformer should be about as good as any other. This was the design philosophy for a long time. It was responsible for "mysterious" failure modes, less-than-optimum efficiency, inordinately high levels of RFI and EMI, as well as other malperformance. Thus, what appeared to be a subtle and unimportant detail of oscillator circuitry, actually was later discovered to greatly affect performance and reliability.

It was previously pointed out that the saturable-core self-oscillating inverter could contain either one or two transformers. The two-transformer version provides best flexibility for the designer and will serve as basis for the interesting comparison of performance between voltage-driven and current-driven feedback arrangements. The plain-driven inverter is not considered because it has no feedback; it has more the nature of an amplifier than an oscillator. Figure 4-62 shows relevant waveforms in a typical voltage-driven inverter working into a resistive load. T1 is an ideal transformer with no leakage inductance.

When the voltage-driven inverter "sees" an inductive load (Fig. 4-63), the primary current in the transformer is altered in both waveshape and timing. Although a purely inductive load would not be encountered in ordinary practice, the assumption of such a load is useful for illustrating problems that are likely to be encountered with more practical partially inductive loads. The changed nature of the transformer primary current produces at least two detrimental operational effects. One is the worsening of the voltage spikes at the transistors. The other is lowered efficiency because of simultaneous current through and the voltage across the transistors. Sometimes, too, such inverters exhibit start-up problems with inductive loads.

Excessively capacitive loads produce similar problems with voltage-driven inverters, although changes in the primary-current waveform and timing are somewhat different. The ordinary voltage-driven saturable-core inverter works best when feeding a resistive load.

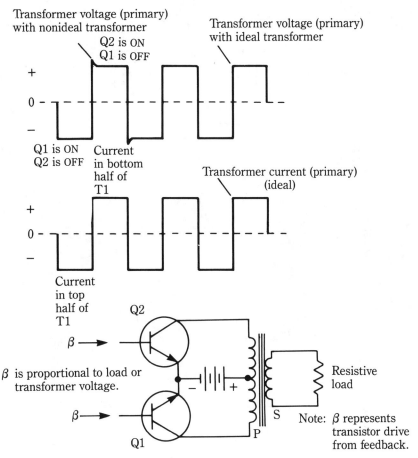

4-62 Resistive load on voltage-driven inverter—no severe problems. Current through T1 primary always finds completion path through an ON transistor. The voltage spike is manageable.

It has long been accepted that inverter transistors can be largely spared destructive voltage-spikes:

- Transformers should be designed for as low leakage-inductance as possible. Toward this end, the use of toroidal cores and bifilar windings are usually desirable.
- Two-transformer inverter oscillators are less prone to spike generation because the output transformer operates in the linear, rather than the saturable, region of its magnetic characteristics.
- Minimal base-drive should be used consistent with acceptable full-load output and efficiency.
- The less reactive the load, the better.

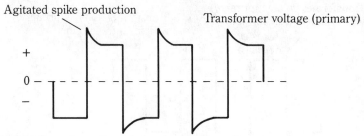

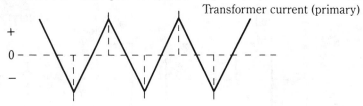

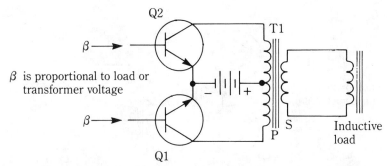

4-63 Inductive load on voltage-driven inverter—a problem arises. Phasing of T1 primary current is now favorable for production of flyback voltage spikes.

- Because they do not store charge in the manner of bipolar transistors, power MOSFETs can be used advantageously. Their faster switching speed also can be helpful in this regard. Special biasing and start-up considerations pertain; because power MOSFETs became somewhat popular after the popularity peak of saturated-core inverters, such circuits have not been common.
- Substitute a current-driven inverter for the more commonly used voltage-driven type.

The last suggestion returns to the topic at hand. The current-driven inverter experiences gentler current disturbances in the primary of the output transformer; this appreciably reduces both spike amplitude and spike energy and relieves the transistors of unnecessary (and often unpredictable) voltage stress. Also, simultaneous transistor current and voltage is reduced so that operating efficiency remains high.

The current-drive characteristic can be achieved, essentially, by inserting an inductor in one dc supply-lead (Fig. 4-64). Notice the interesting situation (with regard to current) in the primary of the output transformer; it remains a near-ideal square wave, regardless of the nature of the load! In essence, the inverter doesn't "care" whether the load is resistive, reactive, or any combination. Such inverters

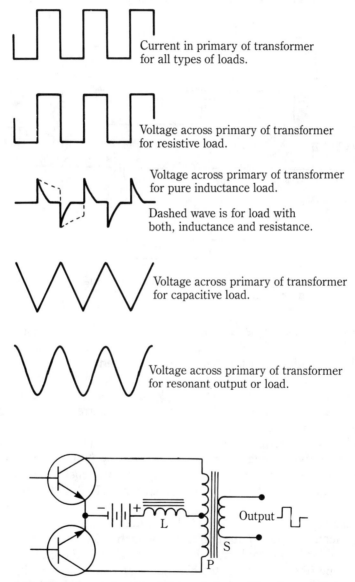

4-64 Waveshapes in current-driven inverter. The shape and timing of the current in the transformer primary is the same for all load situations.

require a minimum load to operate. On the other hand, they continue to oscillate with a short-circuited load and might need protection. They are particularly well-suited as motor power supplies. Interesting too, is the bottom waveform, which is a sine wave when the inverter feeds into a resonant circuit or load.

A more complete circuit of a current-driven inverter is shown in Fig. 4-65. Without the inductor, L, the circuit would operate as the common voltage-driven inverter. Although such inverters are seldom described as being "voltage driven," it is best to do so in order to distinguish them from the current-driven types.

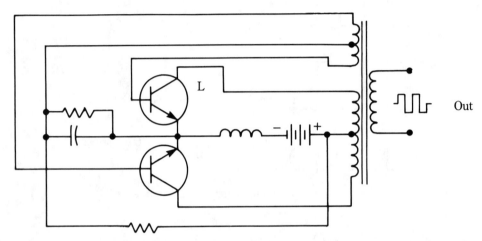

4-65 Basic circuit of a single-transformer current-driven inverter. Insertion of inductor, L, markedly changes inverter operation.

Another circuit technique for obtaining the current-driven operational mode is shown in Fig. 4-66. Here again, it closely resembles ordinary voltage-driven circuits. In this case, the requisite modification to current drive is applied to a two-transformer inverter. Notice that no inductor in the dc supply lead is needed. In order that feedback is positive, as required for oscillation, the phasing of the various windings should be as indicated. In this type of current-driven inverter, it is more evident that there cannot be feedback without a minimum load.

A similar current-driven inverter that uses power MOSFETs is shown in Fig. 4-67. The current-drive characteristic is obtained by interrupting the primary, rather than the secondary, winding of the output transformer. Operation is essentially the same as in the bipolar-transistor circuit of Fig. 4-66. Initially, a little experimentation might be necessary to obtain optimum bias for the power MOSFETs to reliably start up oscillation. Other matters being equal, the MOSFETs should be selected for low R_D to minimize conductive losses. Also, appropriately selected IGBTs should provide meaningful advantages in these inverters, as long as the operating frequency is not too high. For practical purposes, IGBTs can be considered power devices with the input characteristics of MOSFETs and the output characteristics of bipolar transistors. Lower frequency types (up to 5 kHz) perform best, but some types are rated up to 50 kHz.

Improving long-accepted inverter design 243

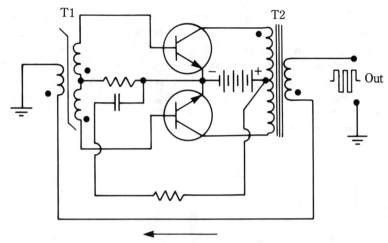

β is proportional to load current

4-66 Circuit of a two-transformer current-drive inverter. T2 is a large nonsaturating transformer. T1 is a relatively small saturating type. Current drive is established by feedback scheme.

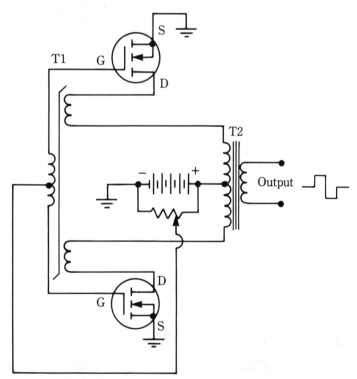

4-67 Current-driven inverter with power MOSFETs. T2 is linear; T1 saturates. Operation is similar to the circuit in Fig. 4-66.

The alluded to current-driven inverter with a resonant load is shown in more detail in Fig. 4-68. The zero-current switching feature, illustrated by the waveform diagram, greatly reduces switching losses. This basic idea has been incorporated in the new resonant-mode power supplies. Resonant-mode operation makes it easier to attain high-frequency operation, and it greatly reduces RFI and EMI problems. Although assuming various topographical forms, resonant-mode inverters and power supplies either switch when the current or voltage is zero. For regulating purposes, frequency modulation, rather than PWM, is generally used.

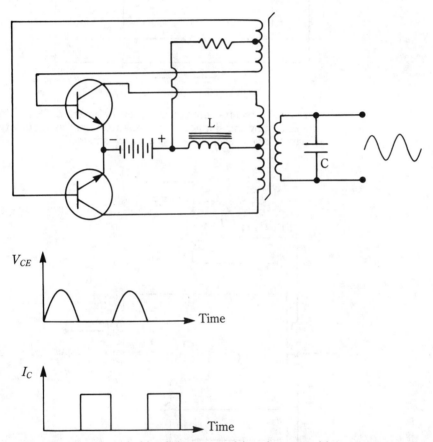

4-68 Zero-current switching is in current-driven inverter with resonant output. This mode of operation largely circumvents switching losses.

Ballparking the series inductor in the current-driven inverter

The function of the series inductance, L, in the basic current-driven inverter of Fig. 4-65 is to maintain uninterrupted dc current through it. This implies a minimum inductance when full power is being delivered to the load. This minimum inductance is not suitable at 50% or 33% load, however. The series inductance will accommodate a three to one variation of load power if the inductance, together with its associated circuitry, behaves as if the effective Q is equal to one (the three-to-one load-power variation pertains to a load-power range from 33% of full load to full load). This criterion might, at first, appear a bit nebulous to apply, but it is quite straightforward and is actually a quick and helpful way to ballpark the value of L.

Thus, $Q = 1$. Rewriting Q as:

$$\frac{X_L}{R_{eff}}$$

provides us with some definite parameters to work with. X_L, for example, is the inductive reactance of the inductor and is given by $2\pi f L$, where f is the switching rate of the inverter.

R_{eff} consists of the dc resistance of the inductor winding plus resistance, R, that the inductor "sees" in the inverter. R is easily estimated. The Ohm's law derived relationship:

$$R = \frac{E^2}{P}$$

uses known parameters. Here, E is simply the dc voltage that is applied to the inverter and P is the full-load power that the inverter is intended to deliver. For our "ballparking" intent, it is OK to assume a 100% efficiency for the inverter and to assume that the dc and ac resistances of the inverter are the same.

Putting everything together so that we can solve for L directly:

$$L = 2\pi \left(\frac{R_L + E^2}{P} \right)$$

Of course, some practical judgment must be exercised to estimate R_L, the winding resistance of the so-far nonexistent inductor. If insight is lacking because of minimal experience with theoretical and practical magnetic-component design, a "first-time-around" approach is to simply assume that R_L and R are equal. Considerable latitude is allowable in ballparking the value of L because such luxuries as minimum weight, maximum efficiency, cost, or production features are not strived for in this one experimental inductor. Also, with little penalty for excessive inductance, L is easier to estimate.

5
Devices and components

THIS CHAPTER TOUCHES ON SOME OF THE DEVICES AND COMPONENTS that have played prominent roles in the modern trend toward more efficient and more compact regulated power supplies. Included, also, are allusions to products that appear destined for serious consideration from designers and manufacturers of supplies in the near future. Comprehensive treatment is not possible because such coverage could easily encompass a book-length format for one device alone. However, sufficient description will be attempted to at least stimulate the interest of those who are concerned with the evolution of regulated power supplies.

Overall, the trend appears to point more and more in favor of switching supplies over linear, and particularly series-pass, designs. However, because of the inherent superiority of the linear supply, with regards to low noise generation and low output impedance, such regulators have not become obsolete. Where their particular qualities are needed, the "switcher" usually is not competitive. On the other hand, where the general drawbacks of switching regulators are not of great consequence, the higher efficiency that is provided by switching techniques is hard to overlook. Not only does the higher operating efficiency entail worthwhile savings in energy consumption, but it is accompanied by an often dramatic reduction in physical dimensions and problems that are associated with heat removal.

Whereas the designer previously had to use garden-variety devices to implement his ideas, he or she now has access to a wide range of items that are specifically intended and optimized for regulator use. This, in itself, is indicative of the long-delayed recognition of the basic importance of regulated power supplies. An example of this turn of events is clearly evident in integrated circuits, where numerous dedicated modules relevantly tailored to the unique needs of the regulated power supply have become available.

Zener diodes

Zener diodes are the solid-state equivalents of the now nearly obsolescent gaseous vr tubes, and are primarily used as voltage references. In turn, the ordinary zener diode now has strong competition from monolithic ICs, which make sophisticated use of field-emission, avalanche, and especially energy-gap "breakdown" mechanisms. Such ICs can provide exceedingly precise performance parameters, as well as new dimensions of operation, such as variable voltage, self protection, and various control options. Nonetheless, the zener diode remains a viable workhorse. Its basic simplicity, low cost, and suitability for numerous applications render it difficult to displace from the designer's arsenal of circuit building blocks. Besides, zener diodes have, in their own right, made considerable progress. The important characteristics of a typical family of 500-mW zener diodes are listed in Table 5-1. It will help to refer to this list in the ensuing section. Trends in dynamic impedance and temperature coefficient are readily discernible. Notice that dynamic impedance is listed for two operating currents, I_{ZT} and I_{ZK}.

The salient operational characteristic of zener diodes is their reverse breakdown behavior. A pn junction diode will sustain only a certain amount of reverse voltage. What current flows as reverse voltage is relatively tiny, until a voltage level (popularly known as the *zener breakdown voltage*) is reached. At that voltage, the current increases to such a great extent that it must be limited or the resulting self-heating will destroy the diode. If the current is limited, however, the voltage that is supported across the diode will remain very nearly constant over a wide range of diode current, as well as applied voltage from the dc source. This is essentially an intrinsic voltage regulating device. Indeed, the zener diode is the simplest of all voltage regulators (regulation, of course, occurs with no external feedback or error-sensing circuitry).

The notion of "zener breakdown voltage" is too simplistic to account for some of the behavior features that are exhibited by these devices. Actually, the physics of the voltage breakdown phenomenon describes two different mechanisms. One is known as *field-emission breakdown*; the other is known as *avalanche breakdown*. Field emission is the predominant phenomenon at low voltages and avalanche breakdown predominates at high voltages. The demarkation between the two breakdown mechanisms is in the vicinity of 6 to 8 V. Field-emission breakdown is essentially a tunneling process. What is observed is evidence that charges will penetrate a barrier field. Such probability exists already at zero volts and increases exponentially as voltage is increased. Thus, it has a gradual or "soft" breakdown transition. Zener diodes with breakdown ratings below 6 V tend to have less-abrupt "knees" than higher voltage diodes.

The avalanche breakdown mechanism is much more abrupt than the breakdown that is premised upon field emission. The difference between the two can be seen in Fig. 5-1. *Avalanching* is a process in which charge carriers are accelerated to high velocities by the applied voltage. These charges collide with valence electrons, which, in turn, impact other valence electrons, and produce a "snowballing" effect that is manifested by a very sudden increase in current flow through the

Table 5-1. Family of high-quality 500-milliwatt zener diodes.

($T_A = 25°$ unless otherwise noted. Based on dc measurements at thermal equilibrium; lead length = 3/8"; thermal resistance of heat sink = 30 °C/W) $V_F = 1.1$ max @ $1_F = 200$ mA for all types.

JEDEC type no. (Note 1)	Nominal zener voltage V_Z@I_{ZT} volts (Note 2)	Test current I_{ZT} mA	Max zener impedance A and B suffix only Z_{ZT}@I_{ZT} ohms	Z_{ZK}@I_{ZK}=0.25 mA ohms	Max reverse leakage current A and B suffix only I_R μA	@V_R volts A	B	Nonsuffix I_R@V_R used for suffix A μA	Max zener voltage temperature coeff. (A and B suffix only) θ_{VZ} (%/°C) (Note 3)
1N5221	2.4	20	30	1200	100	0.95	1.0	200	−0.085
1N5222	2.5	20	30	1250	100	0.95	1.0	200	−0.085
1N5223	2.7	20	30	1300	75	0.95	1.0	150	−0.080
1N5224	2.8	20	30	1400	75	0.95	1.0	150	−0.080
1N5225	3.0	20	29	1600	50	0.95	1.0	100	−0.075
1N5226	3.3	20	28	1600	25	0.95	1.0	100	−0.070
1N5227	3.6	20	24	1700	15	0.95	1.0	100	−0.065
1N5228	3.9	20	23	1900	10	0.95	1.0	75	−0.060
1N5229	4.3	20	22	2000	5.0	0.95	1.0	50	±0.055
1N5230	4.7	20	19	1900	5.0	1.9	2.0	50	±0.030
1N5231	5.1	20	17	1600	5.0	1.9	2.0	50	±0.030
1N5232	5.6	20	11	1600	5.0	2.9	3.0	50	+0.038
1N5233	6.0	20	7.0	1600	5.0	3.3	3.5	50	+0.038
1N5234	6.2	20	7.0	1000	5.0	3.8	4.0	50	+0.045
1N5235	6.8	20	5.0	750	3.0	4.8	5.0	30	+0.050
1N5236	7.5	20	6.0	500	3.0	5.7	6.0	30	+0.058
1N5237	8.2	20	8.0	500	3.0	6.2	6.5	30	+0.062
1N5238	8.7	20	8.0	600	3.0	6.2	6.5	30	+0.065
1N5239	9.1	20	10	600	3.0	6.7	7.0	30	+0.068
1N5240	10	20	17	600	3.0	7.6	8.0	30	+0.075
1N5241	11	20	22	600	2.0	8.0	8.4	30	+0.076
1N5242	12	20	30	600	1.0	8.7	9.1	10	+0.077
1N5243	13	9.5	13	600	0.5	9.4	9.9	10	+0.079
1N5244	14	9.0	15	600	0.1	9.5	10	10	+0.082
1N5245	15	8.5	16	600	0.1	10.5	11	10	+0.082
1N5246	16	7.8	17	600	0.1	11.4	12	10	+0.083
1N5247	17	7.4	19	600	0.1	12.4	13	10	+0.084
1N5248	18	7.0	21	600	0.1	13.3	14	10	+0.085
1N5249	19	6.6	23	600	0.1	13.3	14	10	+0.086
1N5250	20	6.2	25	600	0.1	14.3	15	10	+0.086
1N5251	22	5.6	29	600	0.1	16.2	17	10	+0.087
1N5252	24	5.2	33	600	0.1	17.1	18	10	+0.088
1N5253	25	5.0	35	600	0.1	18.1	19	10	+0.089
1N5254	27	4.6	41	600	0.1	20	21	10	+0.090
1N5255	28	4.5	44	600	0.1	20	21	10	+0.091
1N5256	30	4.2	49	600	0.1	22	23	10	+0.091
1N5257	33	3.8	58	700	0.1	24	25	10	+0.092
1N5258	36	3.4	70	700	0.1	26	27	10	+0.093
1N5259	39	3.2	80	800	0.1	29	30	10	+0.094
1N5260	43	3.0	93	900	0.1	31	33	10	+0.095
1N5261	47	2.7	105	1000	0.1	34	36	10	+0.095
1N5262	51	2.5	125	1100	0.1	37	39	10	+0.096
1N5263	56	2.2	150	1300	0.1	41	43	10	+0.096
1N5264	60	2.1	170	1400	0.1	44	46	10	+0.097
1N5265	62	2.0	185	1400	0.1	45	47	10	+0.097
1N5266	68	1.8	230	1600	0.1	49	52	10	+0.097
1N5267	75	1.7	270	1700	0.1	53	56	10	+0.098
1N5268	82	1.5	330	2000	0.1	59	62	10	+0.098
1N5269	87	1.4	370	2200	0.1	65	68	10	+0.099
1N5270	91	1.4	400	2300	0.1	66	69	10	+0.099

Note 1. **Tolerance**—The JEDEC type numbers indicate a tolerance of ±20% with guaranteed limits on only V_Z, I_R, and V_F as shown in the electrical characteristics table. Units with guaranteed limits on all six parameters are indicated by suffix "A" for ±10% tolerance and suffix "B" for ±5.0% units.

†For more information on special selections contact your nearest Motorola representative.

Note 2. special selections† available include:
1. Nominal zener voltages between those shown.
2. Two or more units for series connection with specified tolerance on total voltage. Series matched sets make zener voltages in excess of 200 volts possible as well as providing lower temperature coefficients, lower dynamic impedance and greater power handling ability.
3. Nominal voltages at nonstandard test currents.

Courtesy Motorola Semiconductor Products, Inc.

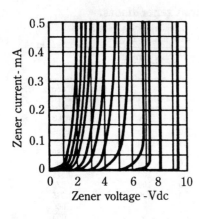
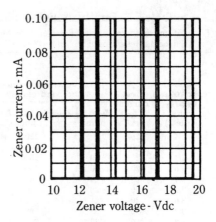

(A) Field emission voltage breakdown.   (B) Avalanche voltage breakdown.

5-1 Field-emission vs. avalanche voltage-breakdown characteristics. TRW Power Semiconductors

diode. Actually, both electrons and holes participate in this phenomenon—it suggests flame propagation once ignition temperature is attained in a chamber of volatile vapor. As previously pointed out, as long as this current "explosion" is limited by the external circuitry, no damaging effects are sustained by the diode.

It is only natural to now ponder the internal physics of device operation. This brief diversion from concern with only the terminal characteristics of the device is that the two breakdown mechanisms cause practical behavior features that have much to do with the relative "goodness" of various zener diodes.

The avalanche breakdown phenomenon causes a positive temperature coefficient for zener diodes with greater breakdown voltages than 6 to 8 V. The field-emission breakdown process (sometimes referred to as *true zener breakdown*), on the other hand, causes a negative temperature coefficient. Thus, zener diodes with voltage ratings much below approximately 5.5 V decrease in voltage, as their operating temperature is increased. Conversely, zener diodes with voltage breakdowns greater than about 6 V will increase in voltage as they become hotter. The 6- to 8-V region is one of transition, in which both temperature coefficients are competitive. As can be suspected, it is more difficult to predict, control, and deal with the effects of temperature in this region than in the clearly field-emission region (low voltage) or the clearly avalanche region (high voltage). The behavior of the temperature coefficient as a function of breakdown voltage in typical zener diodes is shown in Fig. 5-2.

You can take advantage of the competing temperature coefficients. As would be expected, the appropriate design of voltage breakdown in the vicinity of the transition region must result in a diode that exhibits a zero temperature coefficient. In practical devices, this desirable characteristic is attained in the neighborhood of 5.5 V.

Is the 5.5-V zener diode the "best"? Not necessarily, because other parameters are also important. Indeed, either a low or a known temperature coefficient is satisfactory or even desirable. Then, the temperature coefficient of associated

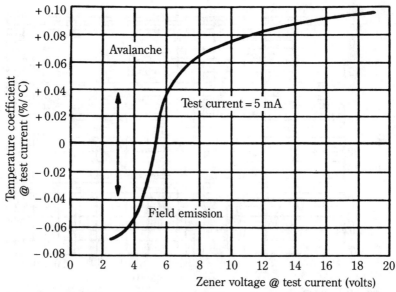

5-2 Temperature coefficient vs. breakdown voltage in typical zener diodes.
TRW Power Semiconductors

components can be used to attain an overall zero or low temperature coefficient. With this in mind, you are more likely to seek a low dynamic impedance. The more abrupt the knee of the current versus voltage curve, the lower the dynamic impedance is and the "tighter" the voltage-stabilizing characteristic is. Minimal dynamic impedance occurs around 6.8 V, at least for low currents. At higher currents, the optimal breakdown voltage is less critical. The dynamic impedance of typical zener diodes as a function of breakdown voltage is illustrated in Fig. 5-3.

Reduced dynamic impedance can be caused by increasing the operating current of the zener diode. Assuming that the increased power dissipation is otherwise acceptable, this technique has limits to which it can be profitably exploited. This is because the increased self-heating from the heavier current is likely to excessively shift the breakdown voltage. Thus, a balance must be decided upon between dynamic impedance and temperature stability. Many factors usually enter the picture here, such as the effectiveness of heat removal, the actual dissipation rating of the zener diode, the temperature coefficient of associated components and devices, and the availability of operating power. Other things being equal, it is generally desirable to operate well within the power rating of the zener diode, to choose the most appropriate breakdown voltage in the face of design constraints, to select a high-quality diode, and to locate and mount the diode with regard to thermal considerations.

The LVA zener diode

From the previous section, it is clear that the avalanche breakdown phenomenon produces more desirable characteristics for zener diodes than does the field-emis-

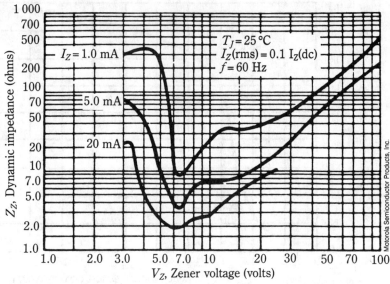

5-3 Dynamic impedance vs. breakdown voltage in typical zener diodes.

sion mechanism (an exception is the manipulation of the two breakdown mechanisms to produce a zero or near-zero temperature coefficient). Particularly, for low-voltage zener diodes, it had long been the objective to sharpen the knees of their current versus voltage curves. Because the soft knees were derived from the intrinsic field emission, any fabrication technique that could enhance the ratio of avalanche to field-emission activity in low-voltage zener diodes would produce more abrupt breakdown and lower dynamic impedance (its practical manifestation).

This, indeed, has been accomplished in the LVA™ (low-voltage avalanche) zener diode. In accomplishing this, the zero-temperature voltage has been shifted down from the nominal 5.5 to 4.7 V. A dramatic side effect of suppressing field emission has been the lowering of operating current. The LVA zener diodes can often be operated at several tens of microamperes. At more ordinary currents (1 to 10 mA, for example), they exhibit low dynamic impedances that are attainable in ordinary zener diodes only at much higher currents. The LVA zener diode merits consideration for low-voltage regulated power supplies, such as the 5-V types that are commonly used for digital and computer equipment. The improved performance of the LVA zener diode is clearly evident in the oscilloscopic display of breakdown characteristics of this diode and of an ordinary zener diode (Fig. 5-4). The abrupt knee of the LVA characteristic, together with the near-vertical slope of its current-versus-voltage curve, reveal inordinately low leakage current and high dynamic impedance. The same circuit was used to find both curves in order to make the comparison valid.

The temperature-compensated voltage reference

The competing temperature coefficients from field emission and from avalanching cancel in the vicinity of 5.5 V in conventional zener diodes. It is, of course, imprac-

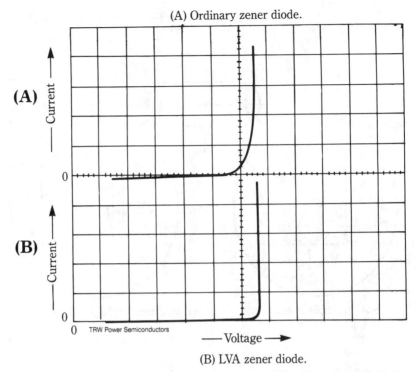

5-4 Breakdown of ordinary 5.6-V zener diode and a like-rated LVA diode.
TRW Power Semiconductors

tical to have to be constrained to one breakdown voltage in order to secure temperature independence. A very satisfactory technique has been devised for producing a reasonable range of breakdown voltages with near temperature independence. Diodes that rely primarily on the avalanche mechanism (above 6 to 8 V) display good breakdown and dynamic impedance characteristics, but are subject to an appreciable voltage drift from a positive temperature coefficient. Conversely, a forward-biased diode has a negative temperature coefficient. An appropriate series combination of reverse- and forward-biased diodes, and a specified operating current can yield a very low temperature coefficient. Such a combination of diodes is packaged in a single container; the resulting unit is known as a *temperature-compensated voltage reference* or a *temperature-compensated zener diode* (Fig. 5-5).

These devices can readily provide temperature coefficients in the vicinity of 0.005%/°C. Careful factory selection and circuit implementation can yield at least an order of magnitude greater stability than this. The need for such precise stability is confined to sensitive and critical scientific instrumentation. Characteristics of a series of 9.4-V reference zener diodes are shown in Table 5-2.

Many designers operate these voltage references in simple series-resistance circuits, as are commonly used with ordinary zener diodes. This is sufficient for many applications. However, in order to advantageously use the temperature stability, these temperature-stabilized references should be operated from either a voltage preregulator or from a constant-current source.

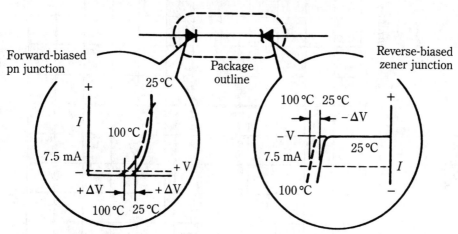

5-5 Principle of temperature-compensated zener voltage reference.
Motorola Semiconductor Products, Inc.

Table 5-2. Characteristics of a series of 9.4-volt zener-reference diodes.
$V_Z = 9.4$ Volts ± 0.4 V (± 0.2 V Suffix "A") @ ($I_{ZT} = 10$ mA)

Type number	Max voltage change ΔV_Z (volts)	Test temperatures °C	Temperature coefficient %/°C	Max dynamic impedance Z_{ZT} (ohms)
1N2163,A	0.033	0, +25, +70	0.005	
1N2164,A	0.086	−55, 0, +25, +75, +125	0.005	
				15
1N2165,A	0.115	−55, 0, +25, +75, +125, +185	0.005	
1N2166,A	0.007	0, +25, +70	0.001	
1N2167,A	0.017	−55, 0, +25, +75, +125	0.001	
				15
1N2168,A	0.023	−55, 0, +25, +75, +125, +185	0.001	
1N2169,A	0.004	0, +25, +70	0.0005	
1N2170,A	0.009	−55, 0, +25, +75, +125	0.0005	
				15
1N2171,A	0.012	−55, 0, +25, +75, +125, +185	0.0005	

Courtesy Motorola Semiconductor Products, Inc.

The zener diode as a transient suppressor

The abrupt turn-on characteristic of the zener diode makes it an excellent transient-suppressing device when it is placed in shunt with a circuit that normally operates at a somewhat lower voltage level than that of the breakdown characteristic. Designers have often used 10- and 50-W units for this purpose. Because of the short time and low-average power of commonly encountered transients, a heatsink is usually not required. For the best protection of vulnerable circuits and components, it is best to use specially designed units that were made specifically for rapid energy absorption. One such unit is the Motorola MPZ5. Actually, the MPZ5 designates a family of zener-diode assemblies with a selection of convenient breakdown voltages. Each assembly consists of several closely matched 50-W zener diodes that are connected in parallel. For transients of a 0.1 ms or so, peak power absorption can be several tens of kilowatts. For longer transients, the rated peak-power absorption is necessarily reduced so that a transient that endures for a full second could dissipate about 500 W.

Figure 5-6 shows a typical MPZ5 transient suppressor. This construction facilitates application to nonpolar circuitry; then, two suppressors are simply bolted together. The two aluminum plates are generally sufficient to remove heat because the occasional, rather than the repetitive, transient usually must be guarded against the most.

Table 5-3 lists the electrical characteristics of the MPZ5 family of zener transient suppressors. The response time is not indicated because of the great dependency of that parameter on lead inductance and on other circuit strays. However, careful installation should enable the full-protective capability of the device to be in effect within 30 ns after its breakdown voltage is exceeded. In the absence of transients, the power dissipation will, for all practical purposes, be zero. This, alone, is a compelling feature, when compared to transient suppressors that have "soft" characteristics (such as selenium and silicon carbide). The suppression effects of the zener-diode transient suppressor are shown for several commonly encountered situations (Fig. 5-7).

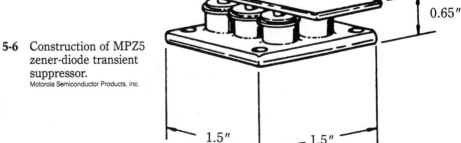

5-6 Construction of MPZ5 zener-diode transient suppressor.
Motorola Semiconductor Products, Inc.

Table 5-3. Characteristics of MPZ5 family of zener-diode transient suppressors.
($T_A = 25\,°C$) ($V_F = 1.5$ V max @ 10 A for all types)

Type	Nominal operating voltage		Maximum device clamping factor $CF = \dfrac{V_Z@I_{Z(PULSE)}}{V_Z@I_{ZT}}$	Minimum zener voltage		Maximum zener voltage pulse width = 1.0 ms		Maximum reverse current $I_{R(MAX)}$ @$V_R = V_{OP(PK)}$ µAdc	Typical capacitance C(typ) @$V_R = V_{OP(PK)}$ µF
	$V_{OP(PK)}$ Vdc	$V_{OP(RMS)}$ V rms		$V_{Z(MIN)}$ Vdc	@I_{ZT} Adc	$V_{Z(MAX)}$ Vdc	@$I_{Z(PULSE)}$ Adc		
MPZ5-16A	14	10	1.25	16	0.4	24	200	50	0.025
-16B	14	10	1.25	16	0.4	20	200	↑	0.025
-32A	28	20	1.25	32	0.2	50	100		0.011
-32B	28	20	1.25	32	0.2	45	100		0.011
-32C	28	20	1.25	32	0.2	40	100		0.011
-180A	165	117	1.14	180	0.03	250	20	↓	0.0012
-180B	165	117	1.14	180	0.03	225	20		0.0012
-180C	165	117	1.14	180	0.03	205	20	50	0.0012

Courtesy Motorola Semiconductor Products, Inc.

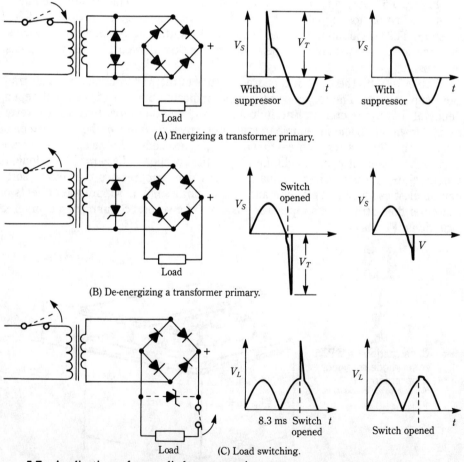

(A) Energizing a transformer primary.

(B) De-energizing a transformer primary.

(C) Load switching.

5-7 Applications of zener diode as a transient suppressor. Motorola Semiconductor Products, Inc.

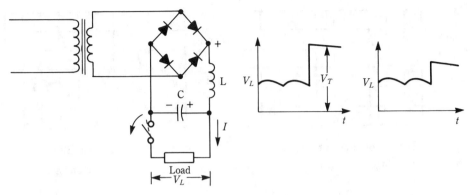

(D) Removing the load from an LC-filtered full-wave supply can create a voltage transient,

$$V_T = \sqrt{\frac{LI_L^2}{C} + V_L^2}.$$

5-7 Continued

The Schottky diode

The trend toward high-frequency switching supplies to provide the nominal 5 V (often at high amperage) for computer equipment has revealed certain shortcomings of conventional silicon rectifier diodes. For example, the popular bridge-rectifying circuit uses two series-connected diodes for each half cycle of the impressed ac voltage. Practical silicon diodes might produce a forward-biased voltage drop on the order of 0.8 V or more. The resultant 1.6-V drop is an appreciable portion of the 5-V output. Thus, the built-in power loss can negate the high operating efficiency in the switching circuitry. Partial remedial measures sometimes involve the use of center-tapped rectifying circuits or even half-wave circuits. Also, in some instances, germanium rectifying diodes might conceivably be used to alleviate the high forward-voltage drop. However, another source of power dissipation exists in conventional silicon and germanium rectifier diodes.

At 60 Hz or even 1000 Hz, both silicon and germanium diodes perform as ideal rectifiers: they fairly faithfully permit passage of one-half cycle of the applied ac waveform and behave as substantially open circuits for the opposite half cycle. At higher frequencies (in the vicinity of 20 or 30 kHz, for example) such rectifying performance no longer exists. Figure 5-8 illustrates what you might see on the oscilloscope. As might be surmised, the impaired rectification at the higher frequency is also a source of additional power dissipation. This malperformance is caused by charge storage during forward conduction and the time needed to deplete the stored charge during the reverse half cycle. Not only does this effect reduce operating efficiency, it also increases ripple current and lowers the effective level of the output voltage. Obviously, in addition to devising more efficient rectifying circuits you also need a more efficient rectifying device.

The Schottky diode is a semiconductor rectifying device, which, unlike conventional pn-junction diodes, depends for its operation only on majority current

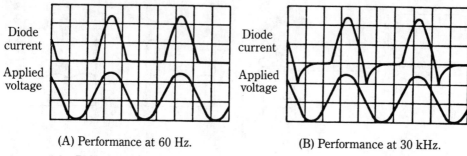

(A) Performance at 60 Hz.   (B) Performance at 30 kHz.

5-8 Behavior of ordinary silicon pn-diode at low and at high frequencies.

carriers. This is because it consists of semiconductor material (usually n-type silicon) that is in contact with metal. We are accustomed to think of such fabrication as nothing more than an ohmic contact. However, by appropriate processing and doping of the semiconductor material, a "half-junction," with essentially unilateral conduction, is produced. Strange as it might seem, this device is reminiscent of the crystal detector (aside from application of modern materials, the chief difference is that the Schottky diode is an area, rather than a point-contact device). Because the Schottky diode does not involve minority carriers in its rectifying performance, it has no charge-storage phenomena. The practical manifestation of this is that it has no interval of extended conduction when the applied forward voltage reverses polarity. You might say that the reverse recovery of this device is instantaneous in terms of specifying parlance. Thus, Fig. 5-8A would not only apply to a Schottky diode at 60 Hz, but also at 30 kHz, and even at much higher frequencies (only internal capacitance limits high frequency performance).

Not only is the Schottky diode a natural choice for many high-frequency applications, but it is also superior to conventional pn diodes in the other important rectifying parameter—forward voltage drop. Figure 5-9 is a comparison of several Schottky diodes and a typical silicon pn diode, with regard to forward voltage versus current. The reduced power dissipation of the Schottky diode (as a result of its lower forward voltage) can be a very significant factor in low-voltage power supplies (in 5-V supplies in particular). Commercially available Schottky diodes are silicon devices that can now be obtained in ratings that extend considerably from those thought to be limiting values just a few years ago. Specifically, current-handling ability has been increased from 20 to 75 A, operating temperature is now specified to 175 °C (rather than 100 °C), and voltage capability has been increased from about 30 to 50 V. Impressive as these improvements have been in a relative way, they nonetheless are, at the same time, suggestive of inherent limitations of the device.

Balancing the Schottky diode's assets of low forward-voltage drop and excellent high-frequency response are its shortcomings of high reverse current and inapplicability to voltages that are much above 50 V. These detrimental features actually go hand-in-hand inasmuch as reverse current is a strong function of reverse voltage. An example of this behavior is shown in Fig. 5-10 for a typical

The Schottky diode 259

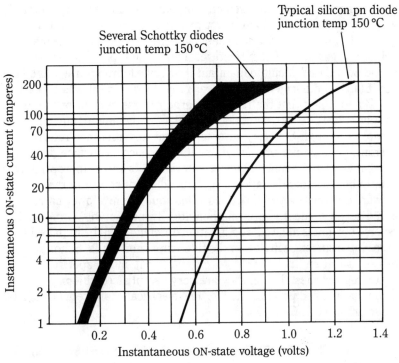

5-9 Forward voltage drop: several Schottky diodes vs. typical silicon pn-diode.

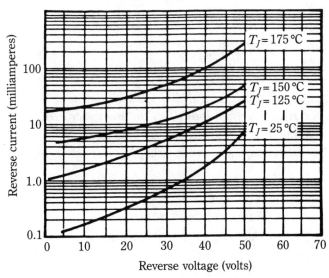

5-10 Reverse characteristics of a typical 30-A Schottky diode.
TRW Power Semiconductors

30-A Schottky diode. Voltage and temperature can be viewed as trade-offs: at elevated temperatures, voltage must give way in order to keep the reverse current down. However, specification sheets generally assign a single maximum operating voltage, regardless of temperature. This makes practical sense; in operation, it would be rather difficult to maintain the device temperature in the vicinity of ambient if (as is ordinarily the case) near-rated load current was being handled.

Do not think that the relatively large reverse current of the Schottky diode is simply a leakage current that is similar to the microampere currents in some diodes. Although Schottky reverse current is sometimes referred to as "leakage current," it does not primarily owe its existence to surface defects or to internal anomalies of the semiconductor material. It is most appropriately explained as simply being a manifestation of enhanced conductivity in the device, both in the forward and the reverse directions. This stems from a lower barrier potential (compared to pn diodes) and is mathematically predictable. Practically, if excessive reverse current is allowed, rectification efficiency will be impaired and inordinately high ripple current will flow in the filter capacitor. Thus, Schottky diodes find their best application in low-voltage circuits and, more specifically, in low-voltage switching supplies. There, they are very useful as rectifiers, and as freewheeling, commutation, and isolation diodes. The electrical characteristics of a 60-A Schottky diode are listed in Table 5-4.

Table 5-4. Characteristics of the 60-A SD-51 Schottky diode.

Maximum Ratings

Symbol	Rating	Limit	Units
V_{RMM}	Repetitive peak reverse voltage	45	V
V_{RWM}	Repetitive peak reverse working voltage	35	V
I_{FM}	Repetitive peak forward current (50% duty cycle)	120	A
I_{FSM}	Nonrepetitive peak one-cycle surge current. 8.35 msec half sine wave	800	A

Electrical Specifications (25°C unless specified)

Symbol	Specification/conditions	Limit	Units
I_{RM}	Maximum reverse leakage current $V_R = 35$ V	50	mA
I_{RM}	Maximum reverse leakage current $V_R = 35$ V, $T_j = 125$ °C	200	mA
V_{FM}	Maximum forward voltage (300 μsec, < 1% duty cycle)		
	$I_F = 60$ A	0.70	V
	$I_F = 60$ A, $T_j = 125$ °C	0.60	V
	$I_F = 120$ A	0.87	V
	$I_F = 120$ A, $T_j = 125$ °C	0.84	V

Table 5-4. Continued.

Electrical Specifications (25°C unless specified)

Symbol	Specification/conditions	Limit	Units
C_j	Typical junction capacitance $V_R = 5$ V	4000	pF
t_{RR}	Typical reverse recovery time $I_f = I_R = 1$ A, $T_C = 125$ °C	50	n-sec
dV/dt	Maximum rate of change of reverse voltage	700	V/μsec

Thermal Specifications

Symbol	Specification	Limit	Units
T_j	Operating temperature range	−55 to +150	°C
T_{sto}	Storage temperature range	−55 to +165	°C
$R_{\theta JC}$	Thermal resistance junction to case	1.0	°C/W

Courtesy TRW Power Semiconductors

A noteworthy practical feature of Schottky diodes is that it is often feasible to parallel similar units to obtain greater overall current capability. With a little experimentation or with consultation of manufacturer's tolerances, this can be readily accomplished with no need for power-wasting ballast resistors. Advantage can also be taken of Schottky-diode assemblies that are composed of two matched diodes in one package (Fig. 5-11). Such units are intended for full-wave rectification with center-tapped transformer windings. With this arrangement, one obtains the low ripple of the bridge rectifier circuit, but without its low efficiency in low-voltage systems (the low efficiency is because the forward voltage drops of two rectifying diodes must be accommodated each half cycle). Yet another practical application of the Schottky diode is in voltage-multiplying circuits. Here, its ability to handle high-repetition-rate square waves merits consideration.

Technological breakthroughs have extended Schottky voltage ratings from the earlier 40-V limit to 100 V, and it is clear that higher ratings are feasible. However,

5-11 Two Schottky diodes in TO-244AB package for a center-tap full-wave circuit. Usually the common-cathode connection goes to the mounting plate. Other arrangements are also available. This package accommodates 200 to 400 A ratings.

this result is an adverse compromise of the forward-voltage drop. The most efficient Schottky rectifying diodes do not much exceed ratings of 40 V.

The LM105 positive-voltage regulator

The National Semiconductor LM105 is a monolithic circuit of a linear regulated power supply. It provides about 18 mA over an adjustable range of 4.5 to 40 V. Although the LM105 has been superseded by more-sophisticated designs, its basic topography pioneered the stage for the more-advanced ICs, many of which established new fashions of acceptance by featuring more "bells and whistles." Instructional objectives are well served, however, by considering the nature and applications of the LM105. Some of the operational modes that are readily attained are:

- It can be made to operate with foldback current limiting. Not only can such limiting be adjusted to occur anywhere between zero and full load, but internal thermal feedback automatically limits output current in the event of a short-circuited load.
- It can be shut off by simply grounding one of the terminals.
- One of the terminals can be connected to an external transistor(s) to boost the current level of the regulated output voltage. Booster transistors can also be used to extend regulation to higher voltages.
- An external frequency-compensation capacitor can be added to provide required stability and transient response for any load situation.
- An external capacitor can be used to bypass the internal voltage reference in order to greatly minimize the noise inherent in breakdown diodes.
- By means of externally connected diodes, the regulator can be protected against the effects of shorted input, from transients in the unregulated supply, against actual reversal of the unregulated-supply polarity, and against reversal of the output-voltage polarity.
- It can be operated as a high-efficiency switching-type regulator.
- With an external transistor, it can be operated as a shunt-type voltage regulator.
- It can be operated as a current regulator.

The regulator circuit

The schematic diagram of this IC regulator is shown in Fig. 5-12. The external connections for basic voltage-regulator operation are shown in Fig. 5-13. It is to explore the general principles of operation, because the familiar configurations of discrete circuitry, though evident in part, involve a number of modifications.

Commencing with the voltage-reference circuitry, breakdown diode D1 (actually the base-emitter diode of a transistor structure) receives constant current from

The LM105 positive-voltage regulator 263

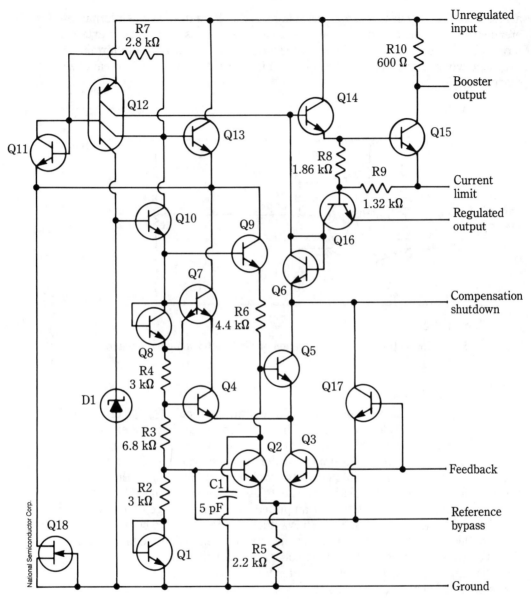

5-12 Schematic of the LM105 voltage regulator.

one collector of multiple-collector transistor, Q12. The voltage that is developed across the breakdown diode is not used directly, but is buffered by transistor Q10, then impressed across the cascaded elements Q8, R4, R3, R2, and Q1. Notice that Q8 and Q1 are diode-connected transistors. The net effect of these series-connected elements is that the temperature coefficient of the voltage that is developed across R2/Q1 is very nearly zero. This voltage is therefore used as the internal reference voltage of the regulator. The reference voltage is also available at external terminal 5. For critical applications, a capacitor that is connected from terminal 5 to ground will greatly attenuate the noise that is inherently produced in breakdown diodes.

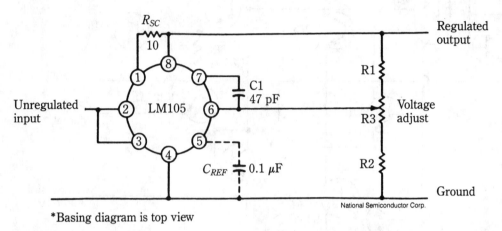

*Basing diagram is top view

5-13 Connections for basic operation of the LM105 voltage regulator.

Voltage comparator

The voltage comparator is composed of the transistor pair Q2/Q3 differential amplifier samples the reference voltage; the other input is available at external terminal 6. Terminal 6, of course, connects to the divided-down output voltage, which is provided by either a fixed resistance divider or a potentiometer (the availability of this "sensing" terminal facilitates many novel applications, such as the stabilization of light intensity, the regulation of a heater, or motor speed control). The error signal from the Q2/Q3 pair is further amplified by a second stage that consists of transistors Q4 and Q5. The usable gain is actually provided by Q5 because the collector of this transistor is fed from a constant current (i.e., high impedance) source, this is one of the collectors of multiple-collector transistor Q12.

The output of Q5 is buffered by emitter-follower Q14, which in turn drives the series-pass transistor, Q15. Transistor Q16 provides current limiting: when the voltage across an external resistor (connected to terminals 1 and 8) increases to turn on transistor Q16, the base drive from Q14 is decreased and the regulator ceases to supply any more load current.

The excellent performance of this regulator is, at least in part, caused by the constant-current sources that are provided by the multiple collectors of transistor Q12. The base bias of Q12 exists only when the regulator is operating. Operation, therefore, would not ensue unless Q12 received initial bias from some other source. In order to make the regulator self-starting, field-effect transistor Q18 is connected to provide an initial current source for the base of Q12. During operation, bias current for Q12 is provided by a lower impedance current source, the collector of transistor Q9.

Transistor Q17 prevents a latch-up condition that might otherwise occur from saturation in transistor Q3, when the external sampling resistances are apportioned for low output voltage from the regulator.

The versatility of the LM105 is indicated in Fig. 6-12, which shows some of the operating modes that are readily achieved by the appropriate external connections. The pertinent characteristics of the LM105 are given in Table 5-5.

Table 5-5. Pertinent characteristics of the LM105 voltage regulator.

Input voltage	50 V
Input-output voltage differential	40 V
Power dissipation*	500 mW
Operating junction temperature range	−55 °C to 150 °C
Storage temperature range	−65 °C to 150 °C
Lead temperature (Soldering, 10 sec)	300 °C

ELECTRICAL CHARACTERISTICS**

Parameter	Conditions	Min	Typ	Max	Units
Input-voltage range		8.5		50	V
Output-voltage range		4.5		40	V
Output-input voltage differential		3.0		30	V
Load regulation†	$0 \leq I_o \leq 12$ mA				
	$R_{SC} = 18\ \Omega,\ T_A = 25\ °C$		0.02	0.05	%
	$R_{SC} = 10\ \Omega,\ T_A = 125\ °C$		0.03	0.1	%
	$R_{SC} = 18\ \Omega,\ T_A = -55\ °C$		0.03	0.1	%
Line regulation	$V_{IN} - V_{OUT} \leq 5$ V		0.025	0.06	%/V
	$V_{IN} - V_{OUT} > 5$ V		0.015	0.03	%/V
Ripple rejection	$C_{REF} = 10\ \mu F,\ f = 120$ Hz		0.003	0.01	%/V
Temperature stability	$-55\ °C \leq T_A \leq 125\ °C$		0.3	1.0	%
Output-noise voltage	10 Hz $\leq f \leq 10$ kHz				
	$C_{REF} = 0$		0.005		%
	$C_{REF} > 0.1\ \mu F$		0.002		%
Standby-current drain	$V_{IN} = 50$ V		0.8	2.0	mA
Feedback-sense voltage		1.6	1.8	2.0	V
Long-term stability			0.1	1.0	%

*For operating at elevated temperatures, the device must be derated, based on a 150 °C maximum junction temperature and a thermal resistance of 45 °C/W junction-to-case or 150 °C/W junction-to-ambient for the metal-can package. For the flat package, the derating is based on a thermal resistance of 185 °C/W when mounted on a 1/16-inch-thick, epoxy-glass board with ten 0.03-inch-wide, 2-ounce copper conductors. Peak dissipations to 1W are allowable providing the dissipation rating is not exceeded with the power averaged over a five-second interval.

Table 5-5. Continued.

**These specifications apply for a junction temperature between −55 °C and 150 °C, for input and output voltages within the ranges given, and for a divider impedance seen by the feedback terminal of 2 kΩ, unless otherwise specified. The load and line regulation specifications are for constant junction temperature. Temperature-drift effects must be taken into account separately when the unit is operating under conditions of high dissipation.

†The output currents given, as well as the load regulation, can be increased by the addition of external transistors. The improvement factor will be roughly equal to the composite current gain of the added transistors.

Courtesy National Semiconductor Corp.

The LM113 energy-gap reference diode

The LM113 represents a practical implementation of the energy-gap voltage principle. The internal circuitry of this novel IC is shown in Fig. 5-14. It uses a pair of transistors similar to that which is described for the basic circuit of Fig. 4-43, but it also contains a buffer amplifier to provide load isolation and to lower the output impedance. The breakdown voltage is 1.220 V. This reference voltage is available at a dynamic impedance of 0.3 Ω from 0.5 to 20 mA. The temperature stability is on the order of ±1% over the wide temperature range of −55 to 120 °C!

The technique of using the LM113 in regulators is the same as for zener diodes. Indeed, the same symbol is used. Two regulators, in which the LM113 is used with op amps, are illustrated in Fig. 5-15. The maximum power dissipation of

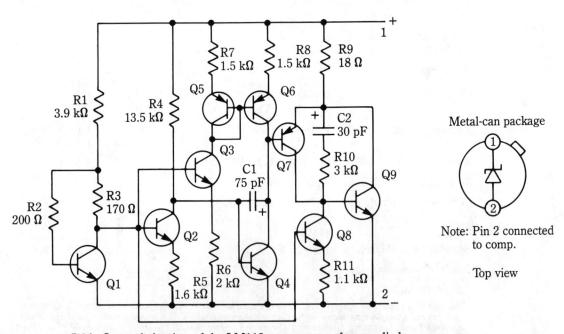

5-14 Internal circuitry of the LM113 energy-gap reference diode. National Semiconductor Corp.

the LM113 at nominal temperatures is 100 mW and its maximum current rating is 50 mA. The salient characteristics of this "synthesized" reference are provided by the performance curves of Fig. 5-16.

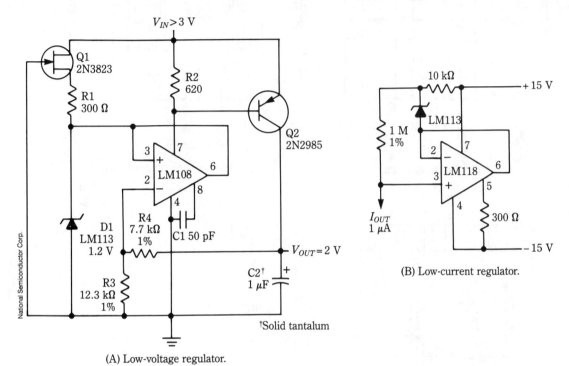

(A) Low-voltage regulator.

(B) Low-current regulator.

5-15 Typical applications of the LM113 energy-gap reference diode.

The three-terminal IC regulator

A very important development in IC regulators is the *three-terminal regulator*. This might seem strange because there also has been a strong trend for IC regulators to have many terminals so that maximum versatility can be achieved. Electronic systems pose distribution problems that partially defeat the desirable characteristics of sophisticated dc supplies. This stems from the resistance and reactance of connecting leads and also from ground loops and "crosstalk" among such leads. To a considerable extent, the use of a regulated supply with remote sensing remedies such troubles. Also, the judicious application of bypassing and filtering is often helpful. However, when a single regulated supply powers multiple loads, these techniques can no longer work so well—indeed, they can provoke problems of their own.

It was realized that point-of-application regulation might solve such distribution problems. That is, instead of relying entirely on a master regulator, each sub-

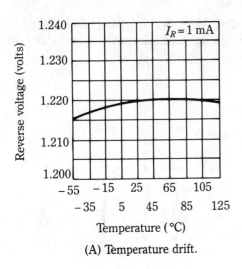

(A) Temperature drift.

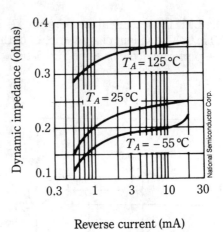

(B) Reverse dynamic impedance.

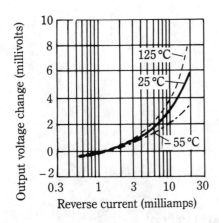

(C) Reverse characteristics.

5-16 Performance curves for the LM113 energy-gap reference diode.

system would contain its own regulator. This appeared feasible because these individual regulators generally could have quite small current capability, compared to that which is required from a single regulator for the entire system. Individual voltage regulators for the various subsystems could also be expected to confer greater decoupling between subsystems than could be readily attained with bypass capacitors. Not only does a good voltage regulator simulate a large bypass capacitor, but the capacitive reactance remains low, right down to dc. No physical capacitor, however large, has this desirable behavior.

What the circuit designer needed, however, was a point-of-application regulator that could be easily mounted on a subsystem circuit board, much in the manner of a bypass capacitor. Not only have the semiconductor companies provided this in the form of the three-terminal IC regulator, but these devices often match the cir-

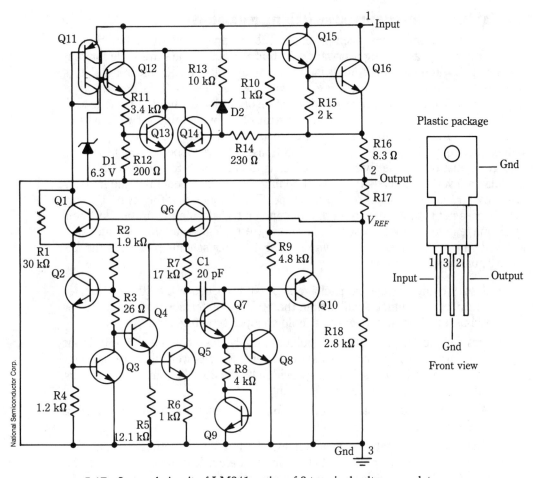

5-17 Internal circuit of LM341 series of 3-terminal voltage regulators.

cuit sophistication of "conventional" IC regulators that have many terminals. Figure 5-17 shows the internal circuitry of the LM341 family of three-terminal fixed-voltage IC regulators.

Despite the easy implementation of this three-terminal IC regulator, it has many desirable features. These features include internal short-circuit limiting and internal thermal overload protection. The internal series-pass transistor has safe-operating-area (SOA) protection. With adequate heatsinking, output currents in excess of 0.5 A are allowable. The three-terminal regulator can be used with external components to provide adjustable voltage or current. For such adjustable-output operation, superior results can be obtained from three-terminal IC regulators that were especially designed for the purpose. One of these ICs is the LM117. Typical circuit applications are described in chapter 6.

Special ICs for inverter-type switching regulators

Considerable attention has been given to the chopper-type regulator. Such a switcher is simple and straightforward now that various ICs are available for controlling the switching transistor. It is also fortunate that the IC linear regulators have performed well when operating as switching regulators. This general technique, whereby either a linear regulator IC or an operational amplifier is made self-oscillatory in order to control a switching transistor, is suitable for a variety of applications. In some instances, an oscillator is used so that the switching rate is independent of variations of operating conditions. However, a better way to achieve voltage regulation via the switching process is to use a high-frequency duty-cycle modulated inverter, instead of merely chopping the dc line. Generally, the inverter is a driven (rather than a self-oscillating) type. This implies that the inverter transformer does not saturate (inverters that depend on transformer saturation for their operation tend to produce worse voltage spikes). A typical switching regulator that is based on a driven inverter is shown in Fig. 5-18. This scheme possesses the following noteworthy features:

- A very compact physical package is possible, because the 20-kHz inverter transformer is much smaller than the 60-Hz transformer that is needed for the unregulated supply of the simple chopper-type regulator.

- Despite the fact that the unregulated supply does not use a 60-Hz input transformer, total isolation exists between the ac power line and the regulated dc output. The isolation is provided by the 20-kHz transformer and by the optoisolator in the feedback circuit.

- A driven inverter is capable of very high efficiencies and no freewheeling diode is needed.

- The ability to operate at a constant switching rate is sometimes of considerable importance in avoiding harmonics and electrical noise that might be particularly offensive to the powered equipment.

- Such an arrangement can operate not only from the 60-Hz line, but also from sources with a wide range of frequencies, including dc.

Although these features have long been recognized, it was not easy to implement this scheme with discrete control devices. Cost and complexity invariably reared their ugly heads when such attempts were made. With the advent of specialized ICs, such as the MC3420, such problems will no longer be encountered. For example, the MC3420 Switchmode Regulator will provide the following basic functions for the system that is depicted in Fig. 5-18:

1. Pulse-width modulator
2. Oscillator, clock, or ramp generator
3. Push-pull drivers for the inverter
4. Voltage reference

Obviously, this multifunction chip resolves much of the complexity of such a regulating system. Moreover, it also provides some important auxiliary functions.

The three-terminal IC regulator 271

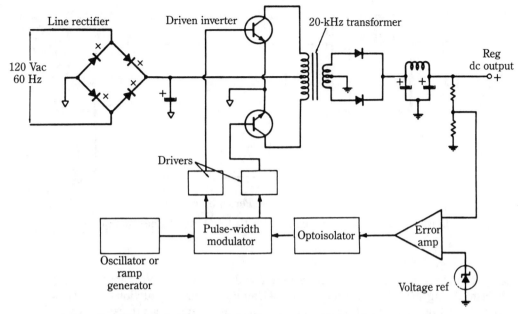

5-18 Simplified switching regulator that is controlled by driven output stage.

The equivalent circuit of the MC3420 is shown in Fig. 5-19. Notice the dead-time provision. This circuitry ensures that the maximum duty cycle of the driven inverter will be as illustrated in Fig. 5-20. It is highly desirable to have some dead time in the waveform to prevent the possibility of having the two inverter transistors simultaneously conduct during switching transitions. Simultaneous conduc-

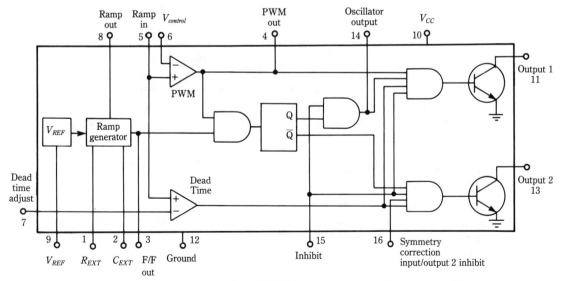

5-19 Block diagram of the MC3420 switch-mode regulator control IC.
Motorola Semiconductor Products, Inc.

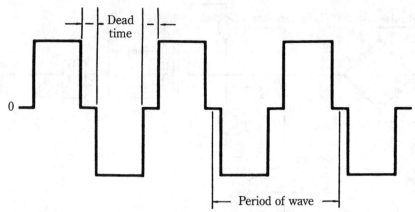

5-20 Drive waveforms for a driven inverter that operates at maximum duty cycle.

tion is the consequence of the time required to deplete the stored charges in the transistor being turned off. Avoidance of such simultaneous conduction diminishes spike production and increases inverter efficiency. You might wish to use the other auxiliary provisions that are included. The symmetry of the duty cycle can be varied by applying a dc voltage to terminal 16. Terminal 15 provides the option of turning off the driven inverter altogether—a useful protective technique.

The waveforms in Fig. 5-21 provide a detailed view of circuit action within the MC3420. Notice that the various timing pulses are available from specific terminals on the 16-pin dual in-line package. This permits extreme flexibility of implementation and allows the ingenious designer or experimenter considerable freedom to apply such techniques as soft start, inrush current limiting, slaving to other regulators, and remote control.

Yet another degree of freedom is derived because the repetition rate of the ramp generator is determined by the values of external resistance and capacitance that are added to terminals 1 and 2, respectively. Figure 5-22 depicts the output frequency of the driven inverter as the function of various values of resistance and capacitance that are associated with terminals 1 and 2. Observe from Fig. 5-21 that the repetition rate of the ramp generator is actually twice the frequency of the driven inverter.

The Silicon General SG1524/SG2524 regulating pulse-width modulator

This monolithic subsystem represents another manufacturer's solution to the control requirements of switching regulators and driven inverters. Figure 5-23 depicts the block diagram of this 16-pin dual in-line IC module. The circuit is similar to that of the Motorola MC3420. However, the ratings and the pin connections for the two devices are not the same.

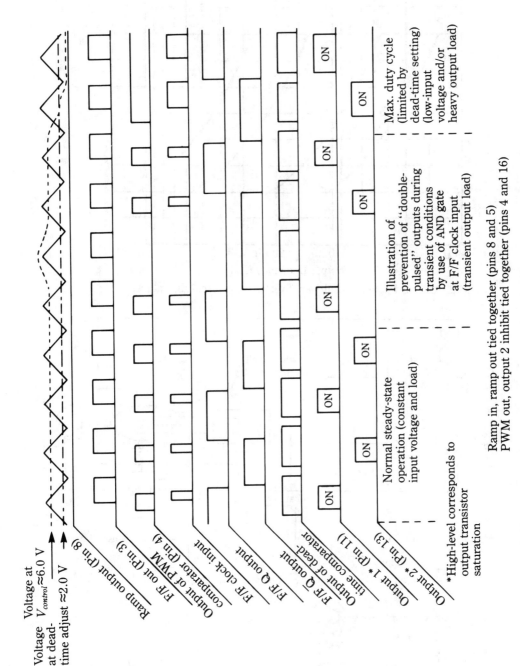

5-21 Waveforms in the MC3420 switch-mode regulator IC. *Motorola Semiconductor Products, Inc.*

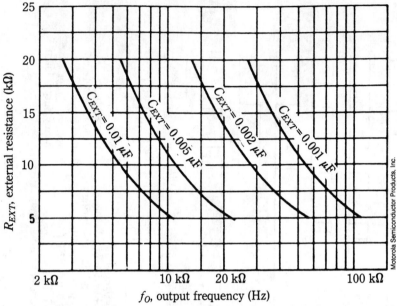

5-22 Output frequency of driven inverter vs. external RC values.

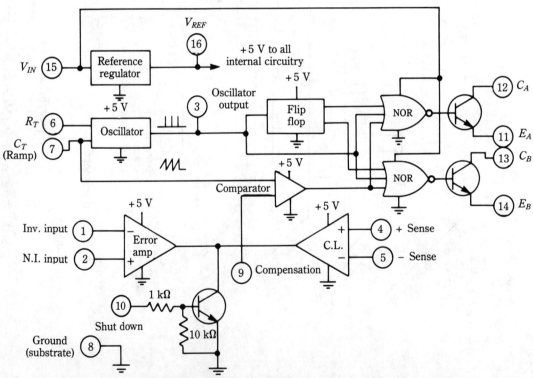

5-23 Block diagram of the SG1524 regulating pulse-width modulator. Silicon General, Inc.

A switching regulator that uses this IC is shown in Fig. 5-24. The performance of this arrangement is generally superior to that of the self-oscillatory regulators, although their circuit configurations are similar. The switching frequency of the regulator in Fig. 5-24 is relatively constant, regardless of line or load conditions and the frequency can be selected to minimize electrical interference or to optimize operating efficiency. Design of the output filter is facilitated by the known switching rate. Tighter regulation can usually be achieved because the control parameter that governs regulation is duty-cycle only, rather than a combination of duty-cycle and switching rate.

The switching rate is easily determined by the values of external resistance and capacitance that are associated with terminals R_T and C_T. The graph in Fig. 5-25 enables you to make a quick selection. Thus, the regulator of Fig. 5-24, with its 3-kΩ resistance and 0.02-μF capacitance combination, has an oscillator period of about 50 μs. The switching rate is then the reciprocal of the oscillator period— 20 kHz, in this case. In single-ended chopper-type regulators, the oscillator and the output frequency are identical. When push-pull switching is involved, such as with most driven inverters, the output frequency is one-half that of the oscillator.

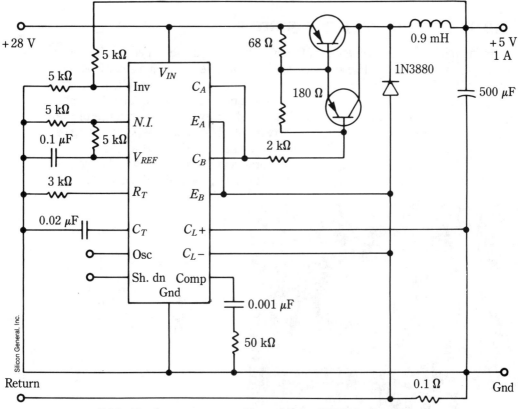

5-24 Dc chopper-type regulator with the SG2524 control IC.

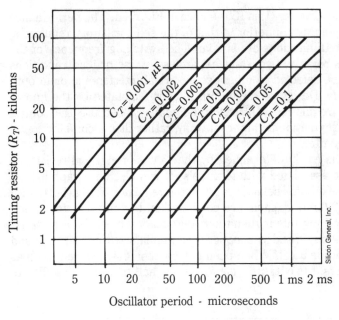

5-25 Oscillator period as a function of external timing RC-network.

Giant Darlington transistors

The modern monolithic Darlington transistor is worthy of mention because it is now a far cry from earlier units. These devices are now capable of yielding superb performance in 20-kHz switching supplies and have voltage ratings that enable use in off-line systems. Unlike their puny predecessors, continuous collector-current ratings are commonly in the neighborhood of 40 to 60 A. An even better evidence of the technological maturity that these devices have attained is the 250-W power rating of the TO-3-packaged MJ10022 and MJ10023 Darlington transistors (Fig. 5-26).

Actually, certain modifications are incorporated in the package, which is somewhat deeper than ordinary TO-3 packages. Not seen by the eye is the liberal use of

5-26 250-W high-voltage high-speed monolithic Darlington transistor.

copper for the package itself and beryllium oxide on the inside. Also, some manufacturers use paralleled chips for the output stage of the Darlington circuit. Although four basic Darlington configurations exist (Fig. 5-27), only those depicted in Fig. 5-27A and 5-27B have been monolithically fabricated as high-power devices. The actual internal circuitry of the MJ10022 and MJ10023 devices is shown in Fig. 5-27E. Notice the base-return resistors and the speed-up and protection diodes. Although both devices carry 250-W ratings, the MJ10022 is a 350-V device and the MJ10023 has a maximum rating of 400 V.

A nice aspect of power devices, such as these giant Darlingtons, is that their ratings pertain to realistic operating conditions in switch-mode power supplies. Thus, their specifications assume reverse-bias assisted turn-off, inductive loading, and a 100 °C case temperature. This practical specification approach, together with the high-voltage and high-speed characteristics of these power devices makes them particularly suitable for off-line power supplies, inverters, and converters. Also, their electrical ruggedness merits consideration for motor-controlled power supplies. Whereas, you would expect to encounter base-drive problems in a 250-W discrete transistor, the high overall current gain of the Darlington configuration greatly simplifies drive requirements.

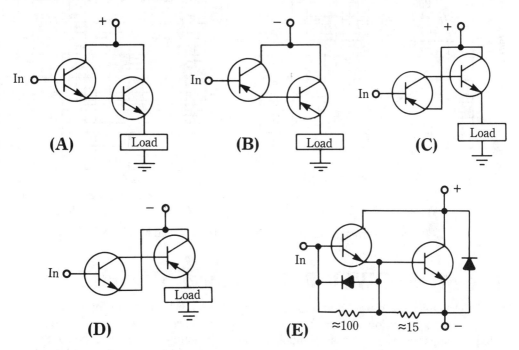

5-27 The four basic Darlington circuits and the 250-W devices. In power systems, the output transistor is much larger than the input transistor.
(A) npn Darlington circuit.
(B) pnp Darlington circuit.
(C) Simulated pnp Darlington circuit.
(D) Simulated npn Darlington circuit.
(E) Internal circuitry of the MJ10022 and MJ10023 devices.

278 Devices and components

The collector saturation voltage of the Darlington arrangement is necessarily higher than in conventional power transistors, but the several volts that are dropped are not of great consequence in off-line operation. It appears that the power Darlington transistor will be a serious competitor to both conventional (single-element) transistors and power MOSFETs. At least, these devices no longer can be ruled out for 20-kHz switching duty as "too slow," electrically nonrugged, or because of low power capability.

The switching characteristics of the MJ10022 and MJ10023 Darlington power transistors are shown in Fig. 5-28. The respectable transition times that are displayed by these curves are for the following parameters:

t_d = delay time
t_r = rise time (from 10% to 90%)
t_s = storage time
t_f = fall time (from 90% to 10%)

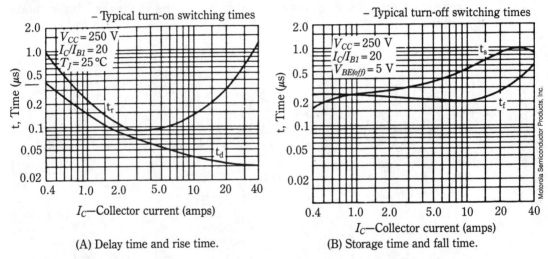

(A) Delay time and rise time.   (B) Storage time and fall time.

5-28 Switching characteristics of the MJ10022 and MJ10023 devices.

The MOSPOWER FET or power MOSFET

MOS field-effect transistors had long been used for VHF tuners and for various "flea-power" applications. Later, these devices became available with current and voltage ratings that rivaled those of bipolar power transistors. Pioneers in the commercial development of this device have introduced broad lines of power MOSFETs with ratings to several tens of amperes and over 500 V. In response to competition from various manufacturers and from bipolar technology, more powerful MOSFETs are continually being introduced. Siliconix cites the following salient features for their MOSPOWER FETs (compare with the characteristics of conventional bipolar power transistors).

- No thermal runaway—temperature coefficient of current is negative.
- No secondary breakdown—no pn junctions are involved.
- No minority-carrier storage time.
- High-speed switching—on the order of 4 ns at 1 A.
- High input impedance—drive power that is suggestive of vacuum tubes.
- Highly linear transfer characteristics.
- CMOS logic-compatible input.
- Intrinsic diode can be advantageously used.
- Gate protected to withstand static discharge (special units only).
- Can be paralleled directly—no power-dissipating ballast resistances are needed.
- Special models can be driven from TTL digital circuits.

It is only natural that this device would be used considerably in switching regulators, even though initial emphasis was focused very much on audio applications. Many fast-switching circuit applications have been worked out and much literature has been written with regard to switching regulators. The basic scheme in Fig. 5-29 applies to many of the switching regulators in this book. In order to advantageously use the fast switching capability of the MOSPOWER FET, the freewheeling diode

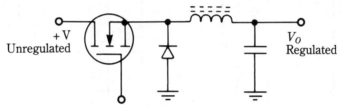

5-29 Basic scheme for using the power MOSFET in a switching regulator. Siliconix, Inc.

should be a fast-recovery or Schottky diode. Switching rates to several-hundred kilohertz and higher will then be feasible if a low-impedance drive is used. The electrical characteristics for a family of MOSPOWER FETs are given in Table 5-6. Accompanying performance curves are illustrated in Fig. 5-30.

Much, if not most of the commercial effort on power MOSFETs has been with the development of n-channel devices. Insofar as concerns dc operating polarities, these devices are similar to npn bipolar transistors. However, p-channel power MOSFETs are also available. Counterparts of pnp bipolar transistors, these devices are often targeted at the audio market for complementary-symmetry stages. However, their applicability to regulated power supplies is inescapable. For example, the bridge-type synchronous rectifier circuit requires two n-channel power MOSFETs and two similarly rated p-channel units. Simplified circuitry for inverters and for bridge output amplifiers will heat up the competition between MOSFETs and bipolars. Semiconductor manufacturers that initially developed p-channel MOSFETs for the commercial market were Hitachi, ITT, and Supertex. International Rectifier Corp. is noteworthy for marketing p-channel power MOSFETs with high current and voltage ratings.

Table 5-6. Electrical characteristics† of several Siliconix power MOSFETs.

	Characteristics	VMP 11 Min.	VMP 11 Typ.	VMP 11 Max.	VMP 1 Min.	VMP 1 Typ.	VMP 1 Max.	VMP 12 Min.	VMP 12 Typ.	VMP 12 Max.	Unit	Test conditions
Static												
1 BV_{DSS}	Drain-source breakdown	35			60			90			V	$V_{GS}=0; I_D=100\,\mu A$
2 $V_{GS(TH)}$	Gate threshold voltage	0.8		2.0	0.8		2.0	0.8		2.0	V	$V_{GS}=V_{DS}; I_D=1\,mA$
3 I_{GSS}	Gate-body leakage			0.5			0.5			0.5	μA	$V_{GS}=15\,V; V_{DS}=0$
4 $I_{D(OFF)}$	Drain cutoff current			0.5			0.5			0.5		$V_{GS}=0; V_{DS}=24\,V$
5 $I_{D(ON)}$	Drain ON current*	1	2.0		1	2.0		1	2.0		A	$V_{DS}=24\,V; V_{GS}=10\,V$
6 $I_{D(ON)}$	Drain ON current*	0.5			0.5			0.3				$V_{DS}=24\,V; V_{GS}=5\,V$
Switch												
7	Drain-source ON resistance*		2.0	2.5		3.0	3.5		3.7	4.5	Ω	$V_{GS}=5\,V; I_D=0.1\,A$
8 $R_{DS(ON)}$			2.4	3.0		3.3	4.0		4.6	5.5		$V_{GS}=5\,V; I_D=0.3\,A$
9			1.2	1.5		1.9	2.5		2.6	3.2		$V_{GS}=10\,V; I_D=0.5\,A$
10			1.4	1.8		2.2	3.0		3.4	4.0		$V_{GS}=10\,V; I_D=1\,A$
Dynamic												
11 g_m	Forward transconductance*	200	270		200	270		170			mΩ	$V_{DS}=24\,V; I_D=0.5\,A$
12 C_{ISS}	Input capacitance		48			48			48		pF	$V_{GS}=0; V_{DS}=24\,V$
13 C_{RSS}	Reverse transfer capacitance		7			7			7			$f=1\,MHz$
14 C_{OSS}	Common source output capacitance		33			33			33			
15 t_{ON}	Turn ON time**		4	10		4	10		4	10	ns	See Switching Time
16 t_{OFF}	Turn OFF time**		4	10		4	10		4	10		Test Circuit

*Pulse test. Pulse test width = 80 μs, duty cycle = 1%.
**Sample test. †25°C unless otherwise noted.

Courtesy Siliconix Inc.

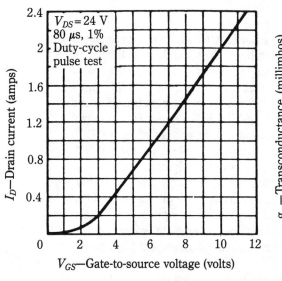

(A) Transfer characteristic. (Courtesy Siliconix, Inc.)

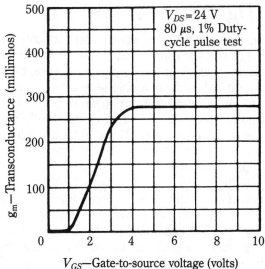

(B) Transconductance vs. gate-to-source voltage.

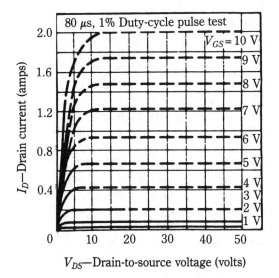

(C) Output characteristics.

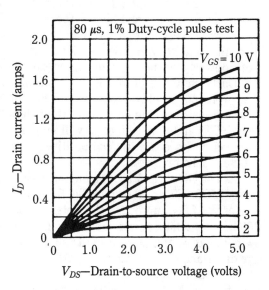

(D) Saturation characteristics.

5-30 Performance curves for the Siliconix VMP-1 power MOSFET.

High-rating MOSFETs

Some notion of the progress in the fabrication of power MOSFETs can be gleaned from two families of devices that are manufactured by the Semiconductor Division of International Rectifier Corp. One of these, the IRF150 series, can be operated with up to 100 V and can handle several tens of amperes. Conversely, the IRF330 series can withstand drain-source voltages up to 400, but at a more moderate current rating (several amperes). In efficient circuits, both series of MOSFETs can reliably control kilowatt loads. Figure 5-31 and Table 5-7 delineate the characteristics of the IRF150 family. Figure 5-32 and Table 5-8 show the same parameter values for the IRF330 MOSFET family. Both types have 150-W maximum power-dissipation rating.

The advent of "giant" power MOSFETs, such as these, has caused a reevaluation of the potentialities of these devices. Thus, designers more or less viewed the power MOSFET as a novelty, with prospects of use in low-power systems. These devices are at least competitive with bipolar transistors in moderate power applications and it doesn't require much imaginative extrapolation to expect high-power levels to soon be commonly supplied by MOSFET designs. This is especially true considering the ease with which these devices can be paralleled and the relatively low driving power that is required in push-pull and bridge circuits. Between the continuing upgrading of the MOSFETs and their applicability to multiple-device circuit configurations, it is probably already true that despite the power level, bipolar systems can be readily replaced by their power-MOSFET equivalents. Interestingly, it is often feasible to replace the bipolar transistors with power MOSFETs (with moderate modification of circuitry and components). At switching rates of several-hundred kHz and higher, the MOSFET is a must.

Despite the manifold virtues of power MOSFETs, it can't be said with any certainty that they are destined to render bipolar power transistors obsolete. Both devices can be expected to produce evolutionary surprises. For example, the Darlington version of bipolar transistors now displays ratings and features that were not readily envisioned a decade ago. Recently, 100-kHz switching power transistors (bipolar) made their debut. What is now clear is that power MOSFETs and power bipolars will mutually motivate their respective advocates to develop devices with competitively inspired features. This motivation, of course, spells progress!

A test circuit and accompanying switching-efficiency curves of the IRF-330 power MOSFET are shown in Fig. 5-33. This might be an eye opener to those who do not yet fully appreciate the potential of this relatively new device. Certainly, 200-kHz switching power supply designs are worthy of serious consideration. The switching efficiency is an overall value, it is composed of losses from forward voltage drops and rise and fall times. Although forward voltage drop is greater than in bipolar transistors, the very fast rise and fall times make the overall switching efficiency of the MOSFET higher at rates that exceed about 75 MHz.

High-rating MOSFETs 283

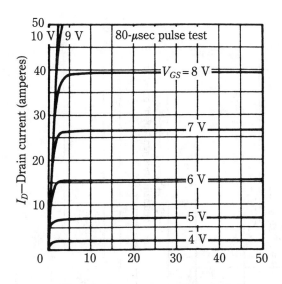

(A) Output characteristics.

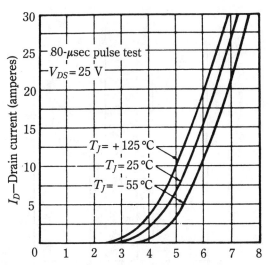

(B) Transfer characteristics.

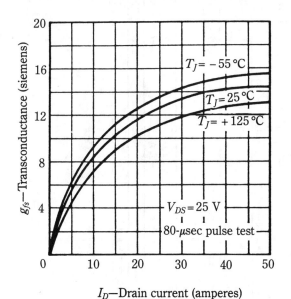

(C) Transconductance vs. drain current.

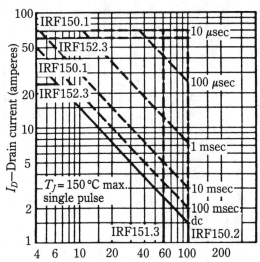

(D) Maximum safe operating area.

5-31 Performance curves for IRF150, IRF151, IRF152, and IRF153 power MOSFETs.
International Rectifier Corp.

Table 5-7. Electrical characteristics of IRF150, IRF151, IRF152, and IRF153 power MOSFETs.
$T_C = 25°C$ (Unless Otherwise Specified)

	Parameter	Type	Min.	Typ.	Max.	Units	Conditions
BV_{DSS}	Drain–source breakdown voltage	IRF150 IRF152	100			V	$V_{GS}=0$
		IRF151 IRF153	60			V	
$V_{GS(TH)}$	Gate threshold voltage	ALL	1		3	V	$V_{DS}=V_{GS}, I_D=1\,mA$
I_{GSS}	Gate–body leakage	ALL			100	nA	$V_{GS}=20\,V$
I_{DSS}	Zero gate-voltage drain current	ALL		0.1	1.0	mA	$V_{DS}=$ Max. Rating, $V_{GS}=0$
				0.2	4.0	mA	$V_{DS}=$ Max. Rating, $V_{GS}=0$, $T_J=125°C$
$I_{D(ON)}$	ON-state drain current	IRF150 IRF151	28			A	$V_{DS}=25\,V, V_{GS}=10\,V$
		IRF152 IRF153	24			A	
$R_{DS(ON)}$	Static drain-source ON-state resistance	IRF150 IRF151		0.045	0.055	Ω	$V_{GS}=10\,V, I_D=14\,A$
		IRF152 IRF153		0.06	0.08	Ω	
g_{FS}	Forward transconductance	ALL	6	10		S(℧)	$V_{DS}=25\,V, I_D=14\,A$
C_{ISS}	Input capacitance	ALL		3000	4000	pF	$V_{GS}=0, V_{DS}=25\,V, f=1.0\,MHz$
C_{OSS}	Output capacitance	ALL		1000	1500	pF	
C_{RSS}	Reverse transfer capacitance	ALL		350	500	pF	
$t_{D(ON)}$	Turn-ON delay time	ALL		40	60	ns	$I_D=14\,A, E_1=0.5\,BV_{DSS}$
t_R	Rise time	ALL		150	200	ns	$T_J=125°C$ (MOSFET Switching times are essentially independent of operating temperature.)
$t_{D(OFF)}$	Turn-OFF delay time	ALL		200	300	ns	
t_F	Fall time	ALL		150	200	ns	
Thermal Characteristics							
$R_{\theta JC}$	Maximum thermal resistance junction-to-case	ALL		0.83		°C/W	

Courtesy International Rectifier Corp.

High-rating MOSFETs **285**

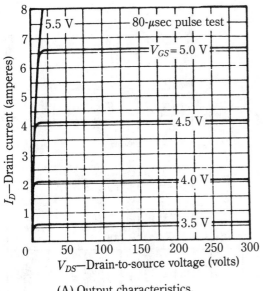

(A) Output characteristics.

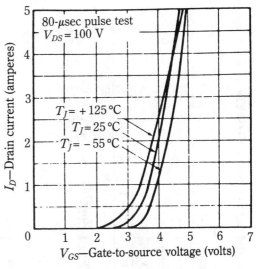

(B) Transfer characteristics.

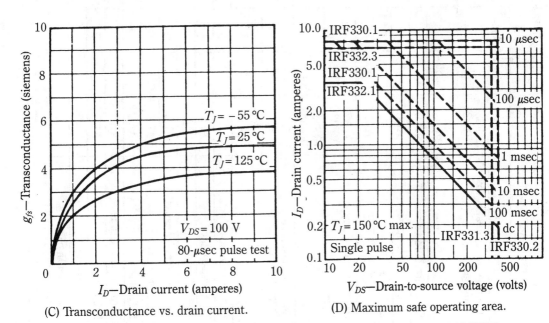

(C) Transconductance vs. drain current.

(D) Maximum safe operating area.

5-32 Performance curves for IRF330, IRF331, IRF332, and IRF333 power MOSFETs.
International Rectifier Corp.

Table 5-8. Electrical characteristics of IRF330, IRF331, IRF332, and IRF333 power MOSFETs.
$T_C = 25\ °C$ (Unless Otherwise Specified)

	Parameter	Type	Min.	Typ.	Max.	Units	Conditions
BV_{DSS}	Drain–source breakdown voltage	IRF330 IRF332	400			V	$V_{GS} = 0$
		IRF331 IRF333	350			V	$I_D = 1.0$ mA
$V_{GS(TH)}$	Gate threshold voltage	ALL	1		3	V	$V_{DS} = V_{GS},\ I_D = 1$ mA
I_{GSS}	Gate–body leakage	ALL			100	nA	$V_{GS} = 20$ V
I_{DSS}	Zero-gate voltage drain current	ALL		0.1	1.0	mA	$V_{DS} = $ Max. Rating, $V_{GS} = 0$
		ALL		0.2	4.0	mA	$V_{DS} = $ Max. Rating, $V_{GS} = 0,\ T_J = 125\ °C$
$I_{D(ON)}$	ON-state drain current	IRF330 IRF331	4.0			A	
		IRF332 IRF333	3.5			A	$V_{DS} = 25$ V, $V_{GS} = 10$ V
$R_{DS(ON)}$	Static drain-source ON-state resistance	IRF330 IRF331		0.8	1.0	Ω	$V_{GS} = 10$ V, $I_D = 2.0$ A
		IRF332 IRF333		1.0	1.5	Ω	
g_{FS}	Forward transconductance	ALL	2.0	3.5		S(℧)	$V_{DS} = 100$ V, $I_D = 2$ A
C_{ISS}	Input capacitance	ALL		750	1000	pF	
C_{OSS}	Output capacitance	ALL		150	300	pF	$V_{GS} = 0,\ V_{DS} = 25$ V, $f = 1.0$ MHz
C_{RSS}	Reverse transfer capacitance	ALL		15	20	pF	
$t_{D(ON)}$	Turn-ON delay time	ALL		30	50	ns	
t_R	Rise time	ALL		50	100	ns	$V_{GS} = 25$ V, $I_D = 2$ A
$t_{D(OFF)}$	Turn-OFF delay time	ALL		50	100	ns	$T_J = 125\ °C$ (MOSFET Switching times are essentially independent of operating temperature.)
t_F	Fall time	ALL		50	100	ns	
Thermal Characteristics							
$R_{\theta JC}$	Maximum thermal resistance junction-to-case			1.67		°C/W	

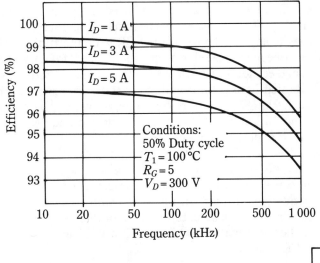

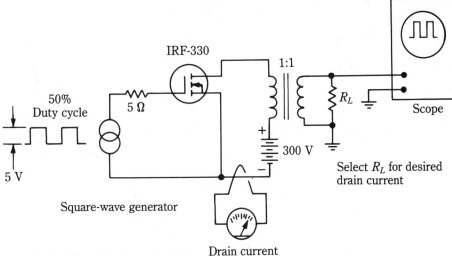

5-33 Test-circuit and switching efficiency curves of the IRF330 power MOSFETs.
International Rectifier Corp.

The gate turn-off silicon-controlled rectifier

The various switching regulators that have been discussed so far have used power transistors as dc switching devices. Many designers and experimenters have wanted to use an SCR for this purpose. The SCR has a very fast turn-on time and, when properly commutated, its turn-off time can also be fast. During the conductive period of the SCR, its internal resistance is low—even while carrying large load currents. Also, minimal energy is demanded from the source that supplies the gate signal to turn the SCR on. "If only another gate signal could be applied to turn the SCR off," has been a common lamentation.

288 *Devices and components*

In the past, circuit techniques have been used to turn an SCR off in a dc circuit. However, such turn-off methods, known as *commutation*, have not been successful for simple dc switching regulators. If the SCR is to interrupt a dc load, considerable energy must be manipulated in a very short time. Ordinary SCRs are not efficient, reliable, and economical when used in this manner. *Silicon-controlled switches* and other SCRs that have gate turn-off characteristics have been marketed for a number of years. However, these devices operate at relatively low power levels. Until recently, no SCR device was capable of being turned ON and OFF in a dc circuit with several amperes or more of current.

RCA has developed the G5001, G5002, and G5003 series 8.5-A gate turn-off (GTO) silicon-controlled rectifiers. These thyristors are turned on in the same manner as are conventional SCRs. Turn-off is produced by a negative 70-V gate signal. Documentation on the device itself is complete and specifications can be readily obtained from the manufacturer. Some of this data is presented here. The basic operating behavior of the GTO SCR is shown in Fig. 5-34. The similarity to a conventional SCR is obvious.

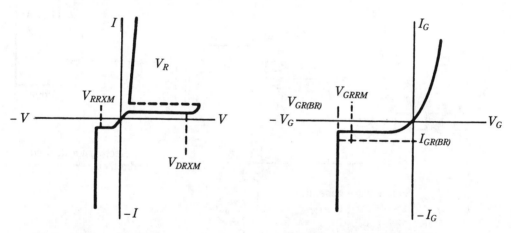

(A) Anode-cathode voltage-current characteristic.

(B) Gate-cathode voltage-current characteristic.

5-34 Operating characteristics of the gate turn-off SCR.

The driven inverter (Fig. 5-35) performs essentially in the same manner as the switching device in a usual switching regulator. This simple arrangement develops approximately 1200 W of chopped dc load power. As shown, the pulse repetition rate is 20 kHz at a 50% duty cycle (of course, in switching regulator service, the duty cycle would be modulated). The electrical efficiency of this simple circuit is in the vicinity of 95%. A *snubber circuit* is connected across the anode-cathode terminals to protect the thyristor from switching transients. Notice that the supply voltage is 400 V. Certain GTOs of this family can block 600 V when in the OFF state.

In experimenting with this device, remember that the negative turn-off pulse must overlap the turn-off interval. Otherwise, the turn-off process will suffer delay and there will be high power dissipation. This requirement should not be difficult to

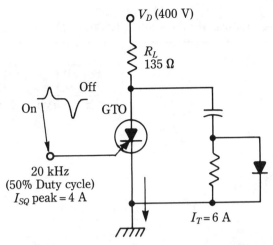

Power dissipation and temperature for 20 kHz.
Resistive load switching with $I_T = 6$ A and $V_D = 400$ V.
1200 W

f (kHz)	T_J (°C)	P_D (On) (W)	P_D (Off) (W)	P_{dc} (W)	P_{total} (W)	T_J-T_C (°C)	T_C (°C)
20	100	10	17.4	7.3	34.7	52	48
20	125	10	23.0	7.0	40.0	60	65

5-35 Driven inverter for demonstrating capabilities of the GTO thyristor.

meet—simply dispel the notion that an extremely sharp negative spike is the proper turn-off signal. Additionally, the impedance of the source for negative turn-off pulses should be low. This is because the gate initially consumes high current. However, by the time turn-off is completed, the gate consumes only a tiny leakage current. The use of a low-impedance drive source for the negative pulses also ensures that the reverse gate voltage does not rise too high during turn-off; then, the gate would be driven into a reverse avalanche condition. In practice, both turn-on and turn-off pulses are likely to be derived from a common source, such as from a pulse-width modulator. That being the case, such a source can display low impedance while delivering both pulses. Turn-on requires about one ampere for a duration of 3 μs. During the application of the turn-on pulse, its amplitude is about 1 volt.

The electrical specifications for this family of gate turn-off SCRs are listed in Table 5-9. To interpret this data, a list of symbols and definitions that are relevant to GTO thyristors is given in Chart 5-1.

The performance curves shown in Fig. 5-36 specify the dc on-state voltage for G5001, G5002, G5003 GTO SCRs. Because this voltage is on the order of 3 V under ordinary operating conditions, the GTO thyristor can operate quite efficiently when chopping high voltages on the order of 100 or more volts. Whereas transistors tend to come out of saturation at high currents, the GTO thyristor does not display this shortcoming. Therefore, its dissipation remains low for heavy loads.

Table 5-9.
Electrical characteristics* for the G5001, G5002, and G5003 GTO SCRs.

Characteristic	Limits — For all types unless otherwise specified			Units
	Min.	Typ.	Max.	
I_{DRXM}: $V_D = V_{DRXM}$, $R_{GK} = 1000\ \Omega$, $T_C = 125\ °C$	—	—	2	mA
I_{RROM}: $V_R = V_{RROM}$, $T_C = 125\ °C$	—	—	2	mA
I_{GRRM}: $V_{GR} = V_{GRRM}$	—	—	300	μA
V_T: For $I_T = 5$ A, $T_J = 100\ °C$ For other conditions	—	1.5	2	V
I_{GT}: $V_D = 12$ V (dc), $R_L = 6.5\ \Omega$, $T_C = 25\ °C$	—	—	2.5	V
V_{GT}: $V_D = 12$ V (dc), $R_L = 6.5\ \Omega$, $T_C = 25\ °C$	—	500	800	mA
I_L: $V_D = 40$ V, $I_G = 200$ mA, $t_P = 50\ \mu s$	—	500	800	mA
dV/dt: $V_D = V_{DRXM}$ value, $V_G = -5$ V, Exponential rise, $T_C = 125\ °C$	500	—	—	V/μs
t_{gt} (t_{ON}): $t_{gt} = t_d + t_r$ $V_D = 100$ V, $I_T = 5$ A, $I_g = 1$ A G5001 series	—	—	$t_d = 1$ $t_r = 1$	μs
G5002 series	—	—	$t_d = 1$ $t_r = 2$	μs
G5003 series	—	—	$t_d = 1$ $t_r = 2$	
t_{gq} (t_{OFF}): $t_{gq} = t_s + t_f$ $V_D = 100$ V for all "A" types, 200 V for all "B," "C," "D," "E," and "M" types, $I_T = 5$ A, $Z_{GS} = 1\ \Omega$, resistive load, $T_C = 125\ °C$: G5001 series ($V_{gq} = -70$ V)			$t_s = 1$ $t_f = 1$	
G5002 series ($V_{gq} = -70$ V)			$t_s = 1.5$ $t_f = 2$	μs
G5003 series ($V_{gq} = -50$ V)			$t_s = 5$ $t_f = 5$	
$R_{\theta JC}$**	—	—	2	°C/W

*At maximum ratings and $T_C = 25\ °C$ unless otherwise specified.
**For temperature measurement reference point, see Dimensional Outline.

Courtesy RCA Corp.

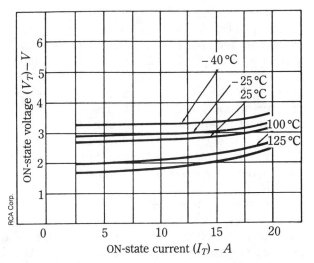

5-36 Maximum ON-state voltage vs. maximum ON-state current for the G5001, G5002, and G5003 GTO SCRs.

Pay heed to the several unique requirements of the GTO device. However, these devices should pose no problems to anyone who has experience with ordinary SCRs. The new device has considerable potential, so anticipate a developmental trend that is characterized by extended ratings and improved parameters.

Another semiconductor maker offers gate turn-off SCRs with gains in the vicinity of 1000. Translated into practical applications, such GTOs enable several amperes of load current to be turned off with several milliamperes of gate current and less than 5 V of negative gate voltage. These GTO SCRs are the Unitrode UGT405 series (12 in all). A unique feature of these devices is the nearly constant gate turn-off power that is needed over a wide range of junction temperatures.

GTO thyristors, as with other semiconductor devices, invariably encounter trade-offs in performance parameters. Such parameters as turn-off voltage, rise and fall times, and reverse-voltage capability are interrelated so that inordinately good ratings in one of them might degrade at least one (possibly more) of the others. However, the availability of these devices with different optimizations enables the alert designer to select one that is best suited to his unique application.

Transcalent power devices

The power capability of any electrical device is ultimately limited by its ability to transfer its self-generated heat to the ambient environment. Thus, conventional power transistors can be permitted to dissipate only the electrical power that keeps the junction temperatures below a certain level (about 200°C maximum). The obstacle is the thermal resistance between the junction and case. No matter how massive the heatsink or how many cubic feet per minute of air are circulated, this thermal resistance limits the practically attainable power dissipation. The use of copper, beryllium oxide, and other thermal conductive materials, has already been optimized. Clearly, a breakthrough in heat transfer is necessary in order for manufacturers to produce transistors with significantly higher power ratings.

Such a breakthrough has been achieved in the RCA group of transcalent power devices (see Fig. 5-37). These use an integrated heat pipe to transfer junction heat to the case and cooling fins. A heat pipe uses the latent heat of a suitable liquid that changes its physical phase while it circulates. Unlike earlier heat pipes, the one used in a transcalent transistor, SCR, or rectifying diode needs no mechanical pumping. Rather, its circulation is caused by screens and orifices that behave as capillary wicks. Junction heat is absorbed by the coolant when it is in its liquid state. It then travels as a vapor to the fins, where it condenses back to its liquid state. In so doing, it transfers considerable heat to the fins. Finally, it again works its way to the junction site, where it repeats the cycle. The effective thermal resistance of the process is very much lower than is obtained from metals or from exotic materials.

Because of this efficient heat transfer, a powerful device can be built with great savings in size and in weight. The devices of Fig. 5-37 can dissipate 500 W when their fins are supplied with 300 cubic feet of air per minute. These devices are used in systems where "money doesn't count," but as with other exotic and initially costly devices, the prices could decline rapidly if market demand picks up.

General considerations for choke and inductor selection

With regard to filter chokes and inductors, the most critical application is in the output circuit of the power switch. It is used here in conjunction with the output filter capacitor and the freewheeling diode to convert the switched waveform to a substantially steady level at the output terminal. The main thing to watch for, once the basic design is otherwise satisfactory, is magnetic core saturation. It would be serious enough if core saturation merely reduced the inductance, which, of course, it does. Such a decrease in effective inductance will increase the output ripple of the supply and will increase the peak current through the switching transistor. The latter effect could penetrate the SOA of the transistor and thereby shorten its life or damage it. However undesirable, the mere decrease in inductance can usually be accommodated by reasonable design safety factors.

More insidious are the effects that are produced at the onset of saturation, particularly if this transition is abrupt. Then, current spikes of high magnitude are produced, which are much more likely to damage the switching transistor—usually via catastrophic destruction. Such current spikes might be periodic, they might occur when the supply is initially turned on, or they might result from a momentary overload from the powered equipment. If not for the practical and economic limitations that are imposed by inordinately large physical size, an air-core inductor would be best for this application. Indeed, this is one of the compelling features of the new generation of very-high-switching-rate supplies (100 kHz and greater), where air-core chokes often are feasible.

For the nominally 20-kHz switching regulator, a magnetic core is needed so that the choke can be of acceptable size and so that relatively few turns of heavy wire can be used. Saturation cannot be altogether avoided, but its effects can be

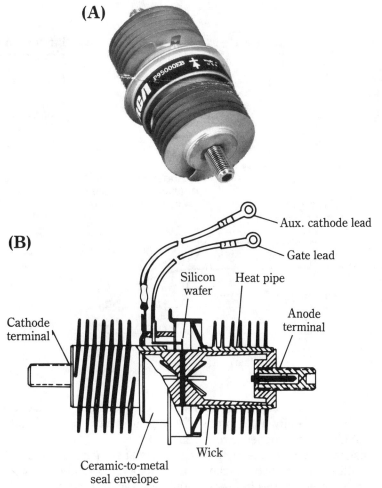

5-37 Transcalent devices pack high power capability in small packages. An internal heat pipe efficiently transports heat to external fins.
RCA Corp.

made more or less benign with an appropriate air gap in the magnetic circuit of the core. Such an air gap can be a physical interruption or it could be the cumulative effect of the large number of magnetic particles that are embedded in the bonding material of composite substances. In either case, the saturation characteristic is considerably more gentle than without such gaps.

An example of the general effect of the air gap on a current-carrying inductor with a magnetic core is shown in Fig. 5-38. Although magnetic saturation of the core still exists with the air gap, considerably more current is required to produce deep saturation and, most importantly, the transition to the saturation region is less abrupt than when no air gap is interposed in the magnetic circuit. The price paid for air gapping is reduced inductance. This detriment can be circumvented somewhat by increasing core area and the number of turns. However, this direction

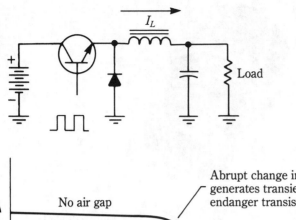

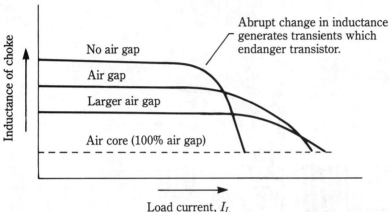

5-38 Effect of an air gap in the magnetic circuit of the filter choke.

can only be taken so far before both design and cost restraints are exceeded. The gapped choke is generally a compromise design in which the core material, air gap, structural size, and copper wire is balanced.

A number of magnetic core materials have been used for inductors and chokes. Laminated silicon iron structures can be satisfactory, but tend to incur high eddy current and hysteresis losses in 20-kHz supplies. Tape-wound cores with varying degrees of nickel content are capable of better high-frequency performance, but are expensive. Molybdenum-permalloy and powdered-iron cores have overlapping characteristics and both have been successfully used in filter inductors. The lower permeability types in these cores are more suitable for filter chokes than those with higher permeabilities. This is because the lower permeability is a trade-off for a larger effective air gap.

Perhaps the designer's favorite material for filter chokes, as well as for other core components that operate at 20 kHz or higher, is ferrite. This name embraces a large family of magnetic ceramics, but most of those that are suitable for chokes are characterized by saturation at relatively low flux densities—usually in the vicinity of several thousand lines per square centimeter. On this basis, ferrite does not compare favorably with the silicon iron that is conventionally used in 60-Hz practice, because such core material saturates in the vicinity of 16,000 to 20,000 lines per square centimeter. Physically more compact chokes result from silicon iron than from ceramic cores.

The core loss of ceramic materials is commonly characterized by the graphical format shown in Fig. 5-39. For many years, this has been the accepted way to depict relative core losses in silicon-iron and nickel-alloy laminations and in tapes. Such graphs show the superiority of ferrite materials for high-frequency switchers (above several, and certainly above 10 kHz, the eddy current and hysteresis core losses of laminated and tape-wound structures pose problems).

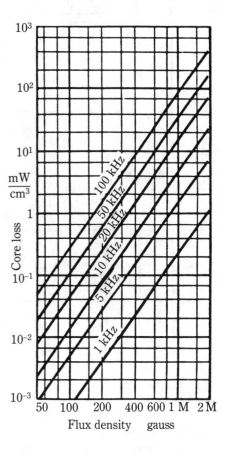

5-39 Core loss vs. flux density in a representative ferrite material.
Amidon Associates, Inc.

Although ferrite materials can have an attractive blend of magnetic, electrical, and physical characteristics, all might not be as well in unfavorable thermal environments. This problem stems from the rather pronounced temperature dependencies that some of these materials exhibit (Fig. 5-40). In Fig. 5-39, the manufacturer tries to minimize core loss over the probable operating temperature (from 20 to 65 °C, for example). The basic magnetic parameters of permeability and saturation flux density can undergo considerable variation as a function of temperature (Figs. 5-40B and 5-40C). As can be surmised, all of these temperature-dependent parameters are interrelated.

In high-power regulators, where high operating temperatures can be involved, remember the *Curie temperature* of the material. This is the temperature at which

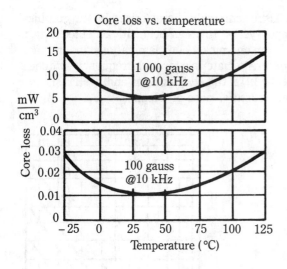

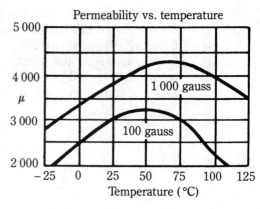

(B) Permeability vs. temperature.

(A) Core loss vs. temperature.

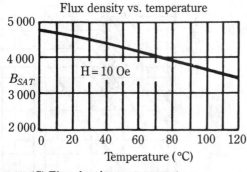

(C) Flux density vs. temperature.

5-40 Temperature dependencies in representative ferrite-core material. Amidon Associates, Inc.

an abrupt transition occurs in the magnetic nature of the material. Specifically, the material changes from a ferromagnetic to a paramagnetic substance. In practice, this action is almost equivalent to reversion to an air core—it is as if the material was physically removed (as far as its magnetic effect is concerned). The Curie point of some ferrites could be as low as 130 °C. The basic idea is to have a safe distance between the Curie point and the operating temperature.

What the hysteresis loop tells us

Manufacturers of magnetic materials generally publish hysteresis loops to depict some of the basic characteristics of these materials. Providing that these curves are plotted on the same scales, they can convey useful comparative information about different materials. For example, Fig. 5-41 shows the hysteresis loops of two ferrite materials. A fat loop (Fig. 5-41A), indicates relatively high hysteresis core

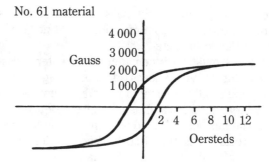

(A) High hysteresis loss, low permeability, "soft" saturation at low flux density.

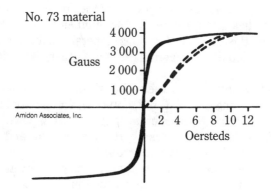

(B) Low hysteresis loss, high permeability, abrupt saturation at high flux density.

5-41 Evaluation of ferrite materials by means of their hysteresis loops.

loss. The loop in Fig. 5-41B not only has a relatively low hysteresis loss, but its permeability is considerably higher than that of the ferrite in Fig. 5-41A. This permeability can be found from the steep slope of its linear section. *High permeability* means high inductance per turn of wire. If the application requires ability to handle high load current, the high permeability probably cannot be used directly because of the early onset of saturation.

What you can do, however, is to use the intrinsic high permeability as a trade-off in order to delay the onset of saturation and to make the transition to saturation less abrupt. As explained previously, this is accomplished by inserting an air gap in the core. The overall result is the modified curve that is shown by the dashed lines.

The effective permeability now is about the same as obtained in Fig. 5-41A, but the hysteresis core loss remains very low and considerable current can be handled before saturation sets in (*oersteds*, the measure of magnetizing force, are directly proportional to ampere-turns). Other things being equal, ferrite material in Fig. 5-41B would appear to better satisfy the requirements of a filter choke than would the material in Fig. 5-41A. Other things might not be equal, however. It might not be feasible to cut the air gap on a production run, the temperature dependencies of the materials might play a decisive role, the switching rate and current capability of the regulator might be such that core loss is not greatly important, the desired core size and configuration might not have the same availabilities in the compared materials, etc.

Although ferrite cores must have considerably greater cross-sectional areas than silicon-iron cores, this drawback loses some of its significance at the higher frequencies (i.e., the popular 20-kHz switching rate). Conversely, however, ferrite cores are inordinately large at 60 Hz and usually give way to silicon iron. Another aspect of this comparison has been alluded to—silicon-iron laminations incur high eddy-current and hysteresis losses at 20 kHz. These core losses in ferrite are very low. Moreover, there is no need to laminate the material—it has a very high resistivity and for practical purposes in switching regulators, this material can be classified as an insulator. Another compelling feature of ferrite materials that are intended for core components in switchers is that they tend to have high permeabilities. This factor implies that a relatively low number of turns are needed for a given inductance. In this respect, ferrite is generally superior to molybdenum permalloy and to powdered iron materials that are otherwise suitable for chokes and inductors.

Manufacturers have made a large selection of standardized ferrite-core configurations available, including E-I, E-E, rod, cup-core, and toroidal types. Even better, some of these shapes have precision air gaps to "slow down" the saturation. One reason for this is that ferrite works better for molding and machining than does molybdenum permalloy or powdered iron.

Of the various shapes, the toroid is capable of the greatest magnetic and electrical operating efficiency because the flux is confined to the shortest magnetic path. This also means that this core merits first consideration, with regard to low EMI, because it is, in principle, self-shielded (in practice, particularly when an air gap is used, the containment of the flux might be less than perfect). The second-best choice is the *cup-core*. It, too, is a very efficient magnetic structure and does not "radiate" much EMI. The outstanding advantage of the cup-core over the toroid is that the winding can be conveniently placed on a removable bobbin.

Magnetically biased chokes and transformers

A basic problem with chokes and transformers is avoiding magnetic core saturation. This is especially true when a net dc current is throughout the ac cycle or when the current being handled consists of ac superimposed on a dc level. The traditional remedy has been to provide an air gap in the core structure. This always allows greater current capability before the onset of magnetic saturation. By preventing abrupt saturation in this manner, you avoid the current and voltage spikes that can destroy transistors. Better filtering action is provided by these chokes when they must handle heavy load current. Unfortunately, you can go only so far with this technique, because the average inductance of transformers and chokes with air gaps is necessarily decreased. Thus, too great of a reliance on air gapping will require inordinate increases in core size, number of turns, and cost.

The Hitachi Magnetics Corporation markets core components in which approximately twice as many ampere-turns can be effectively used as would be the case for a similar device that would use an air gap. The extended performance is obtained by incorporating a biasing magnet in the structure. Figure 5-42 illustrates the operating difference between a conventionally constructed inductor and one with a biasing magnet. The magnetically biased inductor is able to use the entire unsaturated range of the magnetization curve. By contrast, only half of this range is available for current in the conventional inductor. This can be exploited in several ways. For a given current capability, the biased inductor can be made physically smaller. Conversely, for a given size, the biased inductor will enable higher

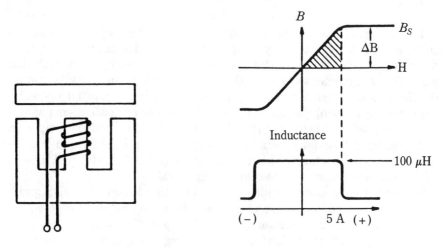

(A) Conventional inductor with air gap only.

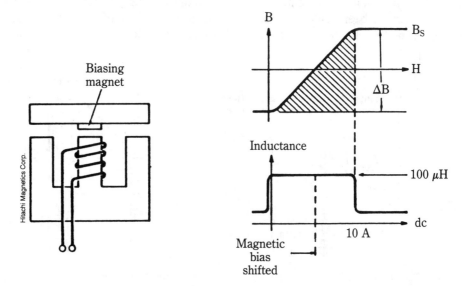

(B) Magnetically biased inductor.

5-42 Operation of conventional and magnetically biased inductors.

currents to be handled before saturation sets in. Generally, the utilization of the full linear range of the magnetization curve indirectly contributes to higher efficiency and reduced EMI.

The magnetically biased inductor is polarized and must be connected in the circuit with regard to direction of dc current flow. On the other hand, the permanent biasing magnet has such a high intrinsic coercive force, that neither ordinary uses or abuses will result in a detrimental amount of demagnetization. This, indeed, is the "secret" of the new device—suitable permanent-magnet materials had not previously been available.

Figure 5-43 depicts the same principle as in Fig. 5-42, except applied to a transformer arrangement. The greater current-handling capability that results from the biasing magnet is clearly evident. An additional bonus that is provided by this operational mode is the attenuation of the "flyback" spike. As with the biased inductor that was previously described, the biased transformer must be connected in the circuit with regard to direction of current flow. Such a transformer will have about twice the volt-ampere capability of a similarly sized conventional unit.

The application of both magnetically biased inductors and the magnetically biased transformer in an off-line switching regulator is shown in Fig. 5-44. Hitachi's line of magnetically biased inductors is called "hicoils" and their magnetically biased transformers are called "hiformers."

The magnetic-biasing principle also enables the design of swinging chokes with very pronounced inductance change as a function of load current. A typical plot that depicts this characteristic is shown in Fig. 5-45. By using such a choke, good filtering prevails at low load current because of the increase in inductance. Conversely, at higher load currents, where low inductance is sufficient for acceptable filtering action, the dynamic response time of the supply is not compromised.

Allusion has been made to the exceedingly good retention of magnetism that is exhibited by the Hicorex rare-earth cobalt magnetic material. This material also

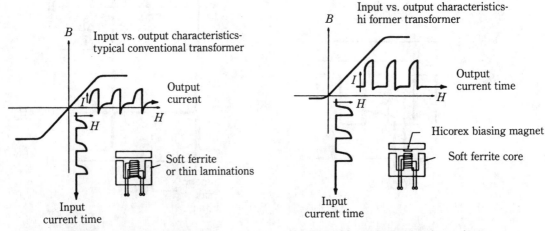

(A) Conventional transformer.   (B) Magnetically biased transformer.

5-43 Operation of conventional and magnetically biased transformers. Hitachi Magnetics Corp.

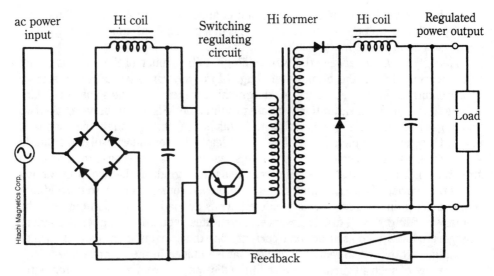

5-44 Use of magnetically biased core components in an off-line switcher.

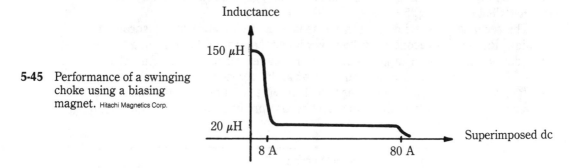

5-45 Performance of a swinging choke using a biasing magnet. Hitachi Magnetics Corp.

All custom designed Hi coil-L's (no standard available).

has a Curie point of 710 °C. Accordingly, ordinary (or even overrated-operating) temperatures are not likely to have any detrimental effect.

The ferroresonant constant-voltage transformer

Dissipation control (in the sense that the series-pass or shunt-regulating device is protected from assuming the burden of high line voltage) can be provided by special regulating transformers. Typical of these is the ferroresonant transformer, known also as the *static-magnetic regulating transformer* or the *constant-voltage transformer*. In this transformer, the flux in the primary magnetic circuit is accommodated by unsaturated core material, as in a conventional transformer. The secondary flux, however, encounters saturated iron; therefore, a change in primary flux produces far less than a proportionate change in secondary flux. Thus, secondary induced voltage remains relatively independent of the voltage that is impressed upon the primary winding (Fig. 5-46).

A *magnetic shunt* between the two windings enables much of the secondary flux to be decoupled from the primary winding. Moreover, capacitor C causes a large reactive current in the secondary winding, which thereby saturates the secondary magnetic circuit. Such saturation increases the magnetic isolation between the two windings because more primary flux then takes the path through the magnetic shunt. Of course, complete decoupling would leave the secondary isolated from its source of energy, the primary circuit. For this reason, and also because the transition from the unsaturated to saturated core region is rather gradual, the regulating action is less than perfect. The regulating action is improved considerably with the addition of a compensating winding. This winding carries ac load current and opposes the primary winding flux. This, in essence, constitutes regulation—higher secondary voltage produces increased ac load current, but this current then causes further decoupling action, which reduces voltage that is induced in the secondary.

The use of such a transformer also limits dissipation from an over-current situation or a short-circuit load. Excessive current demand greatly diminishes the mutual flux between primary and secondary windings, and the secondary voltage rapidly falls to zero (Fig. 5-47B).

The secondary capacitor does not resonate the inductance of the secondary winding. Rather, the secondary voltage is higher than would be indicated by the secondary-primary turns ratio because of the increased secondary flux that is caused by the capacitor current—thus, the term "ferroresonant."

Typical ferroresonant transformers will provide a 10-to-1 improvement over the variations in raw power-line voltage. They also produce surprisingly good load

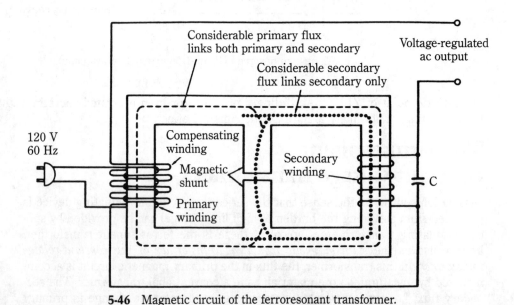

5-46 Magnetic circuit of the ferroresonant transformer.

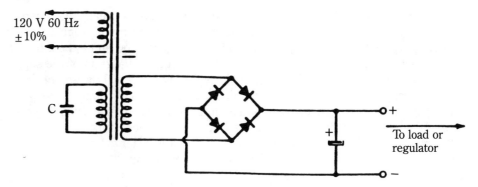

(A) Substitution of ferroresonant transformer in conventional bridge rectifier circuit.

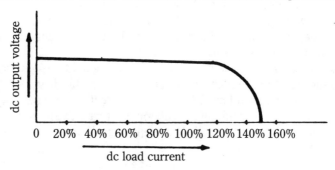

(B) Voltage collapse from excessive load current.

5-47 Dissipation control and preregulation with a ferroresonant transformer.

regulation; ±2% is typical. On the other hand, the ferroresonant transformer is considerably larger and heavier than a conventional transformer of the same power capacity. Their response time is slow (2 to 5 cycles of ac frequency). They are also sensitive to supply-line frequency, but for most applications, this is probably not of consequence. They deliver a waveform that is considerably different from a sinusoid, with a general tendency to approach a square wave. Actually, the harmonics of such a waveform yield readily to filtering. At the same time, rectifying elements receive less stress from peak voltages than when connected to a sine-wave source. A more recent development is the Paraformer transformer, which was discussed in chapter 4.

Filter capacitors

The input filter capacitors of both linear and switching regulators do not present difficult problems. Most people are generally interested in some kind of figure of merit that will provide a lot of capacitance and voltage capability in a small volume at a low cost. Don't be oblivious to ripple current capability, but this does not often pose any stumbling blocks. For such application, the mere fact that the capacitor is intended by the manufacturer to be used in such filter circuits usually takes care of this consideration. Because electronic filtering is provided by the regulator, it is not necessary to design on the premise that "if a little is good, more must be better." Indeed, too much filter capacitance in the unregulated portion of the supply tends to stress the input rectifiers, particularly during the initial turn-on period.

The wrong input capacitor in a switching regulator can contribute to the possibility of oscillation because of the negative input resistance of the switching transistor. Where this problem is encountered, it might be necessary to experiment with small values of resistance in the capacitor circuit in order to lower the Q of the equivalent LC circuit that is "seen" by the switching transistor. Sometimes, an inexpensive electrolytic capacitor with high ESR (effective series resistance) can provide stability in this circuit function whereas a higher quality unit permits oscillation to occur. Of course, the filter capacitor is but one of several components that is involved in this kind of instability. The filter choke (if any), the internal inductance of the capacitor, wiring inductance, and the input characteristics of the switching transistor are all involved.

It nonetheless remains true that minimal filtering action is generally required. Also, EMI and noise production will often be appreciably lower if a high-quality capacitor is used—one with low ESR and ESI (effective series inductance). This pertains, of course, to the switching regulator. However, the high-quality capacitor also pays off in the linear regulator—it more effectively attenuates incoming transients from the power line.

Also, the output filter capacitor of the linear regulator provides few problems and often is not critical in value. This is because very little filtering action, per se, is needed; most is performed electronically by the series-pass (or shunt) transistor. Thus, it is economical to select a high-quality capacitor that is not likely to endanger the transistor if it fails. In those linear regulators that incorporate inordinately high gain in their error amplifiers, the ESR and ESI of the output capacitor might, however, have to be low in order to prevent instability.

The output capacitor of the switching regulator is where all the fuss is encountered, and justifiably so. Here, many operating parameters converge, often with contradictory requirements. Some of these are output ripple voltage, regulating performance, stability, response time, temperature range, ripple-current handling capability, and switching transistor safety. Of course, you must remember such matters as cost, physical size and shape, and terminal provisions. Also, consider the cost and size of the filter inductor, as well as the peak current, which must be handled. The switching transistor depends on the selection of this capacitor. Any notion that a "sufficiently large" electrolytic capacitor should be sufficient is bound to be an exercise in self-deception. It was this approach that previously accounted for the low repute of the switching regulator. Here, more than with linear supplies, trouble lurks at every deviation. More often than otherwise, the output capacitor is the pitfall.

The key to understanding this vital component of switching regulators is to view it from the same vantage point that the switching transistor "sees" it. With such a viewpoint, you must evaluate the capacitor as an equivalent *series-resonant circuit*. The inductance resides in the physical fabrication of the capacitor and in its leads. Because of inherent losses in the elements and dielectric, it has an equivalent series resistance. Summed up, this is a low-Q series resonant tank circuit. Such an analysis would be nitpicking at 60 or 120 Hz. At several kHz, and certainly at 20 kHz, this viewpoint explains capacitor performance much better than mere reference to capacitance alone. Thus, the behavior of electrolytic filter capacitors is most meaningfully inferred from such impedance-versus-frequency plots (Fig. 5-48).

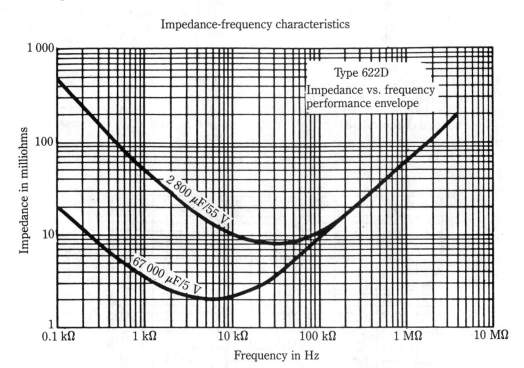

5-48 Typical impedance vs. frequency curves of electrolytic capacitors.
Sprague Electric Co.

Figure 5-49 depicts the series-resonant circuit which, for practical purposes, quite accurately represents the behavior of capacitors and particularly electrolytic filter capacitors. Indeed, if you did not know otherwise, the combined impedance and phase-shift curves of Fig. 5-50 could be properly ascribed to a series circuit that is composed of discrete components, specifically inductance, capacitance, and resistance. The phase-shift curve should provide interest for those who are familiar with stability in feedback amplifiers. In the regulator, the output capacitor is effectively in the feedback path of the circuitry that deals with the error signal. The variation in phase shift from almost 90 degrees negative to almost 90 degrees positive can be the source of "mysterious" tendencies for oscillation and instability to occur. In this regard, it is interesting to sometimes find inferences in the technical literature that a low-ESR capacitor might cause instability. This is true, because of the higher Q of the effective series-resonant circuit formed by the

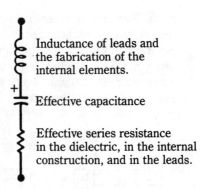

(A) For 60- and 120-Hz filtering applications, the electrolytic capacitor is represented by the above circuit. Often the resistance can be considered negligible.

(B) When the capacitor must handle frequencies from several kHz to several MHz, and higher, the behavior of the capacitor is best represented by a series-resonant circuit.

5-49 What the power supply actually "sees" in the electrolytic capacitor.

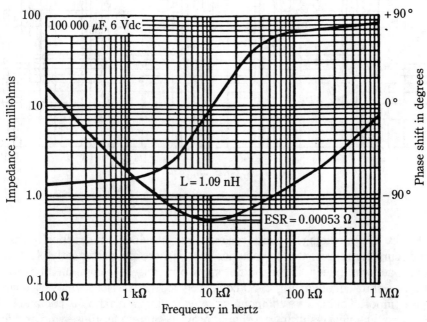

5-50 Accompanying phase shift of a typical electrolytic capacitor.
Sprague Electric Co.

capacitor. Should series resistance, therefore, be sought as a virtue in the electrolytic filter capacitor?

Despite fear of instability, design philosophy should always try to use the lowest ESR capacitor that is economically available. This will result in lower output ripple, the earmark of the well-designed switching regulator (other performance parameters, being equal, of course). The stability problem should be primarily handled in the error-amplifier circuitry by proper attention to frequency response and gain-bandwidth product. Looking at the matter from another vantage point, a capacitor with high ESR cannot function well as a filter capacitor! One of the most rewarding techniques in making practical use of low ESR capacitors is to ascertain that the ESI is also low. This technique requires you to select specially constructed units, often with terminal provisions that are suggestive of radio-frequency techniques. The combination of low ESR and low ESI makes the resonant frequency very high—beyond the error amplifier's tendency to oscillate. At the same time, this combination produces a low-Q resonance without the impairment to filtering resulting when low-Q is attained by a high ESR.

Aluminum electrolytic capacitors are shown in Figs. 5-51 and 5-52. These are specifically intended for service as output filters in high-frequency switching regulators. As such, they merit consideration for switching rates from 10 to 100 kHz and even higher. Even the "workhorse'" types that are shown in Fig. 5-51 have the

5-51 "Workhorse" aluminum electrolytic output filter capacitors.

advantage over general-purpose capacitors in that their parameters are more tightly specified. This helps produce predictable and reproducible power-supply designs and enables the designer to rely less on brute force and "cut-and-try" approaches. The four-terminal (and stacked-foil capacitors, which are not shown) provide much lower ESI values than can be obtained from conventional fabrication. This flattens their impedance curves and makes them particularly useful in high-frequency applications. The stacked-foil construction also helps decrease

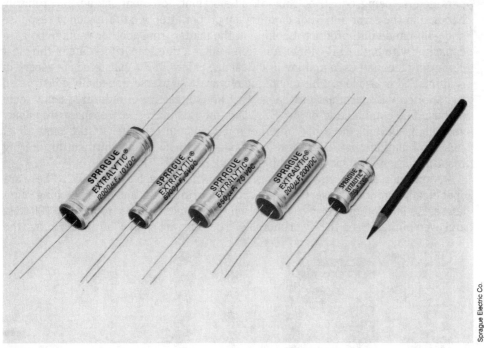

5-52 Four-terminal aluminum electrolytic capacitors.

ESR and enables higher ripple current ratings than do other fabrications; however, stacked-foil types are very costly.

ESR can be estimated if capacitor specifications provide only the dissipation factor, P_D. In such an instance, the approximate ESR is the product of the dissipation factor and the capacitive reactance. One reason this relationship is an approximation is that P_D is generally measured at a low frequency, whereas the capacitive reactance appropriate for most switchers will probably be in the 20 to 40-kHz region, i.e., the switching rate, f. Accordingly, use the relationship depicted below with some discretion:

$$\text{ESR} = P_D \times \frac{1}{2\pi f C}$$

The reason why it is misleading to judge a capacitor by ESR when working at high frequencies is clearly shown in Fig. 5-53. Because of ESL, the impedance curve deviates considerably from ESR.

Tantalum electrolytic capacitors

The resonant behavior of electrolytic filter capacitors (previously discussed) actually dealt with aluminum types. Qualitatively, similar behavior prevails with tantalum electrolytic capacitors. Quantitatively, too, considerable overlap is in the basic

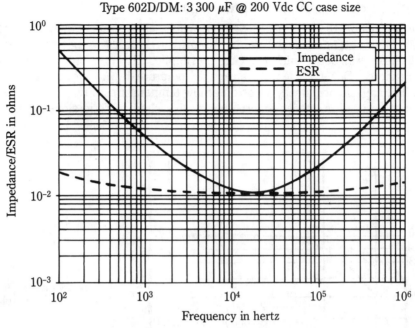

5-53 Typical capacitor curves showing effect of ESI.
Sprague Electric Co.

characteristics of the two types. As with people, however, the little differences are important. Whether or not differences between aluminum and tantalum capacitors are significant depends much on the priorities of the design requirements. In some instances, both aluminum and tantalum capacitors can be considered. In other cases, one or the other will obviously have important advantages.

If you want high volumetric efficiency (high capacitance in small volume), the *wet-slug tantalum capacitor* is generally the best choice. The trade-off here might be cost, however. Also, such capacitors have not been as well optimized for switchers as have the *tantalum-foil capacitors*. Here, you might well find a better blend of performance parameters and cost than in some aluminum counterparts. In this case it must be conceded, however, that high-quality and special-construction aluminum capacitors have generally out-competed tantalum capacitors.

More unique, is the *solid-state tantalum capacitor*. The substitution of "solid electrolyte" for the liquid or gel variety naturally appeals to the senses. Indeed, the solid-state tantalum does exhibit the expected increase in longevity and time stability. A principal drawback is lack of available high capacitances and voltage ratings. They are often used in parallel with larger conventional electrolytic capacitors in order to provide a low-impedance path to the higher frequencies. With higher switching rates destined to be used in forthcoming switching regulators, it is very likely that the solid-tantalum capacitor will assume a more prominent role in filter applications.

Pitfalls in the use of electrolytic capacitors

Once you realize that the electrolytic capacitor behaves as a series-resonant circuit, you have attained much headway toward successfully using this component, particularly as an output filter in switching supplies. However, other factors should be considered. One of these is the strong temperature dependency of these capacitors. The capacitance, itself, drops off at low temperatures (Fig. 5-54). If the supply is to be used in rugged environments, this characteristic alone could cause

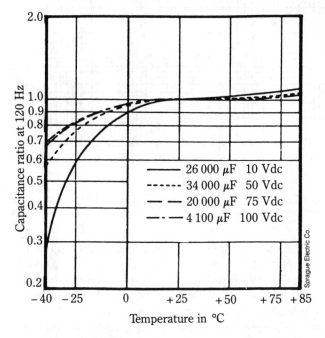

5-54 Temperature effect on capacitance in family of electrolytic capacitors.

supply malperformance. Also, the all-important ESR increases at depressed temperatures and thus aggravates the overall situation (Fig. 5-55). The combined effects of these two temperature-dependent parameters are strikingly depicted in the impedance curves of Fig. 5-56 (temperature has a negligible effect on ESI).

You might believe that it is desirable to operate the capacitor hot. This reasoning is valid to a limited degree. It is probably desirable for the operating temperature of the capacitor to be from 25° to perhaps 45°C. Operation at elevated temperatures is to be discouraged, however. The improvements in capacitance and in ESR then enter a region of rapidly diminishing returns. At the same time (Fig. 5-57), leakage current goes up rapidly and longevity and reliability are bound to be adversely affected. In practice, ripple current often keeps the internal temperature of the capacitor above ambient and, other things being equal, the mounting provision is sufficient for adequate heat removal. However, in tightly packaged supplies that operate at high power levels, forced-air cooling might be as desirable for the capacitors as for the semiconductors.

5-55 Temperature effect on ESR in family of electrolytic capacitors. Sprague Electric Co.

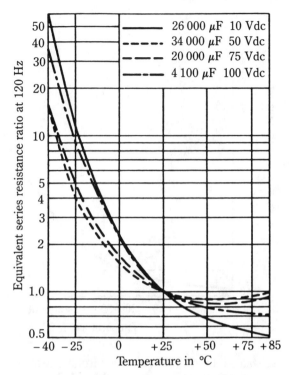

5-56 Temperature effect on impedance vs. frequency curves. Sprague Electric Co.

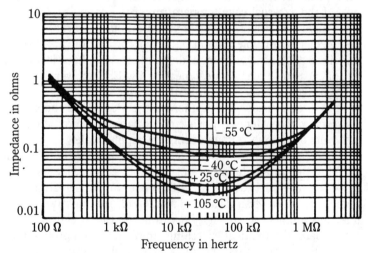

Type 673D: 1 000 μF 20 WVdc JE case

A very important factor is the capacitor's ability to handle ripple current. This capability directly affects self-heating and reliability. Virtually all manufacturers of electrolytic capacitors now provide this rating in their specifications. Because the ripple current rating depends upon frequency, waveform, and dc operating voltage,

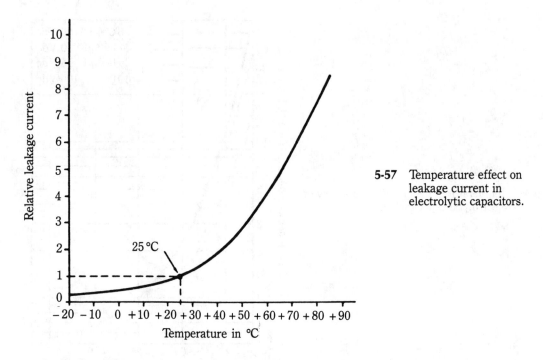

5-57 Temperature effect on leakage current in electrolytic capacitors.

as well as upon the thermal environment of the capacitor, standardization of this rating has been somewhat on the nebulous side, however. In practice, use caution even if the external case temperature of the capacitor is not excessive; the maximum allowable ripple current is premised upon internal temperature. Other things being equal, the ripple current burden of the output capacitor in switching supplies increases as the inductor is made smaller. Also, here is another argument for the use of capacitors that are designed for low ESR; power dissipation from ripple current goes down directly with ESR.

An often-overlooked detriment to the life and reliability of the output capacitor is *reverse voltage*. Generally, electrolytic capacitors do not tolerate more than one or two volts of reverse voltage during the ripple cycle.

The capacitance tolerance of electrolytic capacitors is generally sloppy compared to the parameter values of most other components. Tolerance designations of −10% to +50%, or 0% to +100% exemplify this situation. Also, these capacitance ranges are often measured at 120 Hz; the effective capacitance at higher frequencies might or might not be within the rated tolerance range. Commonly, the two apparently identical switching supplies from a production run exhibit considerable differences in output ripple. The probable reason is capacitor tolerances.

A technological breakthrough in filter capacitors

Modern electrolytic capacitors are greatly improved versions of their predecessors. Although almost every performance characteristic has been enhanced, cost, reliability, and packaging have remained reasonable. Nonetheless, these filter components are often the weak link in the design and operation of high-frequency switching-type power supplies. To appreciate why this is so, consider the common practice of paralleling these capacitors.

The most compelling reason to parallel filter capacitors is not to increase capacitance, but to lower the ESR of the filtering system. The equivalent circuit of a filter capacitor is that of a simple series-resonant circuit. The ESR is then the effective series resistance of such a circuit at the resonant frequency. The ESR represents the effect of power dissipation within the capacitor. When the capacitor is used for filtering, ripple suppression is degraded by the presence of ESR. Notice also the presence of ESI. The ESI is caused by the internal construction and leads of the capacitor. In many applications, the external leads to the capacitor also contribute significantly to the inductance that is "seen" by capacitor C. Paralleling also reduces the ESI, and thereby improves the filtering action at higher frequencies. A third advantage of paralleling is the ability to handle more ripple current. Remember that the heat produced by ripple current is the result of ESR. In other words, such capacitor heating is nothing other than I_2R loss. At best, paralleling is a makeshift remedy for ESR—It would be better to design a capacitor with lower ESR.

A new plastic-film *polypropylene capacitor* with lower ESR has been developed. Figure 5-58 compares the ESR of an electrolytic capacitor and a polypropylene capacitor. Notice that the solid curve represents a medium-grade electrolytic capacitor (aluminum type). The dashed curve represents a polypropylene capacitor. Both capacitors have the same capacitance and both have about the same ESI. However, it can be seen that the polypropylene film capacitor has a lower ESR. The impedance of the polypropylene capacitor is appreciably lower over a frequency range of about 5 to several hundred kHz—the important range for switching regulator operation. Most significant is the greatly reduced impedance—in the vicinity of the 20-kHz switching rate that is used by many switching regulators.

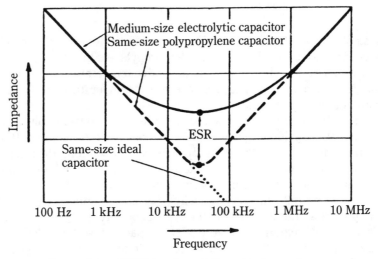

5-58 Comparison of ESR for electrolytic and polypropylene capacitors.

Although the polypropylene capacitor develops less capacitance per unit volume than do electrolytic types, it does not necessarily mean that more space is required for adequate filtering. One-third to, perhaps, 0.1 of the rated capacitance of an electrolytic capacitor is needed for the same ripple-filtering requirement when a polypropylene capacitor is used. This difference is largely because of the lower ESR of the polypropylene capacitor. However, there are two additional reasons why a small polypropylene film capacitor can perform as well as a much larger electrolytic type. One reason is that the true capacitance of electrolytic capacitors decreases rapidly with frequency, whereas the polypropylene capacitor displays nearly constant capacitance over a wide frequency range. Figure 5-59 shows a general comparison of the capacitance versus frequency behavior for the two types of capacitors.

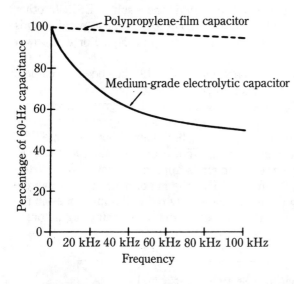

5-59 Capacitance vs. frequency comparison of electrolytic and polypropylene capacitors.

Electrolytic capacitors also suffer a considerable loss of capacitance at depressed temperatures. However, the polypropylene film capacitor maintains a relatively constant capacitance over an extensive temperature range. Figure 5-60 shows a typical comparison of capacitance versus temperature for electrolytic and polypropylene-film capacitors.

Although the different capacitances and types vary considerably for both electrolytic and polypropylene capacitors, it is reasonable to expect 3 to 10 times the ripple-current capability for the polypropylene type. Therefore, the practice of paralleling filter capacitors to increase the ripple-current capability is destined to become less common with the new plastic-film capacitors.

Finally, the overall reliability of switching power supplies should be significantly improved as the new plastic-film capacitors are designed into these supplies. You can anticipate that polypropylene capacitors will also help reduce the ripple and noise outputs of switching supplies. Metallized versions of these capacitors have become commercially available; these exhibit a noteworthy improvement

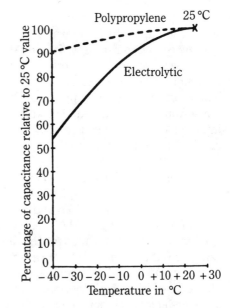

5-60 Effect of low temperature on capacitance.

in capacity-to-volume ratio, compared to some earlier foil fabrications. Such metallized polypropylene capacitors might have tolerances as tight as ±10% in capacity rating. This rating is much better than what usually prevails with electrolytic types, and it is useful in designing and manufacturing supplies with reproducible and predictable characteristics. This type of capacitor is especially well-suited for switchers that operate from 50 kHz to several-hundred kHz.

Thermoelectric heat pump

It is natural to entertain the possibility of interchange between cause and effect. In many cases, no problem is encountered. For example, the same dc machine could serve as either a motor or a generator, depending on whether it is supplied with electrical or mechanical energy. Other transformations appear to be unilateral. Thus, causing heat and light to impinge on a tungsten filament does not produce the hoped for electricity across its terminals. A fortunate situation involves certain metallic and semiconductor junctions. As is well known, such junctions, so-called *thermocouples*, are capable of developing electrical power when exposed to temperature differences. When appropriate materials are used, it is also true that temperature differences between the elements of the junction will occur when an electric current is passed through the junction. The heat that appears at the hot element is extracted from the cold element. This "uphill" flow of heat is commonly described as a pumping action.

This phenomenon is known as the *Peltier effect* and it qualifies among the earliest of observed junction behavior. With modern semiconductor technology, such junctions can be made efficient enough so that they perform as heat pumps (i.e., they remove heat from one locality and transport it to another). A very practical implementation is the "pumping" of heat from a semiconductor package to a heatsink, whereupon the heat energy can be transferred into the ambient environment via convection, radiation, or forced-air cooling. Such a thermoelectric heat pump is an active device and, therefore, it provides much greater cooling effect than would be obtained from merely affixing the heatsink to the semiconductor package. Thus, with no current supplied to the thermoelectric module, the thermal resistance from package to heatsink would be very high—but with current, it is as if this interface thermal resistance becomes negative. In the vernacular, the heat is literally "sucked out" of the package or device that is subjected to cooling, then transported to the heatsink.

Some practical insight into the nature of commercial thermoelectric coolers can be gleaned from the illustrations shown in Fig. 5-61. The vertical dimensions in the sketches of Fig. 5-61B and 5-61C are exaggerated in the interest of clarity. As can be seen in Fig. 5-61A, the thermoelectric module is a relatively thin package, on the order of 0.1 inch for the illustrated models. Thinking of the cooler as a sandwich, the outer layers are ceramic, either *alumina*, or for enhanced thermal efficiency, *beryllia*. The metallic inner layers are usually copper, which is appropriate because of its high thermal conductivity. As shown, alternate P- and N-type semiconductors combine the interfaces with the copper electrodes to form a series electrical circuit. An interesting aspect of thermoelectric cooling is that reversal of the current also reverses the heat-pumping direction. That is, heat would then be extracted from the ambient environment and injected into the device that is attached to the cold plate (which now becomes the hot plate). It is easy enough to see that, because of this reversible behavior, the thermoelectric cooler can be associated with servo circuitry to closely stabilize the device temperature. Thus, voltage references, laser diodes, and crystal oscillators can be so-controlled.

Because the construction of the thermoelectric cooler is suggestive of a pn junction diode or a Schottky diode, it might at first seem strange that the current can be reversed. This confusion stems from unfortunate wordage in physics texts. First, the thermoelectric cooler is not a "diode," such as we are familiar with in electrical work. Second, the so-called "junctions" refer to physical intimacy, not to those of the kind described for conventional pn diodes. It would be better to call the device a *couple*, and to speak of *interfaces*, rather than junctions. The device resembles a thermocouple more than it does a diode. The contacts between the electrodes and the semiconductor material are essentially ohmic in nature, which allows current to flow in either direction.

Greater heat-removal capability is possible with arrays, such as those depicted in Fig. 5-61C. In such a cascade of thermoelectric coolers, heat extraction is as if it was a parallel thermal circuit. Electrically, a series circuit is formed.

In practice, it is often helpful to enclose the thermoelectric module with a thermal insulating material (such as polyurethane) that is an inch or so in thickness. Otherwise, heat from the ambient environment would flow to the cold element and to the device being cooled, which would degrade the cooling effect.

Thermoelectric heat pump

5-61 The thermoelectric cooler.
  (A) Representative models
  (B) Basic structure
  (C) Cascade for greater cooling

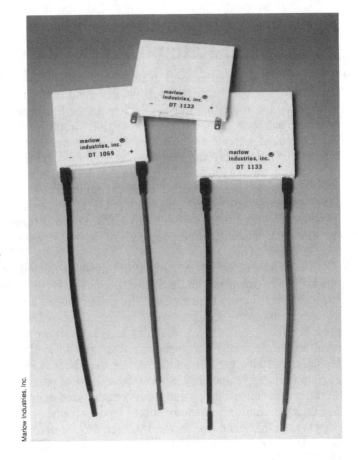

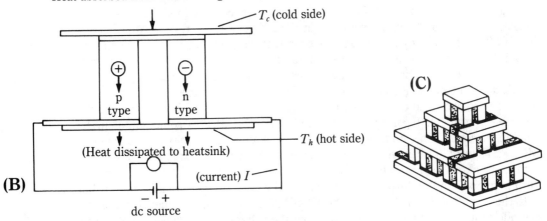

Keep your eye on the IGBT, a new solid-state power device

The power device that is used in regulated power supplies is obviously of prime importance. The technology, itself, was created with the advent of bipolar power transistors and thyristors. These devices made the benefits of solid-state design possible, but they yielded performance that simply could not be duplicated with electron tubes, despite any tolerance of size, weight, or cost. Although thyristors tended to dominate big and brutal power-supply systems, transistors fulfilled more numerous demands for power supplies that were able to provide tens or hundreds of watts, rather than kilowatts. These statements are still more or less true, but fewer areas of overlay exist among competing devices. Also, more factors require consideration than merely power capability. Some of these are frequency capability, efficiency, stability, precision, ease of drive, temperature behavior, reliability, and, of course, cost.

The main contenders in many, if not most, regulated power supplies were the bipolar power transistor, its derivative, the power Darlington, and the power MOSFET. At first, it was thought that the power MOSFET would drive the bipolar devices to obsolescence. However, the very threat of the newcomer inspired and provoked the semiconductor industry to greatly improve bipolar devices. The net result is that designers now recognize that each device family has certain desirable features, but also certain limitations or drawbacks. Power MOSFETs hold the edge in frequency capability, ease of drive, ease of paralleling, and in relative freedom from thermal runaway and from concentrated energy destruction (such as the secondary-breakdown phenomenon of bipolar transistors).

On the other hand, the forward voltage drop or saturation voltage of the bipolar power transistor is less than that as a result of the output resistance, R_D, of the power MOSFET. In low-voltage devices (below 100 V, for example) this statement might have to be tempered in light of the continuing success in lowering R_D. At higher voltage ratings, R_D increases exponentially in power MOSFETs, but remains nearly constant for bipolar. At higher voltages (in the 500- to 1000-V region), R_D of power MOSFETs is annoyingly high and is the source of reduced efficiency from self-generated heat. Even worse, R_D increases rapidly with temperature. In contrast, $VCE_{(sat)}$ of bipolar transistors actually decreases with temperature.

Because of the attributes of the two different power devices, the thought had often crossed the mind of power-supply designers that it would be very nice if the desirable features of bipolar transistors and power MOSFETs could somehow be combined. Toward this end, various combined circuits using bipolars and power MOSFETs were devised with varying degrees of success. For example, the BIMOS composite circuit of Fig. 5-62 enables the power bipolar transistor to exhibit the high-frequency and high-voltage capability of the common-base configuration, but with the easy-drive characteristic of the MOSFET input stage. Another result of this cascade connection of the two devices is that the voltage drop is considerably lower than it would be if a single high-voltage power MOSFET was used. Notice that the

Keep your eye on the IGBT, a new solid-state power device

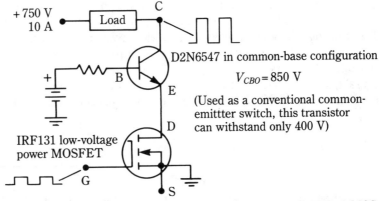

5-62 BIMOS circuit has best features of bipolar transistors and MOSFETs. International Rectifier Corp.

input MOSFET stage only must have a low-voltage rating. Unfortunately, a monolithic version of this circuit never appeared on the market.

Another bipolar-MOSFET combo is shown in Fig. 5-63. Here, the basic idea is to obtain the rapid-rise and rapid-fall behavior of the power MOSFET, together with the relatively low conduction loss of the turned-on bipolar transistor. The MOSFET is turned on first. Next, the bipolar transistor is turned on; it remains on for a bit less than the desired width of the pulse. In so doing, most of the high conduction loss of the MOSFET is bypassed by the lower loss of the conducting bipolar transistor. Finally, the fall time of the pulse is essentially determined by the rapid turn-off characteristic of the MOSFET. Here again, the arrangement provides desirable behavior of the two power devices. However, in practice, it isn't easy to optimally time the gate and base pulses where PWM techniques were needed for regulated supplies. Also, this scheme resisted attempts at monolithic fabrication.

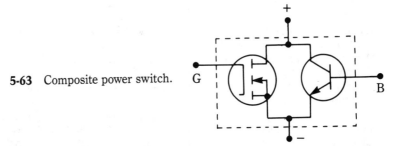

5-63 Composite power switch.

The gist of the foregoing is that a device was long sought that would combine some of the desirable characteristics of power bipolar transistors and power MOSFETs. To a considerable degree, this quest has finally been realized in a discrete device known as an *insulated gate bipolar transistor (IGBT)*. The very name suggests a combined device—a bipolar power transistor with a MOSFET-like gate.

Within certain voltage, current, and frequency areas, the IGBT will be a serious contender for power-supply designs that now use power bipolar transistors, power MOSFETs, Darlingtons, and power op amps. Moreover, the competition will also impact many uses of SCRs, triacs, GTOs (gate-controlled thyristors), and MCTs (MOS-controlled thyristors).

A natural question regarding the IGBT is whether it is a modified bipolar transistor or a doctored-up MOSFET. Strangely, the manufacturers of the new device

Table 5-10. Symbols and definitions for GTO thyristors.

Symbols

Symbol	Definition	Symbol	Definition
dv/dt	Critical rate of rise of OFF-state voltage	t_{gq}	Gate-controlled turn-off time ($t_s + t_f$) comparable to turn-off time (t_{OFF}) for transistors
I_{DRXM}	Maximum (peak) repetitive dc OFF-state current with specified circuit between gate and cathode	t_{gt}	Gate-controlled turn-on time ($t_d + t_r$) comparable to turn-on time (t_{ON}) for transistors
I_g	Pulsed gate trigger current (gate drive current)	T_J	Junction temperature (operating)
I_G	dc gate current	T_L	Lead (pin) temperature during soldering
I_{GM}	Maximum (peak) dc gate current		
I_{gqM}	Maximum gate turn-off current	t_p	Pulse duration
$I_{GR(BR)}$	Reverse gate breakdown current	t_r	Rise time
		t_s	Storage time
I_{GT}	dc gate-trigger current	t_{stg}	Storage temperature
I_L	dc latching current	V_D	dc OFF-state voltage
I_{RROM}	Maximum (peak) reverse current (repetitive), gate open	V_{DRXM}	Maximum (peak) repetitive off-state voltage with specified circuit between gate and cathode
$I^2 t$	Fusing current for device protection		
I_T	dc ON-state current	V_G	dc gate voltage
I_{TM}	Maximum (peak) dc ON-state current	V_{gq}	Gate turn-off voltage
		V_{GR}	dc reverse gate voltage
I_{TGQM}	Maximum (peak) ON-state current gate-turn-off capability	$V_{GR(BR)}$	Reverse gate breakdown voltage
I_{TSM}	Maximum (peak) surge on-state current (nonrepetitive)	V_{GRRM}	Maximum (peak) repetitive reverse-gate voltage
P_D	Device dissipation	V_{GT}	dc gate-trigger voltage
P_{GRM}	Maximum (peak) reverse-gate power	V_R	dc reverse voltage
		V_{RROM}	Maximum (peak) repetitive reverse voltage, gate open
R_{GK}	Gate-to-cathode resistance		
$R_{\theta JC}$	Thermal resistance, junction to case	V_{RRXM}	Maximum (peak) repetitive reverse voltage with specified circuit between gate and cathode
T_C	Case temperature		
t_d	Delay time		
t_f	Fall time	V_T	dc ON-stage voltage
		Z_{GS}	Gate source impedance

have not, themselves, arrived at a common consensus; they thus far seem to disagree about a standard symbol. Even worse, no standard nomenclature appears to exist for the three electrodes of the IGBT. Some claim that it has a gate, source, and drain, as in the power MOSFET. Others believe that it has a gate, emitter, and collector. Awareness of this situation can dispel confusion the first time that this device is used or considered.

Although I see the device as a "doctored-up" MOSFET (with a gate, source, and drain), this viewpoint is not used. The reason is that the International Rectifier Corporation, an outstanding pioneer in the development and marketing of these devices, chooses the base-emitter-collector format. Also, this book uses the IRC's "standard" symbol for the IGBT. At least, all parties agree that the IGBT has a gate, and not a base as input electrode. This, in itself, is important because the gate indicates that the input acts like a capacitor, not like a pn diode ("gate" suggests the input of the MOSFET).

Figure 5-64 compares the cross-sections of an "ordinary" power MOSFET and the IGBT. They are obviously quite similar; the comparison can be summed

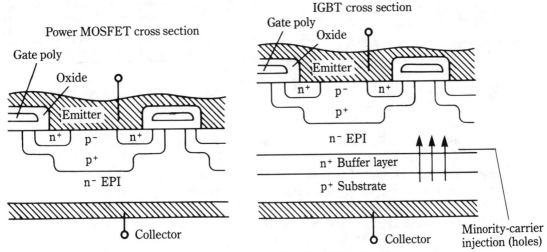

5-64 Comparison of power-MOSFET and IGBT structures.

up by noting that the IGBT has something added—a modified doping profile. The P+ layer injects minority carriers into the high-resistivity N layer of the unmodified MOSFET. This greatly increases current density and lowers the voltage drop. The N+ buffer layer enables the manufacturer to control the lifetime of stored charges so that the frequency capability of the device can be optimized. The unmodified MOSFET has no minority carriers and, therefore, no charge-storage problems in its output section. Thus, its frequency capability is so good.

Because of these added doping layers, the output behavior resembles, but does not duplicate, the output characteristics of a bipolar transistor. What is essentially similar is the very low VCE_{sat} or forward-drop voltage of bipolar transistors. A slight difference is that the IGBT has a 0.7-V offset in its output characteristic. This offset is of little consequence because IGBTs generally operate at high voltages (500 V and up). The family of curves that are depicted in Fig. 5-65 shows this offset for two operating temperatures. It is as if a pn diode was inserted in the collector lead.

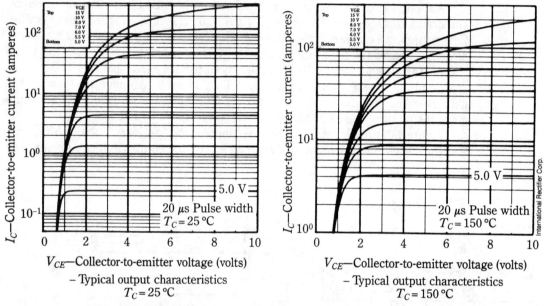

5-65 IRGPC50 output curves showing collector-voltage offset in IGBTs.

Although likening the input of an IGBT to a power MOSFET and the output of an IGBT to a bipolar transistor goes a long way in practical use of the newer device, it has a couple more unique aspects. Unlike the power MOSFET, the IGBT generally shows reverse-blocking ability (Fig. 5-66). This feature is not usually cited as a designed-in parameter. It can be only a few volts in some units, and it varies considerably among units with different voltage, current, and speed ratings. It appears that as the manufacturer has manipulated the other operating parameters, reverse blocking is allowed to assume whatever value makes it easy to obtain the best compromise of the other ratings.

Also, despite the close resemblance to the power MOSFET, the IGBT manifests no external evidence of an internal-body diode. Although this reverse-polarity diode is often useful as a free-wheeling diode or as a transient suppressor, its absence in the IGBT can just as easily be construed a gain as a loss. For one thing, the body diode of the power MOSFET does not have good high-frequency characteristics. Designers often use an external Schottky diode to deliberately cause the body diode to remain passive. Also, in some switching circuits, the body diode is

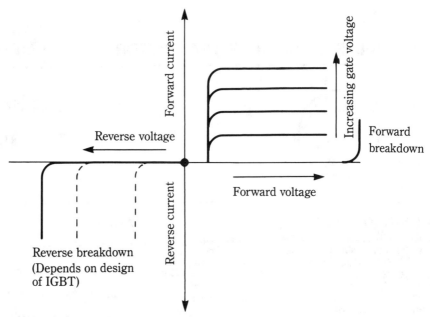

5-66 IGBTs can have considerable reverse-blocking capability. A trade-off is involved with other performance parameters.

actually in the way of intended operation. Finally, even where the body diode is functionally useful, its contribution to the temperature rise of the power MOSFET might be the "straw that breaks the camel's back." With the IGBT, the designer has the flexibility to select an optimally-suited fast-recovery diode or a Schottky diode if circuit voltages allow. Because the maker has many performance tradeoffs to work with, it is important to select the IGBT that is best-suited to a particular application. It is often possible, for example, to find a model with a high reverse-blocking capability and with less spectacular, but adequate speed and/or forward-voltage drop.

Another peculiarity (a beneficial one in most cases) is a lower input capacitance than in power MOSFETs. This problem is not immediately apparent from comparing the actual physical capacitance at the input of the two devices. However, the Miller feedback capacitance from the output can be much less in the IGBT.

Early IGBTs suffered from a tendency to latch up in the manner of a triggered SCR. This problem has been overcome and manufacturers now advertise their products to be "latch-free." Whereas the first IGBTs had several-kHz speed limitations, these devices are now made with 50-kHz ratings. However, 5-kHz specified units can be expected to display superior performance in respects other than frequency capability.

INSULATED GATE BIPOLAR TRANSISTOR

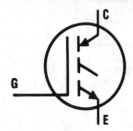

IRGPC50
600V, 55A

600V, 55A, TO-247AC IGBT

International Rectifier's IRG series of Insulated Gate Bipolar Transistors (IGBT) combines the best features of power MOSFET and bipolar transistors. The ease of drive and rugged operation of HEXFETs®, coupled with the low on-state voltage drop of bipolar devices, results in an optimum solution for low-to-medium frequency power converters up to 50 KHz.

The efficient geometry and unique processing of this latch-free design achieves low forward voltage drop and low switching loss.

These devices are ideal for application in ac and dc motor drives, uninterruptible power supplies, welding equipment, power supplies, induction heaters, and high energy pulse circuits.

FEATURES

- Latch-free operation
- Simple drive requirements
- Guaranteed low switching loss
- Low forward voltage drop
- Ease of paralleling

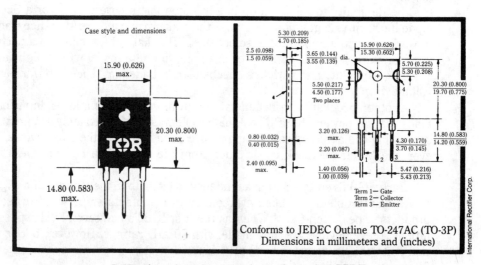

5-67 A peek at the spec-sheet of a typical IGBT.

Table 5-11. Partial specifications for the IRGPC50 IGBT.

Absolute maximum ratings

	Parameter		Units
I_C @ T_C = 25 °C	Continuous collector current	55	
I_C @ T_C = 100 °C	Continuous collector current	27	A
I_{CM}	Pulsed collector current	110	
V_{CE}	Collector-to-emitter voltage	600	
V_{GE}	Gate-to-emitter voltage	±20	V
I_{LM}	Clamped inductive load current	54	A
E_{RV}	Reverse voltage avalanche energy	50	mJ
P_D @ T_C = 25 °C	Maximum power dissipation	200	
P_D @ T_C = 100 °C	Maximum power dissipation	78	W
T_J	Operating and storage temperature range	−55 to 150	°C

Electrical characteristics @ T_J = 25 °C (unless otherwise specified)

	Parameter	Min.	Typ.	Max.	Units	Test conditions
BV_{CES}	Collector-to-emitter breakdown voltage	600	—	—	V	V_{GE} = 0 V, I_C = 250 μA
BV_{ECS}	Emitter-to-collector breakdown voltage	15	22	—		V_{GE} = 0 V, I_C = 1.0 A
$\Delta BV_{CES}/\Delta T_J$	BV_{CES} temperature coefficient	—	0.70	—	V/°C	V_{GE} = 0 V, I_C = 2 mA
$V_{CE(ON)}$	Collector-to-emitter saturation voltage	—	2.2	2.7		V_{GE} = 15 V, I_C = 27 V
		—	2.7	—	V	V_{GE} = 15 V, I_C = 55 A
		—	2.4	—		V_{GE} = 15 V, I_C = 27 A, T_J = 150 °C
$\Delta V_{GE(TH)}$	Gate threshold voltage	2.5	—	5.0		V_{CE} = V_{GE}, I_C = 250 μA
$\Delta V_{GE(TH)}/\Delta T_J$	$V_{GE(TH)}$ temperature coefficient	—	−14	—	mV/°C	V_{CE} = V_{GE}, I_C = 250 μA
G_{FE}	Forward transconductance	13	26	35	S(℧)	V_{CE} = 100 V, I_C = 27 A
I_{CES}	Zero gate voltage collector current	—	—	250	μA	V_{GE} = 0 V, V_{CE} = 600 V, T_J = 25 °C
		—	—	2000		V_{GE} = 0 V, V_{CE} = 600 V, T_J = 150 °C
I_{GES}	Gate-to-emitter leakage current	—	—	±500	nA	V_{GE} = ±20 V

Switching Characteristics @ T_J = 25 °C (unless otherwise specified)

	Parameter	Min.	Typ.	Max.	Units	Test conditions
Q_g	Total gate charge (turn on)	84	110	150		I_C = 27 A, V_{CC} = 480 V
Q_{ge}	Gate-to-emitter charge (turn-on)	8.0	16	23	nC	
Q_{gc}	Gate-to-collector charge (turn-on)	20	48	90		
$t_{D(ON)}$	Turn-on delay time	—	92	—		
t_R	Rise time	—	89	—	nS	
$t_{D(OFF)}$	Turn-off delay time	—	644	—		
T_F	Fall time	—	91	—		
E_{ON}	Turn-on switching loss	—	1.48	—		I_C = 27 A, V_{CC} = 480 V, T_J = 25 °C
E_{OFF}	Turn-off switching loss	—	1.46	—		V_{GE} = 15 V, R_G = 50 Ω
E_{TS}	Total switching loss	—	2.94	3.80	mJ	
		—	4.07	—		I_C = 27 A, V_{CC} = 480 V, T_J = 150 °C V_{GE} = 15 V, R_G = 50 Ω
L_E	Internal emitter inductance	—	13	—	nH	Measured 5 mm from package
C_{IES}	Input capacitance	—	3100	—		V_{GE} = 0 V, V_{CC} = 25 V
C_{OES}	Output capacitance	—	350	—	pF	f = 1.0 MHz
C_{RES}	Reverse transfer capacitance	—	36	—		

Courtesy International Rectifier Corp.

Partial specification data for the IRGPC50 IGBT are shown in Fig. 5-67 and Table 5-11. The following observations are particularly pertinent:

- The IRC symbol is different from several others that appear in the technical literature. Simplified or stylized versions of this symbol are sometimes used; in particular, one or both arrows are dispensed with.
- The device package is small for 600-V 55-A ratings. This stems from the inordinately low conductive loss during ON time. When used within its speed capability, switching losses are also very low.
- $V_{CE(on)}$ is indicated to be in the same ballpark as for bipolar transistors.
- The gate can be driven from CMOS logic.
- G_{FE}, the forward transconductance, is exceedingly high, which implies that a tiny change in the gate voltage can cause a large change in collector current.
- The "ease of paralleling" implies that energy-wasting emitter-resistors are not needed to prevent current hogging.
- The "latch-free" claim overcomes a much-feared shortcoming of earlier IGBTs.
- The 50-kHz rating represents a significant advancement—earlier IGBTs were much slower.

Summed up, IGBTs merit consideration for power supplies that operate at voltages greater than 400, and at high currents, but at not so high a switching rate that "ordinary" power MOSFETs are necessary. Uninterruptible power supplies and motordrive power supplies are good candidates for design around IGBTs. Very large IGBTs trade speed for high voltage and current ratings, such as 1200 V and 300 A. These might be great for welders and traction vehicles. Although not targeted for such use, the IGBT can be advantageously used in some linear-regulated supplies. Providing that power dissipation ratings are not exceeded, the high transconductance would lead to tight regulation in a linear supply. In any regulator, the rectangular SOA curve and the high peak-current capability of the IGBT are bound to be seen as positive features.

6
Linear regulating supplies that use integrated circuits

THE ADVENT OF INTEGRATED CIRCUITS HAS BEEN ONE OF THE MAJOR developments that enabled sophisticated and high-performance regulated power supplies to be produced. Although you could always assemble a large number of discrete transistor or tube stages to accomplish performance objectives, these procedures are inherently costly, wasteful of space, and ill-suited for quantity production. Problems that pertain to thermal and electrical stability are severe enough in any multistage circuit, but in a system that depends upon feedback, such problems become especially formidable. Today, IC regulators are available for modest cost and they possess many advantages over equivalent regulators that use discrete components. These ICs can be used as building blocks for control and programmability that is compliant with the ideas of the most imaginative designer. They can also be used harmoniously with discrete transistors so that the control functions and the power-handling faculties of a regulated power supply can each be implemented for optimum operation.

This chapter investigates examples of linear-type regulated supplies that use integrated circuits. In the next chapter, the application of ICs in switching-type regulators will be presented. However, the physics and fabrication of ICs will not be dealt with here. Instead, refer to treatises that are devoted exclusively to this intricate technology.

Regulators that use IC operational amplifiers

Two methods of using monolithic integrated circuits for regulated power supplies have emerged. With one method, the regulated power supply is designed around an IC op amp, which serves as the error-signal amplifier (also known as the *sensing amplifier*, *input amplifier*, or *comparator*). The series-pass element might simply be

328 Linear regulating supplies that use integrated circuits

the output stage of this amplifier, but more often is a discrete transistor. Additionally, a voltage-reference source is needed. Of course, several resistors and capacitors must also be associated with the arrangement. Figure 6-1 is a block diagram of a voltage-regulated power supply that uses any "work-horse" operational amplifier. The basic design equation for this configuration is:

$$V_o = \frac{A}{1-KA}(V_{ref})$$

where:

V_o is the dc output voltage,
A is the open-loop gain of the operational amplifier,
K is the fraction of the output voltage fed back to the summing junction:

$$K = \frac{R_2}{R_1 + R_2}$$

V_{ref} is the reference voltage.

Of course, if more than the bare essentials of a regulated power supply are desired, such as current limiting, output voltage control, etc., greater circuit complexity is necessary. In this regard, it is often desirable to use a voltage reference that has greater stability than is ordinarily provided by a dropping resistance, in conjunction with a zener diode. In principle, very close regulation is attainable from op-amp regulators if the initial open-loop gain is high. In practice, open-loop gains that are greater than about 50,000 require close attention to the input offset voltage and the common-mode rejection ratio of the IC op amp. Otherwise, the regulation will not be as good as expected from the available gain of the amplifier.

The other method of implementing IC technology in regulated power supplies is to use special ICs, in which virtually all functions that are likely to be needed are

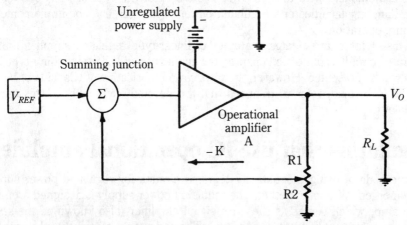

6-1 Block diagram of voltage regulator with an op amp.

processed into a single monolithic chip. Such ICs contain their own error amplifier, voltage-reference source, and various auxiliary features. Depending on the particular IC, such features enable easy implementation of output-voltage control or programming, current limiting, current foldback, remote shutdown, thermal protection, etc. Figure 6-2 is a block diagram for the LM723 voltage regulator. Obviously, the IC regulator represents a very sophisticated approach to power-supply design. However, it is not necessarily preferred. For example, the appeal of having a regulated supply on a chip is somewhat diluted in practice because it is often necessary to use an external series-pass transistor. Some designers feel they have greater flexibility when only the error amplifier is an IC. The combined use of IC regulators with operational amplifiers is increasing. Because operational amplifiers are commonly fabricated as dual or quadrupal units, it is compelling to use them for various auxiliary functions besides the obvious application as error amplifier.

Using either method of implementing integrated circuits for regulated power supplies has pros and cons. For many applications, such schemes represent straightforward and economical ways to construct regulated power supplies. However, if you want a tightly regulated supply and one with very high long-term stability, the "regulator on a chip" might not be the best approach. This is primarily because the thermal interactions within such an integrated circuit might be unfavorably traded off for the convenience of containing all the regulator functions in a single package. Therefore, you are not likely to go astray in the quest for tight operating characteristics by using an operational amplifier to process the error signal and a separate high-stability voltage-reference source.

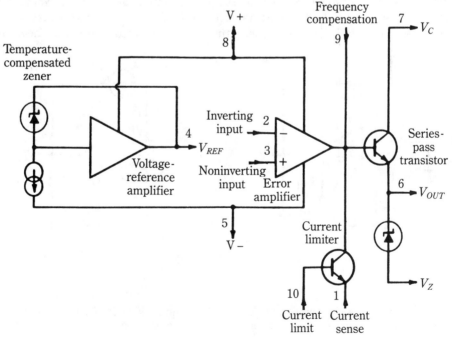

6-2 Block diagram of the LM723 IC voltage regulator.

330 Linear regulating supplies that use integrated circuits

Finally, a common aspect of both op-amp and dedicated IC regulated supplies is the consideration of current or voltage boosting (whether external pass transistors are to be used). Until quite recently, voltages in excess of about 30 V and/or currents that are greater than approximately 0.5 A usually needed a "booster" transistor to relieve the IC from electrical and thermal stress. However, both op amps and IC voltage regulators have become available with greater power capabilities. Designs can now often be implemented to eliminate the booster stage. Although this technique is economical with regard to components, it appears to be best suited to applications where extraordinary regulating performance is not required.

Simple op-amp regulators

The circuits shown in Fig. 6-3 are, despite their simplicity, capable of satisfactory performance where economical voltage regulation is required for loads to several tens of milliamperes. Depending on the type of op amp used, greater load currents can be accommodated. Generally, it is good to use a "booster" transistor with these circuits when more than 50 or 75 mA of current must be supplied to the load.

Additional simplification is achieved by dispensing with the customary dual power supply that is used with IC op amps. Thus, both the op amps and their associated zener reference sources derive operating current from the unregulated power supply. This operation is possible because the output of the op amps need only deliver current of a single polarity. Some op amps, such as the LM741, have terminals that can be used to balance out the effect of the input voltage offset. Not shown in the circuits of Fig. 6-3 is the often-utilized 10-kΩ nulling potentiometer that is connected between terminals 1 and 5, with the wiper connected to the most negative

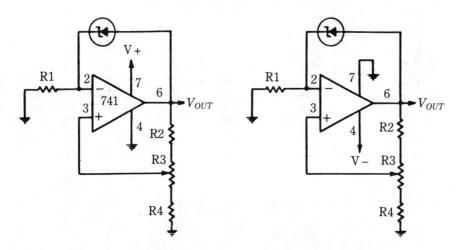

(A) Adjustable positive output voltage greater than zener reference voltage.

(B) Adjustable negative output voltage greater than zener reference voltage.

6-3 Basic configurations of voltage-regulated supplies with op amps.

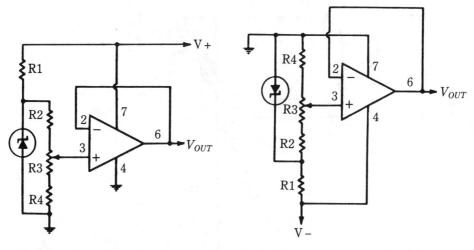

(C) Adjustable positive output voltage less than zener reference voltage.

(D) Adjustable negative output voltage less than zener reference voltage.

6-3 Continued

point of the op-amp supply. For most applications of these simple regulators, such balancing is not needed.

These regulators are often used in conjunction with more complex regulator systems. In such arrangements, the simple regulators constitute the voltage-reference sources for the overall system. This system generally is more satisfactory than the more primitive technique of simply connecting a zener diode in series with a current-limiting resistance to the unregulated dc supply. Basically, the op amp provides a stabilized voltage at a much lower dynamic impedance than is readily attainable from a zener diode alone. Because the zener diode can be operated at relatively low currents in these circuits, thermal drift is considerably alleviated when simple op-amp regulators are used for the voltage-reference source.

Voltage regulators that use the current-mirror op amp

The regulator shown in Fig. 6-4A resembles commonplace configurations, but it incorporates some unique features. Although the LM2900 operational amplifier behaves as a differential-input amplifier, it does not actually contain a conventional differential-input stage (the LM2900 is a quad unit with four such op amps processed into the monolithic silicon chip. The three remaining op amps are available for other uses). This type of op amp is inherently able to work from a single power supply. It can, therefore, be used in certain regulator circuits that would not function with conventional op amps. The regulator circuit of Fig. 6-4A uses a pnp series-pass transistor and the allowable difference between the unregulated dc input voltage and the regulated output voltage is inordinately low.

332 *Linear regulating supplies that use integrated circuits*

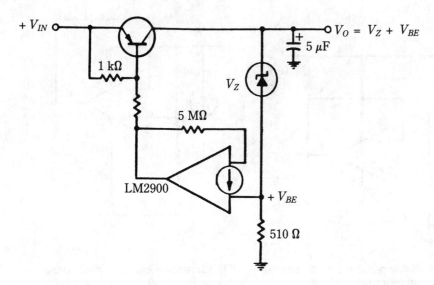

(A) Voltage regulator with the LM2900 employed as the error amplifier.

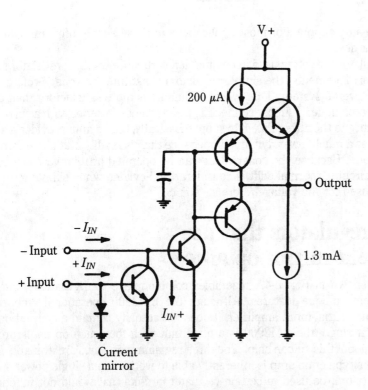

(B) Internal circuit of the LM2900 operational-amplifier IC.

6-4 Simple voltage regulator with a current-mirror op amp.

The unusual circuit of the LM2900 op amp is illustrated in Fig. 6-4B. The current arrows that represent the inverting and noninverting inputs show that the "differencing" between the two signal currents occurs in the inverting input lead. This difference current then flows through the external feedback resistance to produce the output voltage. The overall configuration is relatively simple and unencumbered. Consequently, good all-around op-amp characteristics are available at low cost. For example, the LM2900 is internally frequency compensated for worst-case application, such as unity gain. Also, the output is protected against short circuits.

Op-amp voltage regulators with current-booster stages

A more practical implementation of the op-amp regulator is obtained with the addition of an external transistor to increase the available output current. This technique relaxes the thermal burden of the op amp and indirectly contributes to enhanced overall regulating stability. The circuits illustrated in Figs. 6-5, 6-6 and 6-7 all utilize the popular 741 op amp and an external transistor to supply more current than could be extracted from the op amp alone. These circuits are popular where load currents up to several hundred milliamperes are needed because the booster transistor can be any of numerous signal-level types that are available in either plastic or TO-5 packages. At the higher current levels, a simple "top-hat" or other clamp-on heatsink can be attached to the transistor. Depending on the beta of the transistor and the current output. Base resistor R1 can be considerably lowered or even eliminated.

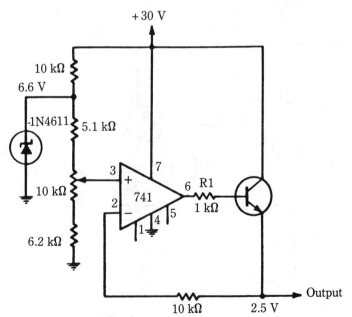

6-5 Current-boost regulator that has an output voltage less than reference voltage.

334 *Linear regulating supplies that use integrated circuits*

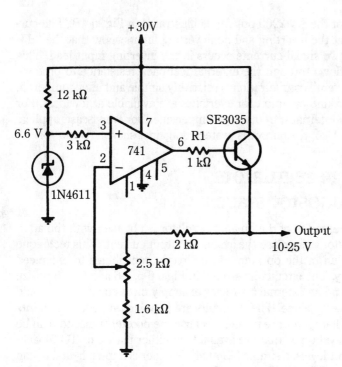

6-6 Current-boost regulator that has an output voltage more than reference voltage.

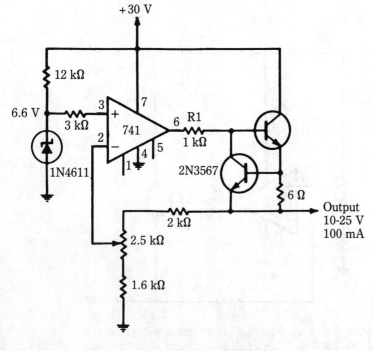

6-7 Current-boost regulator with current-limit stage.

To preserve the salient features of this arrangement, it is probably wise to consider cascaded transistor stages when the application requires more than about 0.5 A of current. In this way, the op amp itself can be spared thermal stress. The usual way of cascading output transistors is to use a Darlington connection with higher-current-rated transistors in the output stages. In many instances, a two-stage Darlington arrangement is sufficient for currents up to 10 A or so. For higher currents, a driver stage and a monolithic two-stage Darlington works out nicely.

The arrangement in Fig. 6-7 incorporates an additional transistor to provide current limiting. As shown, this transistor will deprive the output stage of base drive when the load current is in the vicinity of 100 mA. Thus, even a short-circuited load cannot force the regulator to deliver more than approximately 100 mA. Such self-protection is generally well worth the very nominal cost of the additional transistor and sampling resistance. Not only is it considerably protected from catastrophic destruction conferred to the output transistor, but damage to the load can also be prevented by this simple technique. The scheme is shown as an addendum to the circuit of Fig. 6-6, but it is equally applicable to that of Fig. 6-5.

Op-amp regulator with added features

An op-amp regulator with a zener-diode voltage reference and with a current-boosting transistor to provide practical outputs might be considered as a "bare-bones" regulated dc power supply. Many features that can be added to this basic circuit configuration. The arrangement shown in Fig. 6-8 incorporates two useful features that are lacking in the simpler systems. Instead of using a zener diode and a dropping resistor to develop the reference voltage, the regulator in Fig. 6-8 uses the voltage-reference circuitry of the MC1460G monolithic voltage regulator. This is the only portion of the MC1460G IC that is used. With such a scheme, a much more stable voltage reference is provided for the MC1539G op amp than is economically attainable from discrete zener diodes. Tighter regulation of the overall arrangement is provided than if the MC1460G was used in its entirety. The basic reason for the enhanced performance is that higher gains are generally achieved with external op amps than when the error amplifiers are included on the voltage-regulator IC.

Another feature of the circuit in Fig. 6-8 is the remote-sense provision. No matter how good the regulating ability of a power supply is, such performance is advantageous only if the connecting leads between the supply and the load have very low resistance. In practical applications, this is not always feasible. Remote sensing is the neatest way to circumvent the effect of lead resistance, because the error amplifier then senses the voltage directly at the load terminals. Although the connecting-lead resistance can still cause added power dissipation, tight voltage regulation is nonetheless obtained. The added expense and difficulty of incorporating this desirable feature are minimal because the sensing leads can be of a much smaller gauge than the leads that actually conduct the load current. Sensing leads should be twisted to minimize pick-up of stray fields. The circuit in Fig. 6-8 represents a definite trend in power-supply systems (the combined implementation of IC op amps, IC voltage references, and discrete semiconductor devices).

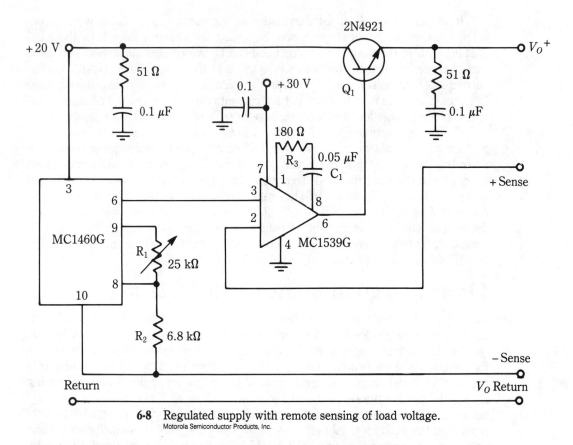

6-8 Regulated supply with remote sensing of load voltage.
Motorola Semiconductor Products, Inc.

High-performance voltage-regulated supply that uses ICs and discrete elements

The one-ampere variable-voltage supply that is shown in Fig. 6-9 follows traditional design concepts, but uses ICs where they can best contribute to overall performance. This includes the error amplifier and the voltage-reference source. The CA3130 operational amplifier has PMOS transistors in the differential-input stage and complementary-symmetry MOS transistors in the output stage. Such an IC op amp is tailor-made for a variable-voltage regulated supply, because it enables adjustment of the output voltage nearly to zero, while maintaining superb performance levels in important op-amp parameters (such as common-mode rejection ratio and power-supply rejection ratio). Also, such an op amp is a "natural" for operation from a single-polarity supply voltage.

The CA3130 (IC1) feeds into a discrete transistor stage, Q4, which, in turn, drives the Darlington-pair output transistors, Q1 and Q2. The actual series-pass transistor (Q2) is a 2N3055. Transistor Q3, in conjunction with the 1-Ω fixed resistance and the 10-kΩ potentiometer, composes the adjustable current-limiting circuit. Current limiting sets in when Q3 receives sufficient base-emitter volt-

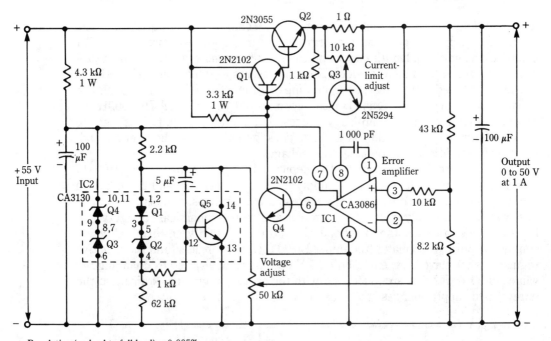

Regulation (no load to full load) <0.005%
Input regulation <0.01%/V
Num and noise output <250 μV RMS up to 100 kHz

6-9 One-ampere voltage regulator with COS/MOS operational amplifier. RCA Corp.

age to render it conductive, which eliminates further increase of the base current in Darlington stage Q1.

The voltage reference source is unique. It has three discrete transistors, Q1, Q2, and Q5, within IC2. Although elements Q1 and Q2 are shown schematically as a forward-conducting diode and a zener diode, they are, in reality, bipolar transistors. The base-emitter sections of ordinary transistors can make fine zener or avalanche-breakdown diodes. The transistors within an IC array often display sharper knees, have lower dynamic impedance, and generate less noise than discrete transistors. Notice that transistor Q5 is deployed as a shunt element, rather than as the more-often-encountered series element. Zener diodes Q3 and Q4 provide preregulation for the voltage-reference circuit, and also provide the dc operating voltage for the CA3130 error amplifier. The performance of this regulated power supply would be both difficult and costly to duplicate with discrete devices only.

Regulators that use power op amps

Once the IC operational amplifier became a technological and commercial success, it was inevitable that serious efforts would be directed toward extending its power-handling capability. This accomplishment has not been easy, for many of the desirable features of low-current op amps are at least partially attributable to the low or moderate temperature rise that is imparted to the chip by the output stage. Also, nonthermal causes of performance degradation are at higher power levels. For example, the gain-bandwidth factor and the slew rate tend to suffer as more current is demanded from these devices. Nonetheless, respectable overall operation is presently attained from such power op amps as the RCA HC2000H, which can deliver a peak current of 7 A and can supply up to 100 W. The RCA HC2000H is a hybrid, rather than a monolithically processed integrated circuit. That is, the entire internal circuitry is not on a single-package chip. From the user's standpoint, however, it is a self-contained IC package.

The dual-tracking regulator that is depicted in Fig. 6-10 uses two LH0041J power op amps to produce a ±15-V supply with one-ampere capability from each section. The ability to dispense with the usual current-boost transistors in such power supplies not only makes for a simpler and more economical configuration, but also tends to reduce the likelihood of electrical instability. In this dual power supply, the positive regulator (the upper LH0041J) is the master, and the negative regulator is the slave. Therefore, both 15-V outputs can be varied by changing the values of R1 or R2 that are associated with the positive regulator. Notice that the unregulated supply voltages can be as low as ±16 V.

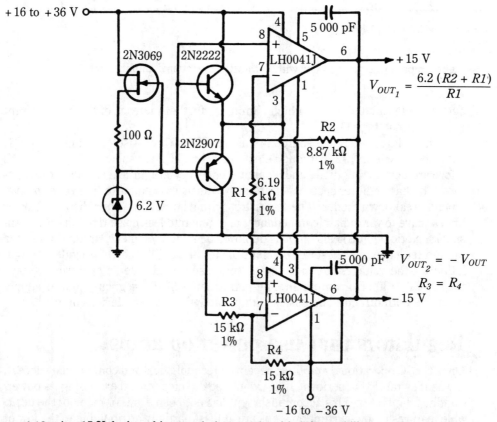

6-10 A ±15-V dual-tracking regulating supply with 1-A capability. National Semiconductor Corp.

The voltage-reference circuit uses a buffered zener diode. Moreover, the zener diode is fed from an FET constant-current source. These provisions enable much greater voltage stability to be developed by the zener diode than from the simpler method of merely operating the zener diode in conjunction with a current-limiting resistance.

A 5-A voltage/current regulator

Many voltage regulators are equipped with current-limiting circuits in order to protect the series-pass element from the effects of excessive loading. Although the load might be short-circuited in such a regulator, the available current is automatically limited to some preset value. When operating in its current-limiting mode, this type of regulator actually performs as a current-regulated power supply. Often, the current regulation is not very tight because the main objective in most regulators that are endowed with current limiting is simply to protect the series-pass transistor from overheating or from catastrophic destruction. However, if sufficient loop gain exists in the current-limit mode and if the current value at which limiting sets in can be controlled, you then have the option of using the regulator to either stabilize load voltage or load current. Such a dual-purpose regulator is shown in Fig. 6-11.

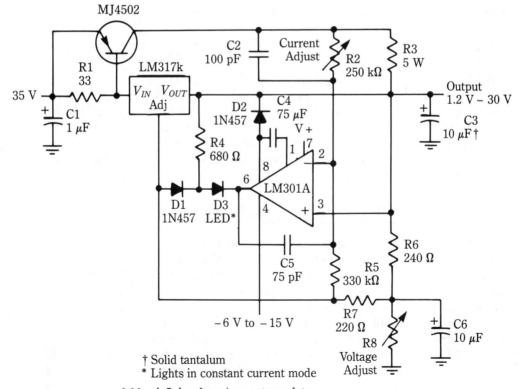

6-11 A 5-A voltage/current regulator. National Semiconductor Corp.

An initial inspection of this circuit might cause some confusion because of its resemblance to traditional configurations. For example, the LM301A op amp is not deployed as an error amplifier and it is not involved in circuit operation when the regulator functions in the voltage-regulator mode. Another possible stumbling block in the analysis of this regulator circuit is the use of the LM317K three-terminal adjustable regulator. A basic understanding of this circuit is aided by considering the LM317k as a "synthesized npn transistor." Viewed in this way, the V_{IN} terminal becomes the "collector," the V_{OUT} terminal becomes the "emitter," and the ADJ terminal functions as the "base." The LM317k functions as the driver for the MJ4502 series-pass transistor. To make the circuit simulation completely "conventional," it is also necessary to know that the voltage-reference source is contained in the LM317k three-terminal regulator.

Current limiting automatically occurs when the voltage drop caused by load current through sampling resistance R3 becomes sufficient to switch the state of the LM301A op amp. This op amp thus functions as a digital voltage comparator and not as a linear amplifier. When the conductive state of the LM301A switches in this fashion, diodes D1 and D2 become forward-biased and the "base" of the LM317k is deprived of any additional drive current beyond that which is sufficient to maintain the op amp in its switched state. In turn, the series-pass transistor can only deliver a limited current to the load. The value of this limited current is controlled by the setting of R2. Because diode D3 is an LED, it becomes illuminated when the regulator operates in its constant-current mode.

Linear regulators that use the LM105 IC

The LM105 positive voltage regulator is described in chapter 5. In general practice, this and similar IC regulators have been used as the "brains" in a linear power supply and external-series or shunt transistors serve as the "brawn." Such "boost" circuits enable almost any practical power level to be achieved. This, indeed, forms the basis for linear-type regulating supplies that use ICs. Figure 6-12 shows a number of such arrangements. The basic ideas can be readily extended for greater boost at higher current levels.

The type 723 voltage regulator

The 723 IC voltage regulator is exceptionally popular. Like the LM105 voltage regulator, it has many operational modes. However, the 723 has a 150-mA output capability and is used more often without current-boost transistors than the LM105. The block diagram of the 723 regulator is shown in Fig. 6-2. Four useful modes of operation are illustrated in Fig. 6-13. Notice in Fig. 6-13A and 6-13B that the resistance values of R1, R2, and R3 can be chosen not only to provide the desired output voltage, but to minimize temperature drift as well.

In all four circuits shown, the V+ and V_c terminals are connected together. This technique is normal because V_c is the collector of the output transistor and V+ is the collector of its immediate driver stage. Connecting the collectors of

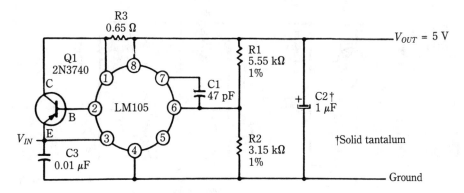

(A) A 0.2-ampere regulator.

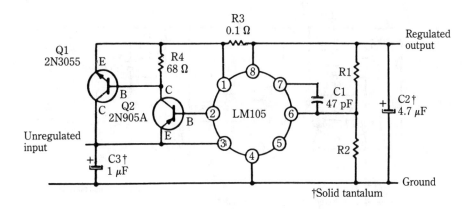

(B) Regulator connected for 2-ampere output current.

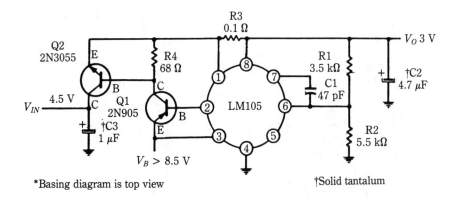

*Basing diagram is top view

(C) Circuit for obtaining higher efficiency operation with low output voltage.

6-12 Boost circuits with the LM105 IC regulator. National Semiconductor Corp.

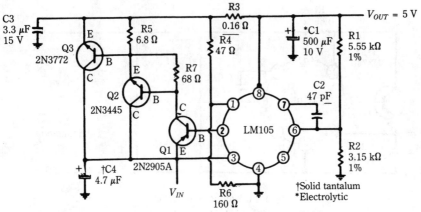

(D) A 10-ampere regulator with foldback current limiting.

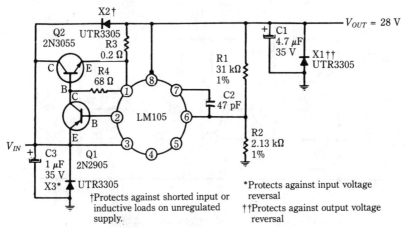

(E) Regulator with protective diodes.

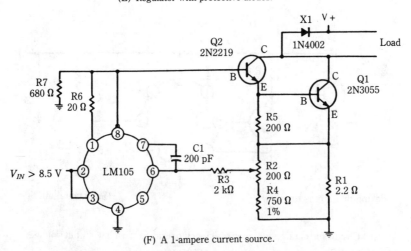

(F) A 1-ampere current source.

6-12 Continued

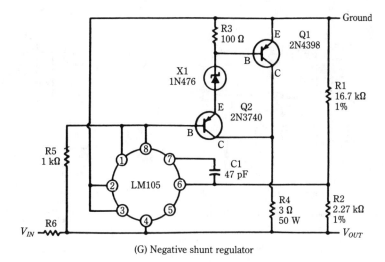

(G) Negative shunt regulator

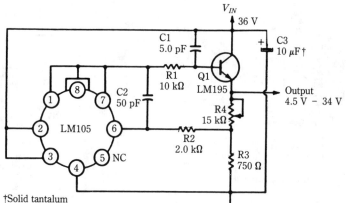

(H) Regulator with the LM195 monolithic power transistor in the output stage.

6-12 Continued

these two transistors together simply makes them a Darlington pair. However, it is sometimes useful to have individual access to the two collector leads. For example, when a pnp boost transistor is used, this option provides the designer with a convenient solution to polarity incompatibility.

In Fig. 6-13D, the V_z terminal, rather than the V_{OUT}, terminal, provides drive for the shunt-regulating transistor. This takes advantage of an internal zener diode to shift the dc level about 6 V closer to ground potential. This circuit technique is better than the alternate approach of using an inordinately high value of base resistance (R3) in conjunction with the V_{OUT} terminal. Such a "brute-force" approach would tend to degrade both the efficiency and the dynamic characteristics of the regulator circuit. The current-sense and current-limit terminals are not used in this scheme because a shunt regulator is inherently immune to damage from overload (it is good to select a wattage rating for R4 that is greater than that which would be sufficient for full-load operation).

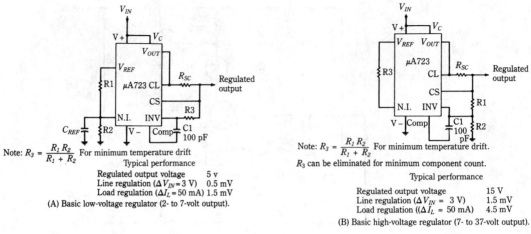
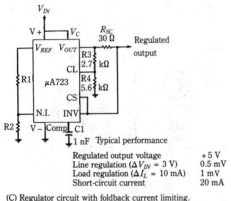
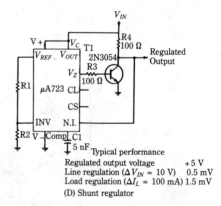

6-13 Four modes of operation of the 723 IC voltage regulator. Motorola Semiconductor Products, Inc.

Three useful regulators that are based on the MC1560/1561 IC

The MC1560/1561 IC voltage regulators have been popular devices because they can provide up to 0.5 A of load current. At the same time, they provide an inordinate number of operational options. This is because of the many internal circuit junctions brought out to external terminals. In addition to V_{IN}, V_{OUT}, and GND, the following terminals are provided: *Electronic Shutdown, Current Limit, Output Sense, DC Shift Output, Output Reference, Noise Filter, and Dc Shift Sense*. The nominal output impedance of this regulator is approximately 0.025 Ω from dc to about 100 kHz. This gain-bandwidth product (200 MHz) is greater than most regulators exhibit. Considerable care is required to realize the benefits of such inherent capability. You must select, mount, and connect the external components according to good high-frequency practices.

The simple scheme shown in Fig. 6-14 requires no other active device than the IC itself. The ratio of R_1 to R_2 can be varied to produce an output voltage that is as low as 3.5 V. A good nominal value for R_2 is 6.8 kΩ. Then, R1 can be adjusted to provide the required output voltage. If output voltages less than 3.5 V are needed, it is only necessary to interchange the connections to terminals 6 and 8. The 1N4001 diode at terminal 4 is generally used with this IC to help suppress any tendency toward instability as a result of the high gain-bandwidth product. If satisfactory operation is obtained with this diode shorted out, so much the better, because its presence increases the value of R_{SC} that is required for a given onset of current limiting.

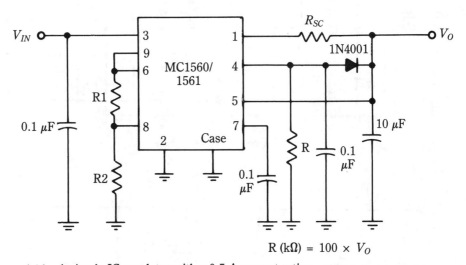

6-14 A simple IC regulator with a 0.5-A current rating. _{Motorola Semiconductor Products, Inc.}

In Fig. 6-15A, an npn power transistor has been added in order to boost the available output current to 5 A. Notice the use of the de-Qing networks that are connected to terminals 3 and 5. Yet another technique for achieving stability is using the PC-board ground plane. Output voltage is determined by resistance R_1. A somewhat different booster technique is used in Fig. 6-15B to provide a 100-V output. Current limiting is produced by Q2 and it occurs at 100 mA with the 5.6-Ω resistance that is indicated for R_{SC}. Two input voltages are required, but the 30-V source for operation of the IC can be derived from the 110-V dc supply via a current-limiting resistance and an appropriate zener diode. In this circuit, the 1N982 zener diode functions as a voltage-level shifter, which enables the IC to operate within its voltage rating and the series-pass transistor operates at a much higher voltage.

A simple ±15-V dual regulated supply

Operational amplifiers and many other ICs require a ±15-V dual power supply for optimum performance. Formerly, the design and construction of such a dual

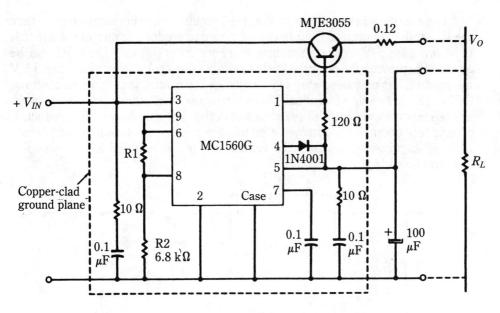

(A) Current-boost technique.

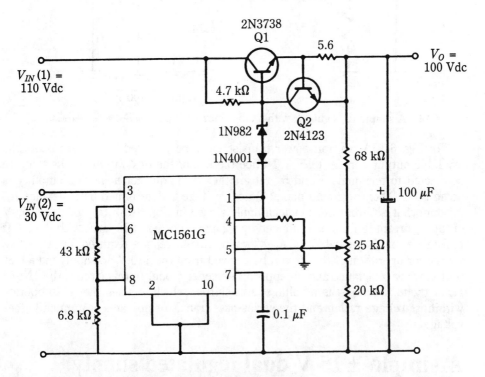

(B) Voltage-boost technique.

6-15 Boost techniques applied to the MC1560/1561 IC regulator. Motorola Semiconductor Products, Inc.

source of regulated dc involved considerable complexity. The likely troubles were more than double those that pertained to a single source of dc. At best, if two single power supplies were used, the parts count was quite high. Although modifying a single source to provide a "centertap" often appeared the easy way out, practical implementation usually involved enough major surgery to self-defeat the practice.

Therefore, it is exceedingly useful to use a low-cost IC that contains two oppositely poled 15-V regulators (such as the MC1568R). A dual voltage-regulated power supply that uses the MC1568R is shown in Fig. 6-16. Notice the extreme simplicity and the minimal number of external parts that are required. An interesting feature of this IC regulator is the separate sense terminal for each of the two outputs. This enables tight regulation at the load and results in high isolation between the two voltage sources.

In the circuit of Fig. 6-16, the unregulated power supply uses a bridge-rectifier configuration. Because of the centertap on the transformer, two 30-V, oppositely poled voltages are produced and each is derived from its own pair of rectifying diodes in the MDA920-3 bridge.

Although the MC1568R is internally set to provide ±15 V, these output voltages can actually be adjusted from ±8 to ±20 V by connecting a potentiometer to a terminal that is not shown in Fig. 6-16. Such an adjustment swings both voltages simultaneously. Thus, the MC1568R is basically a tracking regulator. One version of this IC, the "L" package, provides a special terminal whereby the balance between the output voltages can be adjusted externally.

Each regulator section of the MC1568R can provide up to 100 mA of output current. This current is adequate for many ICs and for many complex systems. If more current is required, the MC1568R can be used with external current-boost transistors.

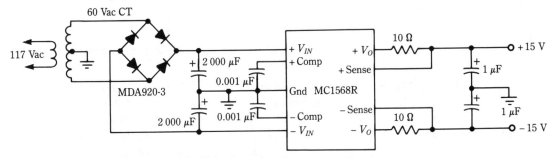

6-16 A simple ± 15-V dual-regulated power supply. Motorola Semiconductor Products, Inc.

Some typical applications of fixed-voltage three-terminal IC regulators

Figure 6-17 shows typical applications for the MC78LXX and MC79LXX families of three-terminal regulators. These examples are typical of the general way in which such fixed-voltage regulators are implemented. Most important is the

348 *Linear regulating supplies that use integrated circuits*

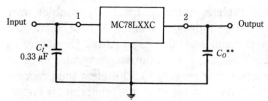

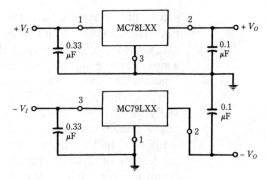

A common ground is required between the input and the output voltages. The input voltage must remain typically 2.0 V above the output voltage, even during the low point on the input ripple voltage.

*C_I is required if regulator is located an appreciable distance from power-supply filter.
**C_O is not needed for stability; however, it does improve transient response.

(A) Point-of-application voltage regulator.

(B) Positive and negative voltage regulator.

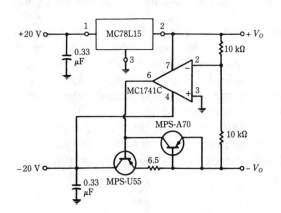

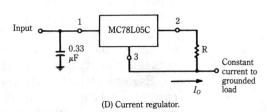

(D) Current regulator.

MC79LXX three-terminal regulators.

(C) ±15-volt tracking voltage regulator.

6-17 Useful applications of the MC78LXX and MC79LXX 3-terminal regulators.

arrangement of Fig. 6-17A in which the bypass capacitors can often be dispensed with. Thus, the compelling feature of these devices is obvious; you can achieve point-of-application regulation with few, if any, additional components to the IC regulator itself. Figure 6-17B shows the dual-voltage version of the supply in Fig. 6-17A. The straightforward system stems from pairing the MC79LXX negative-voltage regulator with the MC78LXX positive-voltage regulator. This dual power supply is not a tracking-type because each regulator independently develops its own output voltage.

The dual supply shown in Fig. 6-17C is a true tracking-type. Here, the negative regulation circuit is slaved to the output of the MC78L15 positive-voltage regulator. Notice that the op-amp (negative) regulator is used with a current-limiting circuit—the MPS-A70 transistor in conjunction with the 6.5-Ω sampling resistance (the three-terminal regulator has its own internal provision for limiting output current).

Figure 6-17D shows just how simply these three-terminal ICs can be used as current regulators. The regulated load current is nearly equal to $\frac{V_o}{R}$. Regulated output voltage V_o is developed across sampling resistance R, rather than across the load, as in more conventional voltage-regulation schemes. Regulated load current is less than $\frac{V_o}{R}$ by the quiescent current of the IC (typically several milliamperes). The quiescent current can usually be found tabulated in the specification sheets.

Rated current is regulated right down to a zero-resistance load. On the other hand, the maximum load-resistance, $R_{\text{Load-max}}$, is the resistance at which rated current or any load current, I_{Load}, can be regulated. The following equation applies:

$$R_{\text{Load-max}} = \frac{V_{\text{IN}} - (V_o + V_{\text{Min}})}{I_{\text{Load}}}$$

where V_{Min} is the smallest voltage differential that can be developed across the series-pass transistor and still have the regulator perform normally. V_{Min} is often called the *minimum input-output voltage differential* and is commonly plotted as a function of output current. Several curves are often depicted that show the variation of V_{Min} at different temperatures. In many IC regulators, it is sufficient to estimate the minimum input-output voltage differential, V_{Min}, in the vicinity of 1.2 to 1.6 V.

Another way to view this situation is that $R_{\text{Load-max}} \times I_{\text{Load}} = V_{\text{IN}} - (V_o + V_{\text{Min}})$ and that both of these quantities define the compliance voltage of the current-regulated supply. If the load resistance is made greater than $R_{\text{Load-max}}$, the supply will not "comply." That is, it will no longer regulate the load current.

Specially designed three-terminal regulators for adjustable voltage applications

Although the output voltage of most three-terminal regulators can be made adjustable, better results are forthcoming from three-terminal regulators that are specifically intended for such service. An example of such a device is the LM117. Four typical circuits in which the adjustable-voltage feature of the LM117 is used are shown in Fig. 6-18. The equation that is associated with Fig. 6-18A is more accurately expressed as

$$V_{\text{OUT}} = 1.25\left(1 + \frac{R_2}{R_1}\right) + I_{\text{ADJ}}(R_2)$$

However, the current that is associated with the ADJ terminal is not only very small in this regulator (on the order of 100 μA), but it is fairly constant. As a result, the term $I_{\text{ADJ}}(R_2)$ can be neglected for most practical applications. With three-terminal regulators not specifically intended for adjustable-output applications, the

350 *Linear regulating supplies that use integrated circuits*

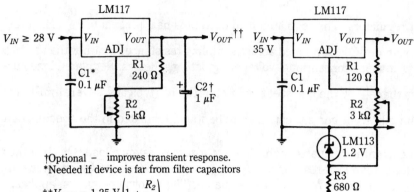

†Optional – improves transient response.
*Needed if device is far from filter capacitors

††$V_{OUT} = 1.25\text{ V}\left(1 + \dfrac{R_2}{R_1}\right)$

(A) Basic voltage-regulator circuit with output adjustable from 1.2 volts to 25 volts.

(B) Modification for making circuit adjustable down to zero volts.

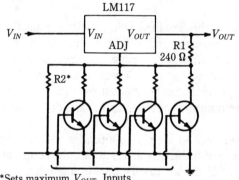

(C) Arrangement for digital selection of output voltage.

*Sets maximum V_{OUT} Inputs

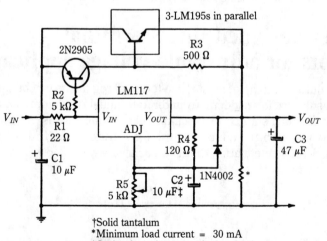

†Solid tantalum
*Minimum load current = 30 mA
‡Optional – improves ripple rejection

(D) Method for boosting output current of adjustable voltage regulator.

6-18 Typical circuits with the LM117 3-terminal adjustable regulator.

current that is associated with the third terminal (is usually designated as the ground terminal) tends to be larger and more variable.

Because of its superb performance in adjustable-voltage applications and because of its useful current capability of 1.5 A, the LM117 eliminates the necessity to stock many fixed-voltage regulators. Not only does it simplify inventory, but it is flexible when deployed as a "workhorse" regulator throughout a system. Its internal circuitry includes current limit, thermal-overload shutdown, and safe-operating-area (SOA) protection. All of these protective provisions remain functional, even if the ADJ terminal is disconnected.

The circuit in Fig. 6-18B is adjustable down to zero output voltage. Although the LM113 is shown as a zener diode, it is actually an IC "reference diode." The energy-bandgap voltage of silicon junctions is used to produce the precise 1.220-V output. The LM113 features extremely low noise because no voltage-breakdown mechanism is involved. Its temperature stability is also much better than can be readily or economically attained with zener or avalanche voltage sources. It is used in the circuit of Fig. 6-18B to cancel the effect of a similar voltage reference source within the LM117 when the output voltage is adjusted for minimum. This is done in such a way that the load current does not pass through the LM113. The adjustable output voltage of the regulator circuit in Fig. 6-18B is 0 to +35 V.

In Fig. 6-18C, the regulated output voltage of the LM117 is controlled by binary signals representing digital data. The transistors, which behave as switches, are driven hard into their saturation regions. When all four of these transistor switches are in the OFF state, the maximum output voltage of the regulator is limited by resistance R_2.

The adjustable-output regulator circuit shown in Fig. 6-18D incorporates a booster to provide increased current capability over that which is attainable from the LM117 itself. Although such a technique is common, there is more than initially meets the eye in this scheme. The three LM195s in parallel are indicated as a transistor. Actually, the LM195 is a special monolithic power transistor. It behaves as a transistor with very desirable characteristics, not the least of which is its paralleling ability. This action is because no emitter-current apportioning resistances are needed, as with conventional transistors. The internal circuit of this "transistor" is shown in Fig. 6-19. The per-unit cost of this recently developed device is usually less than that of ordinary transistors—with comparable voltage and current ratings.

The LM195 can deliver at least 1 A of current and can withstand 42 V between collector and emitter, or between collector and base. It has internal thermal limiting, which provides reliable protection against overload. It also incorporates current and power limiting. These internal protections render the LM195 virtually immune to destruction from any type of overload. Because the device has an extremely high gain, its typical base current is on the order of 3 μA. Despite this rating, the base can be driven by up to 40 V without damage! The switching times for the LM195 are specified as 0.5 ms. Therefore, it is capable of good response when used as the series-pass element of a regulating circuit. The internal circuitries of the LM195 and the LM117 regulator are quite similar, because both devices were developed from similar circuit technologies.

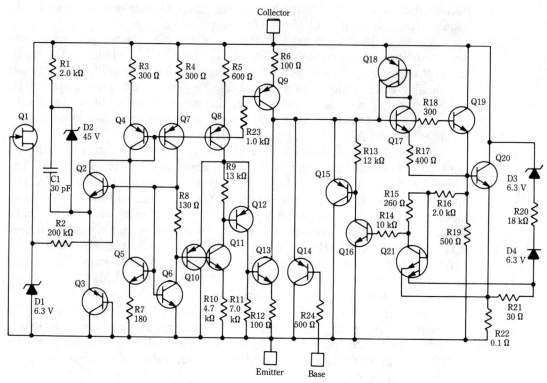

6-19 Internal circuitry of the LM195 monolithic power transistor. National Semiconductor Corp.

The 2N2905 transistor forms a complementary Darlington circuit with the LM195s. Because of this transistor, the npn input polarity of the LM195 is converted to pnp. At this writing, a true pnp version of the LM195 has not appeared on the market.

Figure 6-20 depicts a useful control technique, whereby a number of LM117s (used with individual cards or circuit boards) can be simultaneously adjusted with a single control. When connected in this manner, negligible intercoupling exists

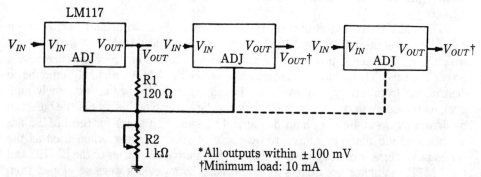

6-20 Control scheme for multiple 3-terminal adjustable IC regulators.

between the regulated loads through the common impedance of control resistance R_2. Such might be the case if the LM117s were not high-performance regulators, with good isolation between the ADJ and the V_{OUT} terminals. In actual practice, you can anticipate a reduction in crosstalk or feedback between various system loads when the technique of Fig. 6-20 is compared to arrangements that use other techniques and devices to accomplish the same result.

Versatility of three-terminal IC regulators

Although three-terminal voltage-regulator ICs are designated as either positive or negative, this represents the manufacturer's endeavor to provide optimum circuit-design convenience. Actually, it is often feasible to use either type, regardless of the desired output polarity. It is just a matter of which line the IC is connected in and floating the unregulated (input) supply. Figure 6-21 shows the various circuit permutations for realizing either output polarity from either type of IC regulator.

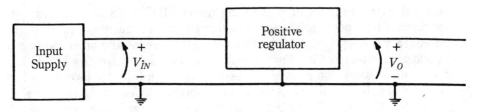

(A) Positive output using positive regulator.

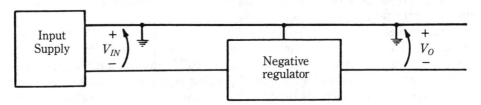

(B) Negative output using negative regulator.

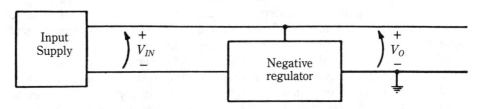

(C) Positive output using negative regulator.

6-21 Required output polarity from either type of 3-terminal regulator.
Motorola Semiconductor Products, Inc.

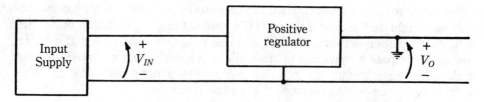

(D) Negative output using positive regulator.

6-21 Continued

Of course, compatibility with the ground status of the load is not necessarily obtained in all situations, which explains the availability of positive and negative regulators.

Power MOSFET battery charger

Almost lost in the enthusiasm of applying power MOSFETS to switching-type power supplies have been their unique benefits for linear regulators. Although the high-frequency switching capability does not apply here, all the other parameter and performance features do. For example, the negligible drive power, freedom from thermal runaway, and immunity to secondary breakdown are favorable when compared to bipolar transistors. Another very worthwhile feature is their paralleling characteristic; generally, no ballast resistors or increased drive are needed if you want to extend the load current capability of a regulator that already uses one power MOSFET. Special consideration is necessary to provide sufficient drive voltage, however, because the gate-enhancement voltage must be about 10 V above the nominal power-supply output voltage.

Figure 6-22 shows a 12.5-A battery charger that uses the VN64GA power MOSFET. The circuit is a basic series-regulator configuration with a low parts count. It is particularly well adapted for charging 12-V storage batteries, but it can be easily modified to serve as a general-purpose regulated power supply. What needs to be done is to increase the capacity of filter capacitor C2 by about a factor of 10, or so. Also, for the sake of stability, a 50-μF 50-V solid-state tantalum capacitor should be connected across the output terminals (when used as a battery charger, the battery supplies the output capacitance). If initially designed as a general-purpose supply, a 30-V center-tapped transformer would be desirable to reduce dissipation in the MOSFET. Notice the use of the CR390 constant-current diode for supplying the zener diode in the voltage-reference circuit.

If an adequate heatsink is provided and care is taken not to short the output, the overall electrical ruggedness of this simple arrangement surpasses that which is ordinarily obtained from bipolar circuits. Current overload protection can be incorporated in a similar manner to the techniques that are used with bipolar series-pass regulators. This supply should be well suited for testing mobile communications equipment.

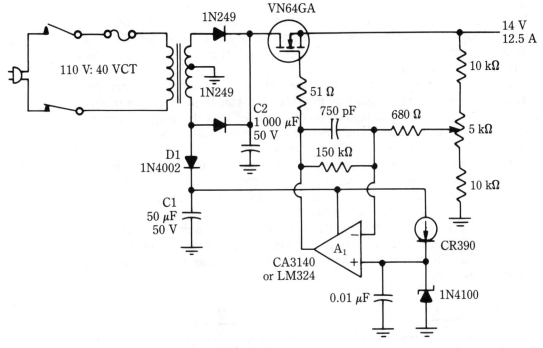

6-22 12.5-A battery charger with a power MOSFET. Siliconix Inc.

Voltage and current regulation for low-voltage and/or low-current circuits

The natural impetus for circuitry and device development in regulated supplies has been for progressively higher power levels at greater efficiencies, lower heat generation, more compact packaging, and, hopefully, at lower costs. These objectives are both commendable and necessary. Another dimension of progress in electronics is the development of circuits, devices, and systems that are capable of operating from several volts or less and often consuming no more than several tens of microamperes of current. Some of the uses for such micropower electronics include electronic-watch circuits, instrumentation, telemetry, electronic jewelry, milliohmmeters (and other portable test equipment), medically implanted devices, A/D and D/A converters, and low-voltage logic supplies.

It is interesting that major semiconductor firms have already found it expedient to make specialized micropower devices available. Included are the LM10, a 1.2-V bipolar op amp, and the ICL76XX family of CMOS op amps, which require only ±0.5 V. Also, the LM4250 programmable op amp performs well from ±1-V sources. Some band-gap voltage references consume a dozen microwatts and provide stable reference voltages of 1.235 V (or double this value). Recently introduced LVA zener diodes display sharp "knees" with as little as 100 nA of current.

The supermatched bipolar transistor pair, the LM194, enables deeper penetration into micropower-circuit functions than could hitherto be readily achieved. Many more micropower devices will make their debut because of ever increasing concern with energy conservation. An immediate benefit will be greatly extended battery life in portable equipment.

These micropower circuits often produce enhanced performance when their supply voltage or current is regulated. Because heat-removal problems are of negligible scope, the regulating schemes generally involve linear techniques. However, the low-power consumption of the powered circuitry should not be overwhelmed by needless dissipation in the regulation device or source.

A number of electronics circuits require a low-current source of either fixed or adjustable biasing voltage that is usually (but not necessarily) negative, with respect to ground. Figures 6-23, 6-24, and 6-25 depict typical examples. Although such biasing voltages are often obtained from nonregulated sources within the system, the use, at least, of a simple regulator is worthy of consideration. This is more precise, reproducible, and is likely to remedy certain noise-injection problems. Because of the miniscule current requirement, the added cost is likely to be trivial in most cases.

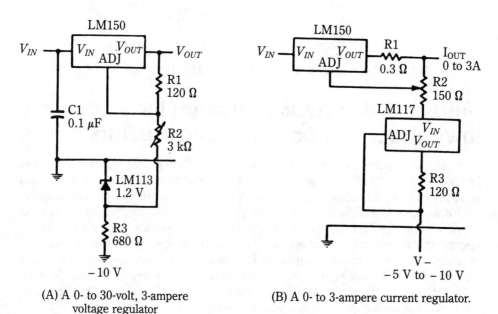

(A) A 0- to 30-volt, 3-ampere voltage regulator

(B) A 0- to 3-ampere current regulator.

6-23 Negative-bias techniques to enable adjustment of regulator to zero.
National Semiconductor Corp.

In Fig. 6-23, the output levels of the two regulators cannot be adjusted to zero unless a negative source of bias voltage is connected to the circuits. This requirement is often realized belatedly and might be considered a nuisance to be quickly resolved in typical design projects. However, even though the bias voltage does not need to be regulated, just "any old source" of 5 or 10 V can lead to noise injection or mutual-impedance interaction with other circuitry. On the other hand, a simple regulated source of such bias voltage can justify its inclusion by its contribution to overall good performance.

The RCA CA3097E IC (shown in Fig. 6-24) contains a number of useful devices that are isolated from one another by reverse-biased pn junctions. Thus, substrate pin 10 should be biased with a higher negative voltage than any other pin. Here again, the notion that any stray or residual voltage in the system is suitable might lead to unanticipated malperformance. Such a voltage should maintain its nominal value in the face of changing system operating conditions. The use of a regulated source is a good idea and it might be helpful in circumventing difficult-to-interpret misbehavior, such as inadvertent triggering of the thyristors.

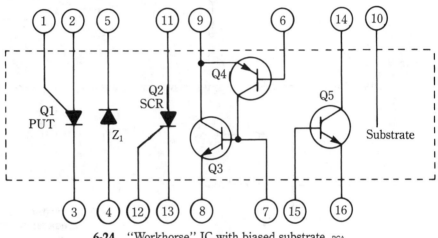

6-24 "Workhorse" IC with biased substrate. RCA

Switching supply transistors are often investigated for their rise and fall times before being approved for actual use. Figure 6-25 shows a simple setup. The negative bias should be known and it best derived from a small shunt-type voltage regulator. Bias that is too high can lower the SOA capability of the power transistor and cause its destruction.

Small voltage-regulated supplies with adjustment potentiometers are extremely useful for tuning in passive LC circuits and in oscillators that use varicap diodes or varactors. Three such voltage-dependent capacitors are shown in the IF bandpass filter of Fig. 6-26. Although some consumer products have attempted to deploy this kind of "electronic tuning" without benefit of bias-voltage regulation, such penny-pinching designs have been plagued with tuning-stability problems. In communications receivers, where relatively high-Q resonant circuits are involved, the quality of the bias supply is as important as the characteristics of the tuned circuit, itself. Fortunately, high-performance IC regulators and precision potentiometers are both available at minimal expense. Indeed, the degree of "tracking" between stages that incorporate voltage-tuned resonant circuits is surprisingly good. This effect is largely attributable to the readily obtainable IC regulator.

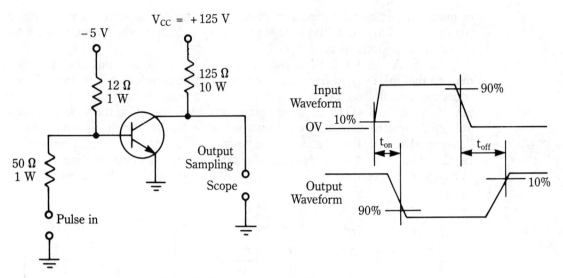

6-25 The use of negative bias to test a switching transistor. Delco Division of General Motors.

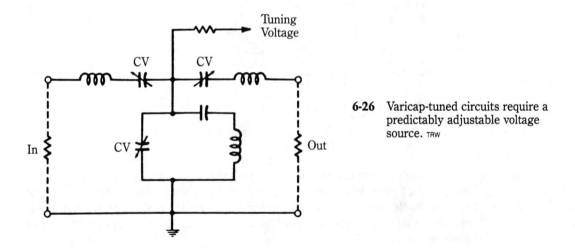

6-26 Varicap-tuned circuits require a predictably adjustable voltage source. TRW

A 1.2-V, 200-µA voltage-regulated supply

The "flea-power" voltage-regulated supply (Fig. 6-27) can be valuable in the design, testing, and implementation of various micropower circuits. Actually, the 200-µA capability endows it with "brute" status, with regard to many of these circuits that often consume no more than several tens of microamperes.

The base of the 2N4250 series-pass transistor "sees" a reference voltage that is produced by a constant current through its 100,000-Ω base-emitter resistor. The constant current is provided by the LM134 adjustable current source, in which the current level is set by the potentiometer-diode network, which is connected to its R terminal. The diode, by virtue of its temperature coefficient, makes possible an

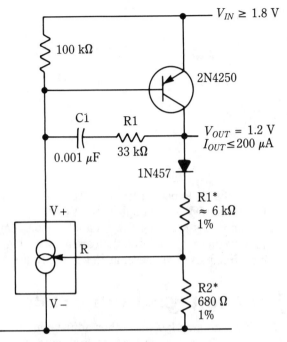

6-27 1.2-V, 200-μA regulated power supply.
National Semiconductor Corp.

*Select ratio of R1 to R2 for zero temperature drift.

overall near-zero temperature coefficient of the supply when R1 and R2 are properly chosen. This supply will operate from as low as 1.8 V from the "unregulated supply"—most appropriately, two AA penlite cells.

The presently available micropower ICs are so improved with respect to leakage currents, specification precision, and parameter tolerances, that ultimate performance is more readily attained when operated from regulators, rather than directly from batteries.

Low voltage from an LED

The scheme shown in Fig. 6-28 is suggestive of a simple zener-diode shunt-regulator circuit. However, this is a forward-biased light-emitting diode. A stabilized output of about 1.4 V is developed. Although not commonly exploited, the forward-conduction characteristic of the LED displays a much sharper knee than is ordinarily obtained from forward-biased silicon junction-diodes. Also, the temperature coefficient is lower than in common silicon devices. Thus, the requirements for voltage regulation are inherent in this device—even though major emphasis understandably focuses on its use as a "solid-state lamp." LEDs are basically gallium arsenide, but different types also contain various other dopants—primarily to produce different colors.

Generally, satisfactory regulating action begins when the LED is just visible in the dark. Some experimentation is in order, depending on the LED type and the

360 Linear regulating supplies that use integrated circuits

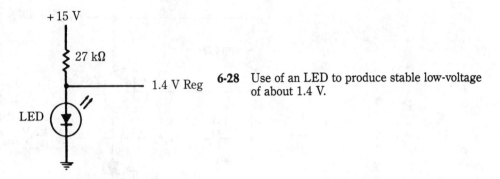

6-28 Use of an LED to produce stable low-voltage of about 1.4 V.

current that is expected from the regulated output. This setup is an economical reference for 5-V logic supplies. In its own right, it is useful for powering micropower circuitry. By reducing the value of the current-limiting resistance, much lower source voltages can be used than are shown. In general, the same design and operating principles apply here, as in zener circuits.

Low voltages from Zener regulators

Many bench supplies and other adjustable sources of regulated voltage will not operate below a 6, or at best, several volts. Although lower voltages are available from a number of band-gap devices and from sophisticated ICs, a quick (and often very satisfactory) solution to this problem is likely to be closer—perhaps even from that abundant arsenal of odd-ball components, the junk box. Figure 6-29 exemplifies a method of producing low voltages from two zener diodes. The output voltage of this regulating circuit is simply the difference in breakdown voltages between the two zener diodes. The technique is especially applicable when two temperature-compensated reference diodes are used. Also, the LVA zener diodes, because of their sharp knees and low current requirements, can develop low dynamic impedance at minimal power dissipation. You can, for instance, make a nominal 1.5-V regulated supply from LVA zener diodes (with breakdown voltages in the vicinity of 3.3 to 4.7 V) and operate it from a 6-V battery or other source.

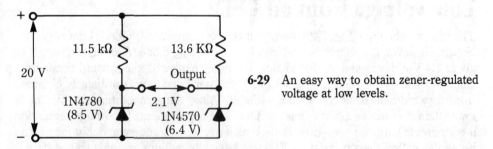

6-29 An easy way to obtain zener-regulated voltage at low levels.

When using the circuit of Fig. 6-29 with a rectifier-filter source of input voltage, make sure that there are no ground conflicts—either the rectifier-filter circuitry or the load should be "floating."

Low voltages from reverse-biased base-emitter diodes in transistors

Except where overdrive occurs, the reverse-bias characteristic of small-signal silicon transistors usually does not receive much attention. However, some of these devices display excellent "zener" characteristics (as low as approximately 3.5 V). Moreover, junk box transistors, which are otherwise defunct, will often perform well in this mode. A rapid way to evaluate a number of transistors for this quality is with an ac source. Then, with an oscilloscope, you can quickly discern the clipping levels as the base-emitter diode undergoes sequential forward and reverse conduction. Thus, those units with high reverse-breakdown voltages can be eliminated from consideration as low-voltage "zeners."

As shown in Fig. 6-30, no connection is made to the collector because ordinary transistor action is not involved. This scheme is a good low-voltage source for the experimenter. In factory production, difficulties could arise from the rather wide tolerances in the reverse breakdown characteristic. This problem could defeat the economics of what is otherwise a satisfactory low-voltage regulating technique for many applications.

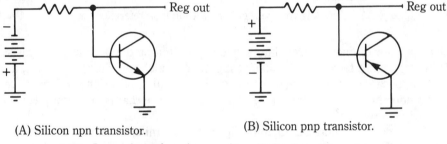

(A) Silicon npn transistor. (B) Silicon pnp transistor.

6-30 Low voltage from base-emitter diode of small transistors.

The ultralow voltage regulator shown in Fig. 6-31 has a unique voltage-reference source. The diode-connected transistor (2N797) is a germanium type, whereas the 2N2222 is silicon. The reference voltage that is developed by the pair is the differential energy-gap voltage between the two materials (0.5 V). The LM111 is no ordinary operational amplifier, but it is a voltage comparator that is specially designed to have extremely low input current and to provide output down to zero volts when powered from a single 5-V supply. With the circuit configuration of Fig. 6-31, the minimum adjustable output is approximately 500 mV. The temperature coefficient of the overall regulator circuit is very low.

Miscellaneous applications of linear IC regulators

The linear IC regulators that have been covered thus far were selected to provide reasonable coverage of many application areas. Unfortunately, a really comprehen-

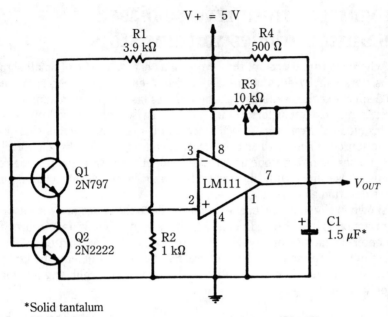

*Solid tantalum

6-31 Ultra-low voltage regulator with unique reference source.

sive overview of linear IC regulators cannot be attempted in a treatise of this size. However, before terminating this investigation of such regulating schemes, a few more applications are covered. Although these final applications will be more or less randomly selected, each will contain some feature that is new, unique, or particularly interesting.

The negative voltage regulator (Fig. 6-32) is similar to the positive regulator shown in Fig. 6-12H. The interesting feature here is the combination of the LM104, an early IC which has survived because of its high performance, and the more recent LM195 monolithic power transistor. Although the LM104 is the negative version that corresponds to the LM105 positive regulator, the scheme illustrated in Fig. 6-32 is not a "mirror image" of the circuit shown in Fig. 6-12H. Functionally, however, the arrangement of Fig. 6-32 is a true negative voltage regulator. No special grounding techniques need be followed, as is the case when a positive regulator is used to provide negative output voltage.

The special feature of the 0- to 250-V regulator (Fig. 6-33) is that no technique for shifting voltage level is required for the circuit. This feature is feasible because the MC1466 and MC1566 ICs are designed to function as floating regulators (for an example of the conventional approach to high-voltage regulation, refer to the circuit in Fig. 6-15B, where voltage-level shifting is accomplished with a 1N982 zener diode). In Fig. 6-33, the auxiliary 25-V source must float with the IC regulator and must be electrically isolated from the 275-V unregulated supply. However, the unregulated supply and the regulated output voltage use the same grounding system. The circuit has an output-current rating of 0.1 A.

Miscellaneous applications of linear IC regulators 363

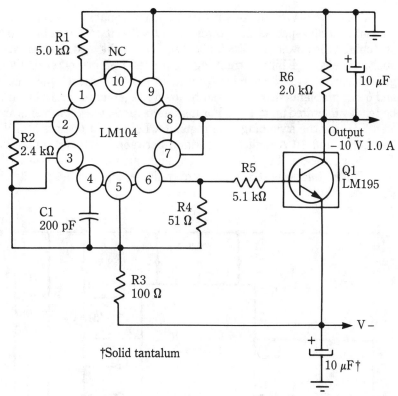

6-32 A negative voltage regulator with the LM195 monolithic transistor.

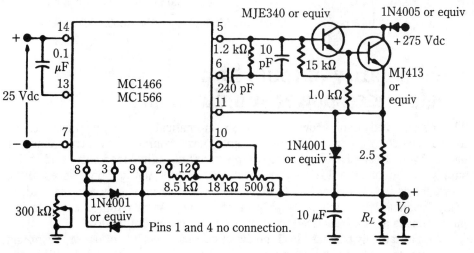

6-33 A 0- to 250-V 1-A regulator that uses the MC1466 and MC1566 ICs.
Motorola Semiconductor Products, Inc.

A negative 5-V regulator with 10-A current capability is depicted in Fig. 6-34. The interesting aspect of this power supply is that it performs in an area where the switching-type power supplies (to be discussed in the next chapter) are used. Obviously, a low-voltage high-current regulator must become increasingly inefficient as greater currents are demanded from it. In most practical situations, the thermal and other problems that are posed by low-voltage high-current linear supplies can be readily tolerated up to 10 A. For higher currents, these problems are best alleviated by using the switching-type regulator. Loads that operate at 5 V and require between 1 and 10 A invite competition between the two regulating techniques. The arrangement in Fig. 6-34 is a good candidate for those who favor linear regulation. This regulator is straightforward, reliable, and, because of its current limiting, safe.

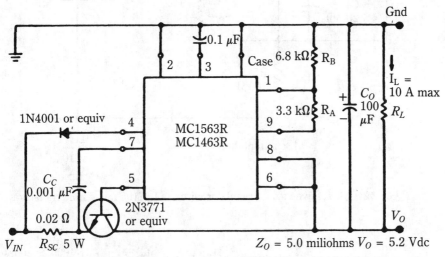

6-34 A 5-V 10-A regulator that uses the MC1463R and MC1563R ICs.
Motorola Semiconductor Products, Inc.

Dual-polarity outputs from single-ended power supplies

The self-powered circuit shown in Fig. 6-35 automatically maintains the center tap for loads (such as op amps) that require split power sources. This arrangement is useful if you already have a regulated supply with a single-ended output. If such a supply can be made to float, with respect to the load, then it is only necessary to insert the active divider of Fig. 6-35 between the floating supply and the load. In many instances, the regulated supply satisfies the floating requirement if its output terminals are disassociated from the ground terminal of the load.

With a statically balanced load, no direct current flows into or out of the ground (center-tap) terminal. Such a condition corresponds to equal current demands by the loads, but it will not be the usual dynamic operating mode. For example, if the load consists of a number of op amps that perform as class-A amplifiers, the signal-rate

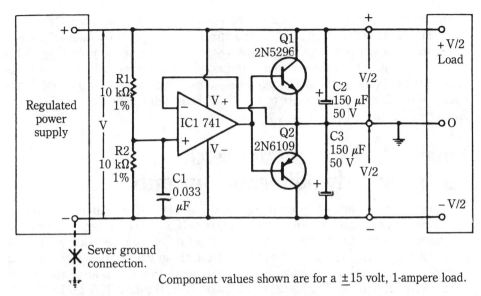

Component values shown are for a ±15 volt, 1-ampere load.

6-35 Conversion of single-ended supply to dual-polarity output voltage.

will not be balanced—even though balance prevails on the average, that is, statically. In any event, whether imbalance of load current is considered for static or dynamic conditions, such imbalance will destroy the equality of the two load voltages. However, before any appreciable deviation can occur, the feedback loop of the circuit will restore the equality by appropriately varying the conduction of output transistors Q1 and Q2. Indeed, these transistors compose the shunt elements for the positive and negative regulator circuits. The error signals are not in conflict just because the two shunt transistors are controlled from a single command; only one of these transistors is ever active at any given time.

Accurate voltage halving will prevail not only for wide variations of the two loads, but also over a wide range of input voltages from the regulated supply. In principle, a "dead zone" will exist because of the base-emitter potentials of the transistors. In practice, the high open-loop gain of the op amp renders this zone negligibly narrow, if not undetectable. The accuracy in voltage splitting depends on the two resistances, R1 and R2, which establish the reference voltage. Transistors Q1 and Q2 should have dissipation ratings of $1/2$ VI, where V is the single-ended voltage that is obtained from the regulated supply and I is the worst-case difference in load current. This will usually correspond to the situation where one load is maximally ON and the other load is OFF.

If so desired, resistors R1 and R2 can have other than the indicated one-to-one relationship. This difference will enable dual-polarity output voltages with unequal magnitude to be obtained. As with center-tap operation, such output voltages will maintain their relationship over a wide range of operating conditions.

The active divider (in conjunction with a regulator) provides a unique operational mode for split-polarity loads. Although a dual-polarity supply (Fig. 6-17B) independently regulates each of its two output voltages, it is under no compulsion to preserve the equality of the output voltages if one of them should deviate or drift from the effects of load variation, temperature, or aging of their internal voltage references. On the other hand, a tracking-type dual-voltage supply (Fig. 6-10) maintains the equality of the output voltages, but is under no compulsion to preserve the total value. For many purposes, these three modes will prove interchangeable. However, awareness of the differences will enable optimum specific load requirements.

Dual-polarity regulator with independently adjustable outputs

The dual-polarity regulator shown in Fig. 6-36 is similar to the arrangement of Fig. 6-17B in that both systems utilize three-terminal ICs. However, the dual-polarity regulating scheme of Fig. 6-36 has the added feature of independently adjustable output voltages. This feature is feasible because these ICs are designed for such operation, whereas the IC regulators that are used in Fig. 6-17B are intended for fixed-voltage operation. The LM317 is a member of the LM117/LM217/LM317 family of adjustable three-terminal IC voltage regulators and has been described in reference to the circuit arrangements of Figs. 6-18 and 6-20. The LM137/LM237/LM337 family are more recent developments and are negative, rather than positive, regulators. In other respects, they are "mirror images" of the earlier positive regulators. Thus, if you view the LM317 as an elegant npn transistor, then the LM337 qualifies as a matched pnp transistor. In any event, much better results are

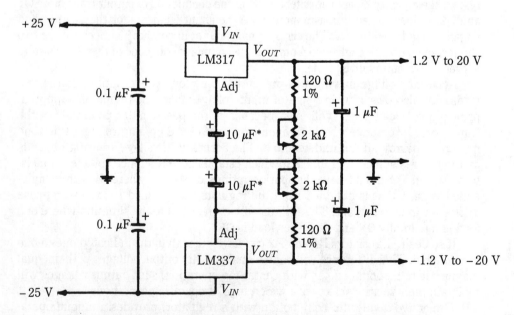

*The 10-μF capacitors are optional to improve ripple rejection

6-36 Dual-polarity regulator with independently adjustable outputs.

forthcoming from these ICs than can be attained from fixed-voltage three-terminal ICs, which have sometimes been forced to provide adjustable outputs. The salient characteristic of the adjustable ICs is that the current required at their "base" or ADJ terminal is very small.

Automobile voltage regulator with dedicated IC

The solid-state version of the electromechanical voltage regulator for automotive use has had extraordinary success. When properly designed and implemented, the inherent maintenance problem that pertains to the older type is completely overcome. Several solid-state automobile voltage regulators are described in chapter 1. Inevitably, a specialized IC module (such as the Motorola MC3325 voltage regulator). It is intended for use in conjunction with an external power stage and a dozen or so passive components, mostly resistors.

The schematic circuit of an automobile voltage regulator that uses the MC3325 is shown in Fig. 6-37. The pin-out of the MC3325 14-pin dual-in-line package is shown in Fig. 6-38. There, you can see three options for the battery sense connection. These options enable you to select three temperature coefficients. The nominal temperature coefficients will be -9 mV per degree C, -11

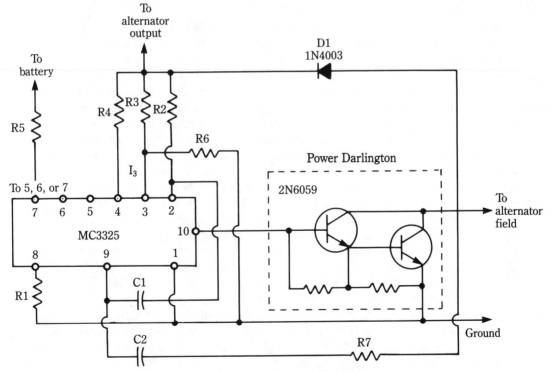

6-37 Automobile voltage regulator that uses a dedicated IC. Motorola Semiconductor Products, Inc.

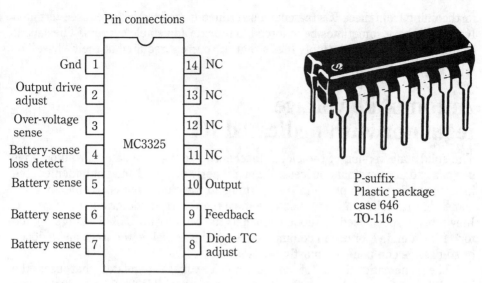

6-38 Pin-out for the MC3325 automobile voltage-regulator IC. Motorola Semiconductor Products, Inc.

mV per degree C, and −13 mV per degree C for pin 5, pin 6, and pin 7, respectively. Control of the temperature coefficient is also provided by R1, but it is the manufacturer's intention that the current into pin 8 should be within the range of 0.5 and 1.0 mA.

The nominal values of the associated components and their basic functions are depicted in Table 6-1. Experimentation will probably be required with regard to the best value for R5 in a particular application. This resistor governs the value of the regulated voltage that is maintained across the battery. On the other hand, resistors R3 or R6 can be changed in order to vary the battery voltage level at which the regulator cuts out. The idea is to supply field current (and, therefore, battery charging current) only when the battery terminal voltage drops below a predetermined level. The amount of charging current that is delivered to the battery could be controlled by resistor R2, which controls drive current to the power Darlington transistor. This, however, is not the manufacturer's intention.

The power stage is intended to operate as a switch; when the regulator senses a lower-than-predetermined battery voltage level, fully available charging current is delivered. Once the battery voltage builds up to this predetermined value, charging current is turned off. Accordingly, R_2 should be low enough that the Darlington transistor is driven into saturation whenever the regulator turns on. Because the current gain of a Darlington power transistor can vary widely from unit to unit, sufficient drive must be available from the regulator IC. The collector-emitter saturation voltage of the 2N6059 Darlington power transistor is in the neighborhood of 2 to 5 V. No "rheostat" field-current control occurs. High rates of charging current result from a relatively large difference in voltage between the alternator output and the battery. As the battery charges, its terminal voltage increases, which reduces the current that it can accept from the alternator. However, the 2N6059 always remains saturated.

Table 6-1. Parts list for automobile voltage regulator that uses a dedicated IC.

Part	Value	Circuit Function
R1	13 kΩ	Limits current into pin 5, 6, or 7 to vicinity of 0.5 to 1.0 mA
		Also controls temperature coefficient.
R2	1.2 kΩ	Controls drive current to output Darlington transistor.
R3	1.5 kΩ	In conjunction with R6, determines the overvoltage sense level.
R4	2.2 kΩ	Enables regulator to shut down in event of loss of battery sense.
R5	3.3 kΩ	Determines the level of output regulated voltage.
R6	1.5 kΩ	In conjunction with R3, determines the overvoltage sense level.
R7	3.3 kΩ	In conjunction with C2, inhibits self-oscillation.
C1	0.01 μF	Helps promote dynamic stability.
C2	0.01 μF	In conjunction with R7, inhibits self-oscillation.
D1	1N4003	Protects output Darlington transistor from inductive transients.
Power Darlington	2N6059	Delivers alternator field-current on an on-off basis.

All resistors 5% 1/2 watt composition.

Courtesy Motorola Semiconductor Products, Inc.

Figure 6-39 shows the required field-winding connections. Some alternators already are connected this way. Such an alternator is said to have a "pulled-up" field. In some alternators, two noncommitted field-winding connections are available on the alternator housing. In such a *floating-field alternator*, it is easy enough to make

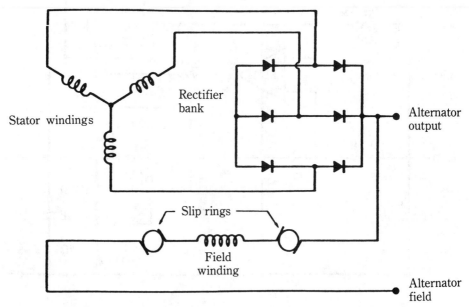

6-39 Required field-winding connections for the regulator of Fig. 6-37.

370 *Linear regulating supplies that use integrated circuits*

external connections to comply with the required circuit of Fig. 6-39. In a third group of alternators, one field-winding connection is internally grounded. In some cases, it is feasible to modify the situation so that the field winding can be connected (Fig. 6-39). On other "grounded-field" alternators, the necessary modification cannot be readily made and the regulator circuit of Fig. 6-37 cannot be used.

The power Darlington transistor should be heatsinked so that the metal box that houses the overall regulator can transfer heat to the surrounding air. This is necessary because of the several amperes of field-winding current and the relatively large saturation voltage of Darlington power transistors. The resulting 8 or 10 W of power dissipation could develop a high temperature in a confined compartment. This heat, in turn, could act on the temperature coefficient of the IC in an unintended way.

Paralleling regulator ICs

To extend the power handling capability of three-terminal adjustable regulators, it is only natural to think of paralleling them. When this is attempted by merely connecting together similar terminals, the scheme generally cannot be optimally exploited because of unequal current division. If you have a number of such ICs, it is possible to obtain acceptable results experimental selection for near-identical units. However, a better technique (Fig. 6-40) is where two LM350 regulator ICs can be combined to produce a total load current of 6 A, with 3 A shared by each

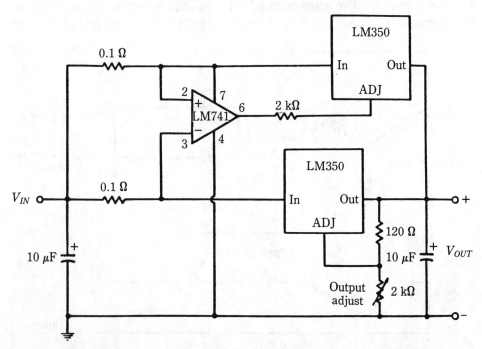

6-40 Paralleling technique for 3-terminal adjustable regulator ICs.

LM350. Indeed, such operation is forthcoming—even if the two LM350s are at the opposite extremes of manufacturers' tolerances.

In Fig. 6-40, the lower LM350 might be considered as the master regulator because its ADJ terminal determines the output voltage of the arrangement. The upper LM350 is a "slave," being forced to supply one-half of the load current. Such current division is quite accurate, being determined primarily by the matching of the two 0.1-Ω resistances.

Additional LM350s can be similarly connected to the output of the LM741 op amp through their individual 2-kΩ resistances. Modification of one of the current-sense resistances is then required, however. The basic idea (Fig. 6-41) is that three LM350s equally share the load current. Notice that additional 0.1-Ω resistances are incorporated as current ballasts to help maintain equal current division between the slave LM350s. This design approach is powerful, because a respectable current capability can be readily attained with minimal parts count. With the LM350s, a tenfold adjustable voltage range is feasible—from 3 to 30 V. To keep down internal power dissipation, however, the input voltage should be as low as

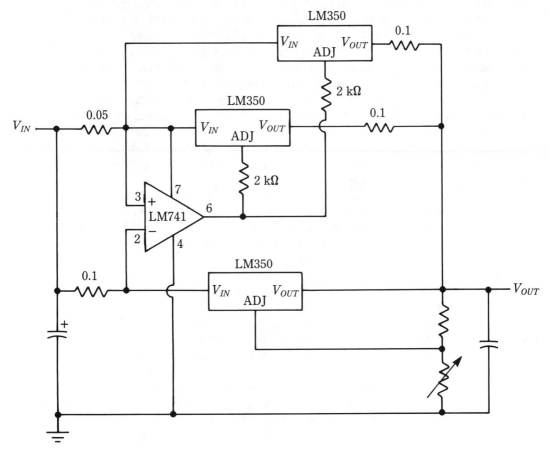

6-41 Method of adding additional current-sharing slave ICs.

possible. This, together with effective heatsinking, can result in a voltage regulated supply that is appreciably simpler and considerably less costly than "conventional" designs. Also, the protective circuitry and the voltage-reference source are already embodied within the LM350s.

Distributed power systems

In chapter 5, *point of application* was discussed as an alternative to one large central power supply for distributing dc power throughout a system. Point-of-application distribution simply uses a number of small power supplies to provide localized dc power to the individual system units. This distribution scheme has advantages of cost and efficiency for systems that draw overall currents of up to several tens of amperes. For larger systems, where total current demand is on the order of several tens to several hundreds of amperes, a new distribution method has been gaining favor.

The so-called "point-of-application" method of distribution generally used individual regulated dc supplies and these often were configured around the three-terminal linear voltage regulator. The newer larger-system power distribution scheme is somewhat similar in basic concept, but the individual dc sources are dc/dc converters that operate as switch-mode supplies. Thus, in a 5-V, 100-A system, actual power distribution might be via a 50-V 10-A dc distribution cable. Figure 6-42 illustrates an example of a high-efficiency power distribution system.

From a practical standpoint, it is much easier to implement 10-A cable than it is to provide wire that is capable of carrying 100-A in the previous example. Also, it is less costly and much easier to limit voltage drop and power dissipation in the distribution system. Individual dc/dc converters provide high isolation for the various subsystem units; this is not the case with the point-of-application scheme, in which ICs (such as three-terminal regulators are used).

Some might say that a 500-W centralized power source would still be needed. This is true. However, the performance demands on this main supply are extremely lax, because it is to deliver 50-V at 10-A, rather than the more difficult 5 V at 100 A. Moreover, it can be a simple brute-force supply that does not even have to be regulated. Because the individual dc/dc converters are switch-mode circuits that can operate at high frequencies and tolerate wide changes in dc-voltage input, the overall system can be implemented with low weight and space penalties, and still operate efficiently and reliably.

Distributed power systems have been recognized as the architecture of the future by semiconductor firms, some of whom have taken the lead by marketing designated-control ICs for the dc-dc converters that are used in such systems. Not only does this building-block approach ease the designer's problems, but it confers compelling benefits to the user. A system predicated on such a distribution technique is likely to attain higher overall efficiency and better reliability than one that is thrown together from the more-ordinary power supplies.

The Siliconix Si110 high-voltage switch-mode controller can be used to implement commonly used switcher topologies (such as flyback and forward circuits)

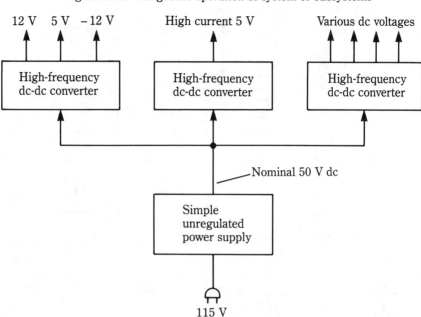

6-42 Example of high-efficiency dc power distribution system. Relatively small-gauge distribution cabling is required and the exact voltage of distributed dc power is not important.

and can accommodate switching-rates up to 1 MHz. This fact alone, resolves much of the size, weight, and cost objections of using ordinary 20- to 40-kHz power supplies or dc-dc converters in the distributive supply-system. The IC is designed to require minimal external components. For example, the switching rate is set by a single external resistance, and an internal start-up circuit is incorporated. The overall operating efficiency (as a dc-dc converter) can be as high as 80 to 84% in typical applications. This efficiency stems largely from the fast speed and low dissipation of the CMOS control circuitry of the Si9110, and also that its input power does not need to be delivered at low-voltage and high current—it can accommodate up to 120 V of dc input (where the input dc is accompanied by transients, 50 V or so would be a safer level and would allow "headroom"). Aside from transients, the IC will operate efficiently throughout the 10- to 120-V range of dc input.

The block diagram of the Si9110 is shown in Fig. 6-43. Both current-mode and voltage-mode control are featured. The current-mode sensing technique makes possible tight line regulation and fast response to line-voltage variations. The frequency of the oscillator, which is twice that of the output switching rate, can be easily synchronized to an external rate by pulse injection at pin 8. Synchronization occurs when the injected pulse rate is 10 to 20% higher than the natural operating frequency of the internal oscillator. Synchronization is often a good idea in systems

374 *Linear regulating supplies that use integrated circuits*

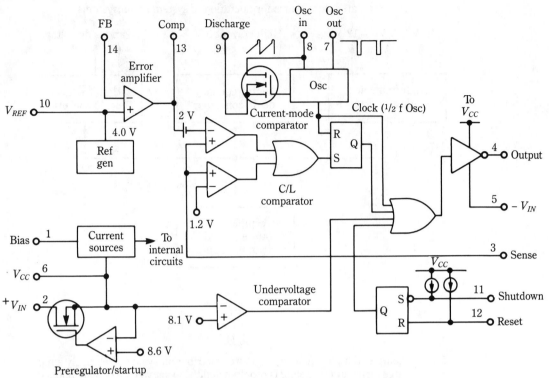

6-43 Block diagram of the Si9110 switchmode regulator. This control IC is well suited for use in the individual power supplies of a distributed-power system. Siliconix, Inc.

that use multiple power supplies so that the levels of noise on the various dc outputs are reduced. Multiple switch-mode supplies with synchronized switching rates produce minimal intermodulation products and simplify filtering.

The dc-dc converter of Fig. 6-44 delivers 5 W output and operates at a switching rate of 100 kHz, although its internal oscillator produces twice this frequency (200 kHz). As shown, it is intended to work from an output dc source of 25 to 50 V (it could be made to accommodate dc input voltages as low as 10 V). The maximum duty-cycle of its switching waveform is 50%. As shown, complete galvanic isolation is between its 5-V outputs and the dc-input source. Both voltage-mode and current-mode feedback are used to obtain fast transient response and close line regulation. As should be obvious, the parts count is inordinately low and the circuitry is quite simple. Not readily apparent from initial inspection is the incorporation of start-up circuitry in the control IC. As previously mentioned, this control IC is implemented with MOS technology to achieve high-frequency capability and very low quiescent power dissipation. The IC also has provisions for shutdown, reset, and overload protection.

The salient feature of the converter in Fig. 6-44 is its suitability for use in distributed power systems. This stems primarily from the ease with which this 5-W

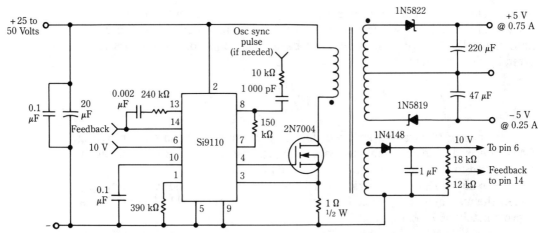

6-44 Experimental dc-dc converter for use in a distributed power system. This is a 5-W 100-kHz circuit. Basically, the same topography can be implemented for operation to 50 W and 500 kHz. Siliconix, Inc.

design can be scaled up to power levels as high as 50 W. This upscaling can be accomplished via the following modifications:

1. An appropriate power MOSFET is substituted for the 2N7004 shown.
2. The 1-Ω current-sensing resistance is reduced in inverse proportion to the higher power level that is desired. Thus, this resistance would become 0.5 Ω for a 10-W design, 0.25 Ω for a 25-W converter, etc. The power rating of this resistance varies directly with the power capability of the converter. This rating implies a 1-W rating at 10-W, a 2-W rating at 20 W, etc.
3. The rectifier diodes and filter capacitors of the dc output section must be changed to accommodate the higher currents involved. Because of the somewhat unpredictable effect of internal impedance on ripple voltage, the sizing of the larger filter capacitors is best left to experimental determination. Acceptable output ripple voltage at full load is the criterion for qualifying the filter capacitors.
4. The transformer must be changed to accommodate the new conditions that are imposed by higher-power operation. Be concerned only with the inductance of the primary winding, $L_{p!}$, because this portion of the transformer is most intimately associated with the basic operation of the flyback converter. The other transformer-design details, such as core selection, secondary turns, insulation, etc. are amenable to ordinary transformer theory. Techniques that are known to promote low leakage-inductance (including toroidal cores and bilateral or trilateral winding-formats) should be used. The Ferroxcube Corp. markets ferrite cores of various sizes that are dedicated to the needs of inverters, converters and switch-mode power supplies.

Three quantities must be known in order to calculate the all-important primary inductance, L_p. These three quantities are the minimum input voltages that are expected for the converter's operation V_{IN}, the peak current through the primary (and through the semiconductor switching device), and the maximum ON time of the PWM switching wave. The basic connection of these quantities is given by the equation:

$$L_p = \frac{V_{IN} T_{ON}}{I_{pk}}$$

For sake of illustration, suppose that a 20-W version of the converter (Fig. 6-44) is desired. The new converter will, again, operate at 100 kHz, and will, again, be expected to accommodate a 25- to 50-V dc input source.

V_{IN} is 25 V. T_{ON} is one period of a 100-kHz wave times the maximum duty-cycle. The maximum duty-cycle is given by the maker of the Si9110 control IC as approximately 50%. Accordingly,

$$T_{ON} = \frac{1}{1 \times 10^5} (50\%) = 5 \times 10^{-6} \text{ s} = 5 \text{ }\mu\text{s}.$$

Now, find the value of the peak primary current, I_{pk}. First, find the average current, I_{av}, as follows: assume an overall operating efficiency of the converter of 80%. Then, the input power, P_{IN}, of the 20-W converter will be $20 \div 0.8 = 25$ W. Knowing P_{IN}:

$$I_{av} = \frac{P_{IN}}{V_{IN}}$$

where again V_{IN} is the worst-conditions input voltage (25 V). Therefore, $I_{av} = 25$ W $\div$ 25 V = 1.0 A.

With one more step, I_{av} can be converted into I_{pk} for use in the original equation that was set down in our quest to find the primary inductance, L_P.

You can obtain I_{pk} from the relationship:

$$I_{pk} = \frac{(2)I_{av}}{D}$$

where D is the maximum duty-cycle (at low input voltage and at full-load output). Then,

$$I_{pk} = \frac{(2)(1)}{0.5}$$

Having now solved for V_{IN}, T_{ON}, and I_{pk}, you can finally obtain L_p for your 20-W converter:

$$L_p = \frac{V_{IN} T_{ON}}{I_{pk}}$$

Be aware of the possible penalties for deviating from the design-value of L_p. The design value of L_p is certainly not as critical as the calculated inductance for a resonant circuit. However, too small a value for L_p will drive up the peak current I_{pk}. This will engender higher losses and might endanger the power switching device. Another adverse effect could be driving the transformer core into its saturation region, which, in turn, would result in even-higher peak current.

Too high a value of L_p will increase copper losses in the transformer, which will increase leakage inductance. Leakage inductance not only causes voltage spikes at the power switch, but it degrades the cross-regulation between output windings. It can upset the converter's ability to properly regulate, because the feedback path receives false information about the sensed output voltage. A flyback converter with excessive primary inductance in the transformer will generate excessive interference because of the voltage and current spikes. Yet another possible effect of too-much primary inductance can be a failure to respond to the duty-cycle variations that are imposed by the PWM control unit.

Summarizing, distributed power systems benefit from the use of high-frequency dc-to-dc converters, not only because of their superiority in efficiency over linear regulators, but also because of their unique ability to perform as dc transformers. This feature enables the use of small-gauge interconnecting cables with negligible penalty of distribution losses.

A unique backup source of dc power

In the past, a storage battery, such as the lead-acid type that is used in automobiles has been likened to a giant capacitor of many millions of farads. This comparison is valid in the sense that the two devices display many common features as energy sources. In particular, both can be electrochemical devices and they share similar, if not identical charge-discharge characteristics. Whereas the use of various types of batteries as primary or backup sources of dc power for electronic systems has a long history, it has not been so with capacitors.

Although the concept of a capacitor as a temporary source of dc power rests on firm theoretical ground, practical implementation has long eluded designers. Ordinary capacitors tend to be too bulky, too leaky, too expensive, and generally, a combination of these drawbacks. Recent developments have overcome these objections for certain applications and have even given capacitors the edge over battery backups in many instances. The need for backup power is especially urgent in computer and logic systems that use volatile memory where power failure wipes out the stored data. What is often desired is a backup source that is capable of maintaining the integrity of volatile memory for periods of hours or days. Even though much progress has decreased the current consumption of volatile logic, a very large capacitor is needed; it was previously considered impractical.

Capacitors of one or more *farads* have become available in small enough packages to present virtually no problem as an add-on to a PC board. Although these capacitors are electrochemical in nature, they cannot be properly defined as electrolytic types. Unlike electrolytic and other types of capacitors, this newer type, sometimes called an *energy-storage device (ESD)*, has no dielectric! Capacitance is achieved via a layer of opposite charges that are formed at the interface of a carbon electrode and a solid-state *electrolyte*. It is reminiscent of a pn-junction, except that the opposing charges are completely mobile in this device. Thus, transportable ions (as in a battery) and a double-layer of charges (as in pn semiconductor devices) are encountered (pn-junction ions cannot move to the electrodes).

In addition to the intrinsically high capacitance, this device displays an internal leakage-current order of magnitude that is below that of electrolytic capacitors. Thus, the charge retention or "shelf-life" makes it appear as a battery for practical purposes. Its voltage rating is seldom above 6.5-V, but this imposes no limitation on 5-V logic systems.

Practical insight into the energy-storage capability of these capacitors can be gleaned from considering a 1-F unit. Let one of these be charged to 5-V. This capacitor can then deliver power to a 100-kΩ load for 25 hours before dropping below 3 V. Alternately, it can provide power to a 1-MΩ load under the same constraints for 10 days. Such loads simulate microelectronic circuits, low-dissipation logic, watches, liquid-crystal displays, and other useful devices. Indeed, application areas where these capacitors can compete not only with batteries, but with electronic power supplies exist. Almost any voltage-limited charging circuit can be used and the number of charged/discharged cycles is essentially unlimited. It appears, too, that ordinarily encountered rates of charge or discharge are not likely to affect the life-span. You must, however, observe this device is polarized, like an electrolytic capacitor or a battery.

A typical 1-F unit can be contained in a 1-cubic-inch package. The volumetric efficiency is, accordingly, an astounding 5,000,000-μF V per cubic inch, based on a 5-V working level. The rated working voltage is actually 5.5 V, with an allowable surge to 6.8 V.

It is only natural to ponder the use of such a capacitor to attenuate the noise and ripple on 5-V switch-mode power supplies. However, the ESR, though not given in data that is ordinarily supplied, is too high for such a filter application. You must be content with the several novel features of this minuscule power source and not demand too many exotic ratings.

An interesting application, which is realizable with available technology, is the association of ESD modules with micropower linear-regulating supplies. Such regulators use MOS elements and are specifically designed to minimize quiescent current and to optimize efficiency. From such a combination, you could obtain constant voltage or current with an acceptably small penalty on the duration of the regulated output.

7
Switching-type regulators that use integrated circuits

VACUUM TUBES PERFORMED VERY WELL IN DISSIPATION-TYPE LINEAR regulators—they were (and still are) used in both series-pass and shunt regulators. However, these tubes are rarely encountered switching-type supplies. A basic reason for this is that vacuum tubes are generally a far departure from an ideal switch.

Transistors provide a much closer approach to ideal switching action—infinite OFF resistance and zero ON resistance. Thus, recent years have had an accelerated development of switching-type power supplies. Steadily, the goals that were promised by the theory of switching have become attainable. Essentially, these objectives are high efficiency, small physical size, and a dramatic relaxation of thermal problems. Such upgraded performance was not realizable overnight because the switching element not only has to turn on and off as a switch, but ideally it must do so in zero time. The early power transistors simply were not fast enough to avoid serious dissipation regions when undergoing their switching transitions. Such a situation partially cancelled the calculated advantages of accomplishing regulation via switching.

Even after suitable switching transistors became available and significant improvements were made in other components (such as core materials, capacitors, and diodes), a problem still remained in the actual implementation of switching supplies. This problem involved stability and noise. Because very high frequencies are involved in switching transitions, a switching supply should be constructed more like a VHF tuner than an audio amplifier. Otherwise, you can expect both internal and external problems to arise from radiation and stray coupling. Unfortunately, the physical layout of the control circuitry of a switching supply was inevitably the weak link when discrete transistor stages were used to control the duty-cycle or repetition rate of the switching transistor.

ICs for switching regulators

The preceding discussion naturally brings you to *integrated circuits*. All of the attributes conferred on linear regulators by ICs also apply to switching regulators. Actually, because of the high frequencies that are involved in the switching process, the benefits provided are even more profound. Even if no other IC than the op amp was available, the switching-type power supply would still be practical. Actually, a number of specialized ("dedicated") IC modules now on the market promise additional impetus in the proliferation of switching power supplies. It is now relatively easy to construct a reliable high-performance switching regulator with ICs. Figure 7-1 shows the trend of the packaging power density as a function of the switching rate for medium-power outputs. At zero-frequency (a linear power supply), the power density would be on the order of 0.3 to 0.6 W per cubic inch.

A surprising aspect of the application of ICs to switching supplies is that very good results can be obtained by simply using linear IC regulators. Thus, popular IC regulators (such as the LM105, the 723 types, and the Motorola and RCA ICs that were initially intended for linear service) all function well in switching circuits.

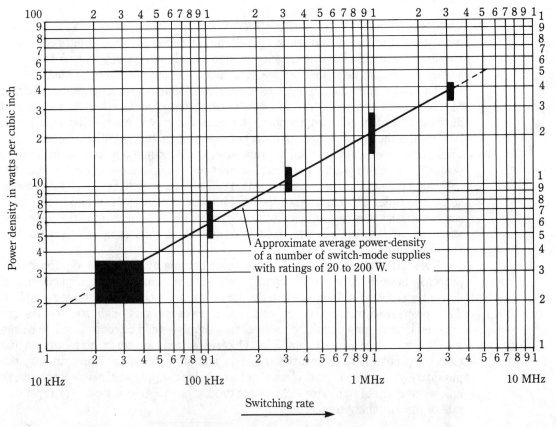

7-1 Power density vs. switching rate.

Sometimes the circuit modifications are quite benign. Three components are generally required to convert a linear regulator to a switching type: an inductor, a capacitor, and a freewheeling diode. Otherwise, the configuration of the switching regulator remains quite similar to that of the linear circuit. Such switching regulators are generally self-oscillating and tend to contradict the notion that switching supplies are inherently complex circuitries.

In addition to the simple self-oscillating regulators are more-sophisticated systems. Some of these systems use the specialized ICs that were previously mentioned. In any event, the phobias regarding RFI and EMI have lost much of their validity. The reduction of electrical noise is more easily manageable when the devices are physically small and interconnections are minimal in length and systematic in routing. Although the residual noise in a switching regulator still exceeds that of a good linear power supply, this factor is often of little practical significance. For example, if the switching supply is used for a digital load, as is so often the case, the prime source of "switching noise" can easily be the load, not the supply.

The IC switching-type regulator

When the load requires high current and system conditions are such that the difference between the dc voltage that is applied to the regulator and to the load is also high, the linear mode of regulator operation becomes very inefficient because an inordinate amount of power is then dissipated in the pass element itself. Under these operating conditions, the switching-type regulator can produce excellent results (the efficiency is often in the 80% to 90% range). This is because the "pass" transistor (now more appropriately called the *switching transistor*) is either hard-driven ON or is in an OFF state of conduction; neither state ideally produces I^2R loss. Practically, the ON resistance is not zero, the OFF resistance is not infinite, and the transition between the two states is not instantaneous. Some power loss is thereby incurred. This loss, however, can be a small fraction of the loss that might attend linear operation under the same conditions of applied dc voltage and of load voltage and current.

Circuit details and operation

Consider, for a moment, the situation that is depicted in Fig. 7-2A. For simplicity, suppose that Q1 is an ideal element and that the rectangular wave that is impressed at its base causes switching of the applied dc voltage, V_{IN} with no power loss in Q1. Inductor L and capacitor C make up a low-pass filter. The output voltage, V_{OUT}, that is derived from this filter will be essentially constant, being equal to the average value of the switched waveform. That is, $V_{OUT} = V_{IN}\left(\dfrac{T_{ON}}{T}\right)$. Notice that V_{OUT} is independent of load current and that the value of V_{OUT} can be changed by varying the duty-cycle of the switching wave. Also, in order to enable this switching circuit to deliver current to a load, diode X1 must be included. During the time that Q1 is in its OFF state, the energy that has been stored in L and C

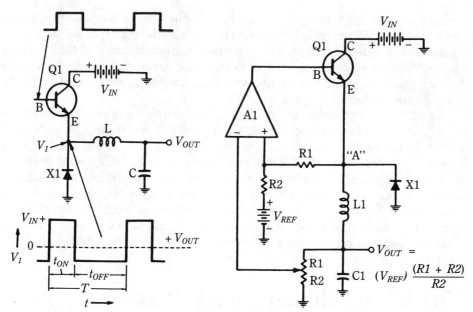

(A) Simple circuit with switch Q1, low-pass filter LC, and freewheeling diode X1.

(B) Circuit of (A) associated with operational amplifier to form self-oscillating system.

7-2 Basic concepts of the switching rate. RCA

during the previous ON state of Q1 becomes the source of load current. The inductor current must then have a return path, which is the function of the diode. This diode is often referred to as a *freewheeling diode*.

Such a circuit might be associated with an operational amplifier (Fig. 7-2B) to produce the basic switching-type regulator. The base of Q1 no longer needs to be driven by an external-wave generator because the entire system now becomes a *self-excited oscillator* (positive feedback results from a small sample of the voltage at A, which is applied to the noninverting terminal of the operational amplifier). Additionally, there is simultaneously dc negative feedback (just as in a linear regulator), which is caused by the sampling of the dc output voltage by the inverting terminal of the operational amplifier. The waveforms involved are shown in Fig. 7-3.

The net result of the simultaneous feedbacks is that the system oscillates at a repetition rate that is determined by L and C, as well as by other parameters of the system. As a result, the noninverting input of the operational amplifier is driven equally above and below the reference voltage. The circuit has an equilibrium in which the steady value of the supplied reference voltage and the sampled output voltage are equal. This system is not unlike the error-signal nulling that occurs in a regulator that operates in the linear mode. However, here the nulling action occurs as the result of variations in the duty-cycle of the switching wave. Thus, if the output voltage drops, the switching transistor will remain in its ON state longer in order to produce the correction that is necessary to bring the sampled output voltage back to equality with the steady value of the reference voltage. The converse mode occurs if the output voltage increases.

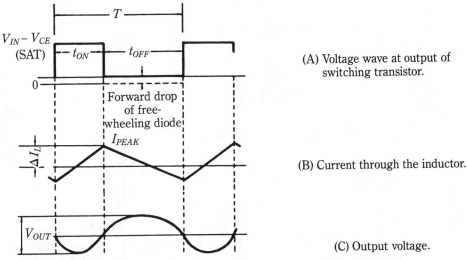

7-3 Waveforms in the switching circuit of Fig. 7-2.

Because the reference voltage, as seen by the noninverting input, is alternately increasing and decreasing, ripple at the switching frequency is superimposed on the dc output voltage. With a proper output filter and with high amplifier gain, this ripple can be reduced to generally acceptable levels. However, the salient feature of the switching regulator is neither extremely close regulation nor infinitesimal output ripple. Rather, it is high operating efficiency—even under the condition of high input voltage, high output current, and low output voltage (the very conditions that would necessitate an inordinately large series-pass transistor and a massive heatsink in a linear regulator). In order to change the dc output voltage of the switching regulator, the steady value of the reference voltage must be changed, or the sampling network, $\frac{R1}{R2}$, must be adjusted.

In Fig. 7-4, this principle is applied to the LM105 IC. The beauty of the technique is the ease with which the linear-oriented circuit is converted to a switching-type regulator.

Determination of the output-filter inductor and capacitor

To best utilize the unique characteristics of a monolithic module (such as the LM105) in the switching mode of regulator operation, it is necessary to use a "ballpark" approach in choosing the value of the output-filter components. This is because during the time that the transistor switch is in its OFF state, the load receives its current from the energy stored in the filter inductor and capacitor. Also, because of the self-excited nature of the switching oscillation, the switching frequency is largely determined by the size of these two reactances. Satisfactory performance can be obtained from selecting the inductor and capacitor by "cut-and-try" methods or if the values that you selected produced a "brute-force" filter. However, it is much better to start from an engineering estimate that is derived

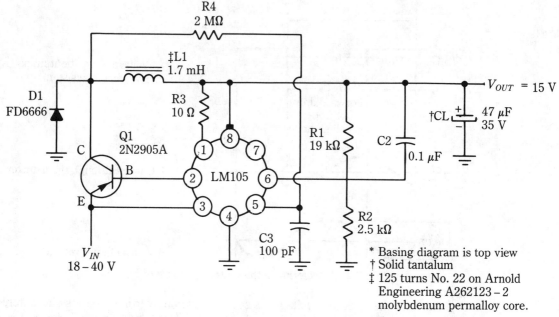

7-4 Connections for switching operation from the LM105 linear regulator.

from basic operating theory and then, if desirable, to resort to empirical techniques to arrive at final values. Otherwise, you are likely either to have unnecessary physical bulk in the filter or to generate dangerously high peak currents in the switching transistor and in the freewheeling diode.

First, you should decide the nominal frequency at which the switching is to occur. Whereas discrete-circuit switching regulators are often operated in the range from several hundred hertz to several kilohertz, the monolithic version can usually be operated at higher switching rates. For example, the LM105 is capable of high-efficiency performance at switching rates of 20 to 100 kHz. Other things being equal, these higher rates can be very advantageous, because the size, weight, and cost of the filter components can then be drastically reduced. Also, post-regulator LC filters can often be used to decrease ripple without appreciably degrading voltage regulation. However, at switching rates that are greater than about 5 kHz, serious attention must be given to the switching transistor and the freewheeling diode. Ordinary power transistors and rectifier diodes simply will not follow the higher switching rates, but their performance will be degraded by dissipating inordinately high power. Special transistors and diodes are available that store relatively little charge during forward conduction and that can be turned off very quickly in response to fast switching rates. In the ideal switching regulator, the switching transistor, and the freewheeling diode are either in their ON or OFF states and they consume zero time in the transition between the states.

After selecting a nominal switching frequency, f, the time that the switching transistor spends in its two conductive states is readily computed with the aid of the nominal dc input voltage, V_{IN}, and the desired dc output voltage, V_{OUT}. The formulas are as follows:

$$T_{ON} = \frac{V_{OUT}}{V_{IN}} \times \frac{1}{f}$$

$$T_{OFF} = \frac{1 - V_{OUT}}{V_{IN}} \times \frac{1}{f}$$

where:

T_{ON} is the time in seconds in the conducting state,
T_{OFF} is the time in seconds in the nonconducting state,
V_{IN} is input voltage in volts,
V_{OUT} is output voltage in volts,
f is frequency in hertz.

Next, an arbitrary decision must be made concerning how much the peak ripple current (ΔI_L) will exceed the maximum dc load current. Referring to Fig. 7-3, the ensuing considerations apply: zero ripple current requires an infinitely large inductor. On the other hand, as the inductor is made smaller, the ripple current through the inductor becomes progressively larger, which thereby forces the switching transistor and the freewheeling diode to carry peak currents beyond the steady maximum dc current that is required by the load, T_{OUT}. Peak currents in excess of 150% of I_{OUT} begin to impose difficult demands on the switching transistor and the freewheeling diode. Because these elements do spend time in their "linear" regions, high peak currents quickly degrade efficiency by causing high power losses during switching. This is particularly true at higher switching frequencies. It appears that many practical designs use ripple currents between 115% and 135% of I_{OUT}. A value of 125% is very reasonable and it is applicable to perhaps the majority of regulators. Remember that this number does not delineate the output ripple voltage; rather it is an intermediate value that must be used in calculation of the inductor, L, which can now be determined:

$$L = \frac{(V_{OUT})(T_{OFF})}{(2I_{OUT})(k-1)}$$

where:

L is in henrys,
V_{OUT} is in volts,
T_{OFF} is in seconds,
I_{OUT} is in amperes,
k is equal to peak

Inasmuch as we have decided to use $k = 125\%$, the formula for L becomes:

$$L = \frac{(V_{OUT})(T_{OFF})}{(2I_{OUT})(1.25-1)} = \frac{(V_{OUT})(T_{OFF})}{(0.5)(I_{OUT})}$$

The inductor should be wound on a core material in which magnetic saturation occurs gently, rather than abruptly. Molybdenum-permalloy toroidal cores are excellent for this purpose. Ferroxcube and other ferrite cores with or without air gaps can also yield excellent results. Ordinary silicon-steel laminated cores are satisfactory, but if the switching frequency is above a few hundred hertz, it is best to use audio-frequency, rather than power-frequency grade. Deltamax, Orthinal, Permenorm, and similar core materials with abrupt saturation characteristics are generally to be avoided because of the possibility of destructive transients that can be generated in inductors that are wound on such cores. With suitable air gaps, however, even these core materials often yield satisfactory performance. At very high switching rates, air cores are sometimes feasible.

Next, is the output capacitor. To calculate C, it is necessary to have some idea of the maximum peak-to-peak ripple, ΔV_{OUT}, that can be tolerated from the power supply. The characteristics of the circuitry or of the system to be powered by the switching regulator. If the maximum allowable output ripple is unknown, a value of 25 mV is a reasonable choice for many applications. Here, the low output ripple is not the salient feature of the switching-type regulator. For critical applications, where very nearly perfectly smooth dc is required, the linear-type regulator is better suited for the purpose. However, the output ripple voltage of the switching regulator can be reduced to satisfactory levels for many applications by appropriately choosing the switching frequency and the filter components. The output filter capacitor is calculated as follows:

$$C = \frac{(T_{ON})^2 (V_{IN} - V_{OUT})}{2L(\Delta V_{OUT})}$$

where:

C is in farads,
T_{ON} is in seconds and is obtained from the relationship,

$$T_{ON} = \left(\frac{V_{OUT}}{V_{IN}}\right)\left(\frac{1}{f}\right)$$

L is in henrys,
V_{OUT} is the postulated peak-to-peak output ripple voltage,
V_{IN} is the nominal voltage applied to the regulator from the unregulated supply,
V_{OUT} is the desired dc output voltage from the regulator.

The formulas and procedures that are outlined represent a practical method for determining the output filter for the self-excited switching regulator. A more rigorous mathematical attack would not only be laborious, but it would defeat its purpose because of parameters that are not easily specified or conveniently measured. You would need a precise knowledge of the rise and decay times of the voltage that is across the switching transistor, the voltage hysteresis of the comparator, the volt-ampere characteristics of the unregulated supply (V_{IN}), the actual inductor behavior (with respect to frequency and direct current), and the dissipation factor and high-frequency behavior of the nebulously specified electrolytic capacitor.

The practical validity of the described approach stems largely from the unimportance of the exact switching frequency. In a typical example, you might calculate an inductor of 17 mH and a capacitor of 120 µF. Because of capacitor availability, and because of the inherent economy of increasing inductance by the addition of a few turns (L increases as the square of the number of turns, N), a capacitor of 100 µF is selected. The inductor is increased to 20 mH. Because the LC product remains approximately the same, the switching frequency has not changed greatly due to the modification. Packaging problems are relaxed by using a single, readily available 100-µF capacitor. In all probability, core-window space is available for the few extra turns on the inductor (also, it should be easier to obtain a 20-mH, rather than a 17-mH inductor).

Switchers that use other popular ICs

Figures 7-5 and 7-6 depict self-oscillating switching regulators that are configured around Motorola and RCA linear regulator ICs. These schemes are very similar to the switching regulator that uses the National Semiconductor LM105 (Fig. 7-4). All three arrangements use pnp switching transistors. An alternative switching regulator, which uses an npn output transistor, is shown in Fig. 7-7. Here, transistors Q1 and Q2 compose a complementary-symmetry Darlington stage.

The 3-A switching regulator shown in Fig. 7-8 is designed around the LM317 three-terminal IC regulator. The overall arrangement is simple and inexpensive, yet it allows the regulated output voltage to be adjusted over the range of 1.8 to

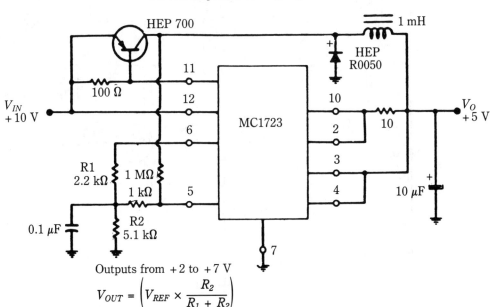

7-5 Switching regulator that is designed around the Motorola MC1723 IC.

388 *Switching-type regulators that use integrated circuits*

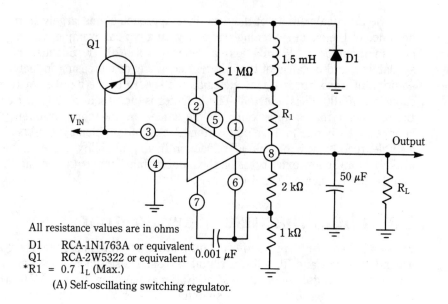

All resistance values are in ohms
D1   RCA-1N1763A or equivalent
Q1   RCA-2W5322 or equivalent
*R1 = 0.7 I_L (Max.)

(A) Self-oscillating switching regulator.

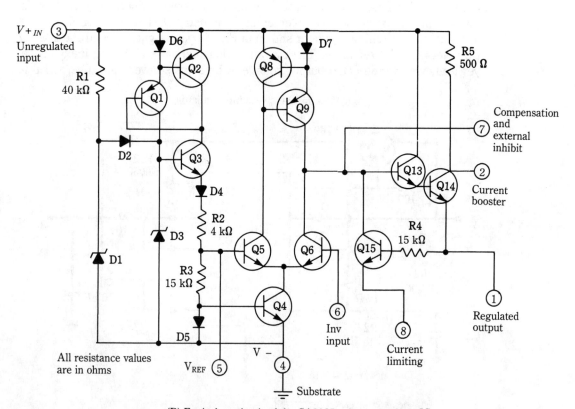

All resistance values are in ohms

(B) Equivalent circuit of the CA3085 voltage regulator IC.

7-6 Switching regulator that is designed around the RCA CA3085 IC.

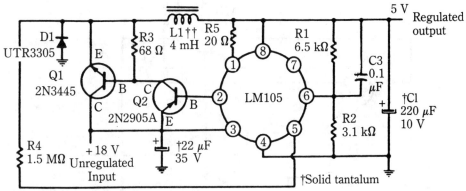

†† 160 turns no. 20 on Arnold Engineering
A-548127-2 molybdenum permalloy core
* Basing diagram is top view

7-7 A switching regulator that uses an npn transistor for current boost.

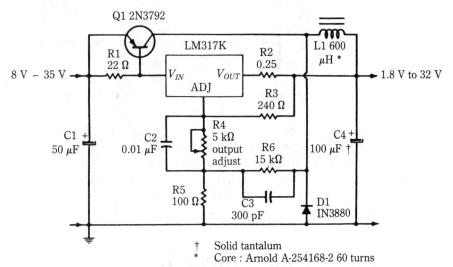

† Solid tantalum
* Core : Arnold A-254168-2 60 turns

7-8 A 3-A switching regulator that uses the LM137 3-terminal IC.

32 V. The positive feedback path is through R6 and C3. Resistor R2 causes the LM317 to operate in a current-limit mode in the event that some malfunction would cause an excessive current demand. Thus, the LM317 is self-protecting. However, switching transistor Q1 is subject to the effects of overloading, because R2 does not monitor the actual load current. To overcome this disadvantage, the modified circuit (Fig. 7-9) can be used. The three paralleled LM195s extend the current capability of the regulator to 4 A and also provide automatic overload protection. Such protection is an intrinsic characteristic of the LM195 monolithic output transistor.

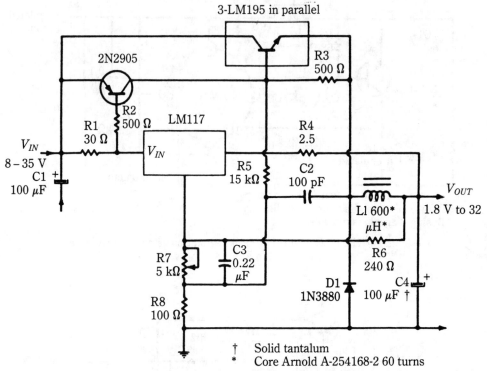

7-9 A 4-A switching regulator with LM195 monolithic output transistors.

Resistor R4 in this scheme produces current limiting in the LM117 IC in the event of a malfunction. Resistance R_4 is higher than its counterpart in Fig. 7-8, R_2. This difference is because much less drive current is required from the three-terminal IC than is the case in Fig. 7-8. These two regulators demonstrate one of the noteworthy advantages that the switching regulator has over linear regulators—the ability to operate efficiently, despite a great disparity between the dc input and output voltages.

Negative switching regulators

By at least two ways a negative-regulated voltage can be produced with switching regulators. A positive-voltage-regulator IC can be utilized in the manner that is illustrated in the example of Fig. 7-10. With minor differences in circuit configuration, the type 723 positive voltage-regulator IC can operate from an unregulated dc source that has its positive side grounded. The regulated negative voltage is obtained with respect to this ground. This approach is generally satisfactory, although it sometimes is difficult to use all of the auxiliary options that are available on the IC module.

Another, and often more satisfactory, technique is to use a linear-regulator IC that is specially designed for use in negative voltage regulators, such as the

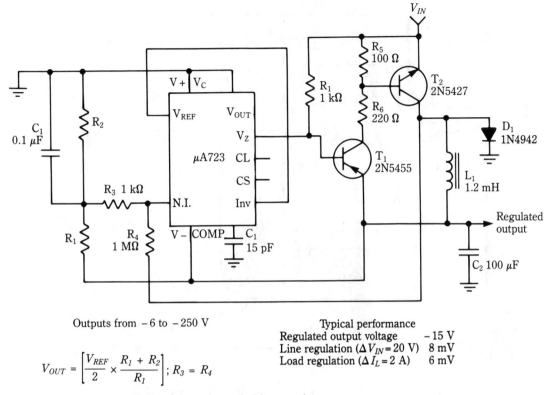

Outputs from -6 to -250 V

$$V_{OUT} = \left[\frac{V_{REF}}{2} \times \frac{R_1 + R_2}{R_1}\right]; R_3 = R_4$$

Typical performance
Regulated output voltage $\quad -15$ V
Line regulation ($\Delta V_{IN} = 20$ V) $\quad 8$ mV
Load regulation ($\Delta I_L = 2$ A) $\quad 6$ mV

7-10 A negative switching regulator. Fairchild Semiconductor Corp.

LM304. A negative switching regulator that uses the LM304 is shown in Fig. 7-11. Current limiting is available in this self-oscillating switching regulator by virtue of the connection that is made to pin 6. A unique feature is that the output voltage can be programmed by a single resistor, R2. The programming relationship is stated by the equation: output voltage $= \dfrac{R_2}{500}$. Thus, with R_2 selected to be 2500 Ω, this regulator circuit delivers 5 V (negative).

The constant-current switching regulator

The preponderance of circuit information on regulators has involved voltage-regulated power supplies. Relatively little information is found with regard to current regulation—especially with *switching-type regulators*. This situation is understandable because the operation of most equipment and systems requires constant voltage. However, many laboratory measurements and certain industrial processes are best made with constant current. If you wish to determine the breakdown voltage of a rectifier, the drop-out current of a relay, or the capacitance of a large electrolytic capacitor, a constant-current regulator best serves these needs. Also, a constant-cur-

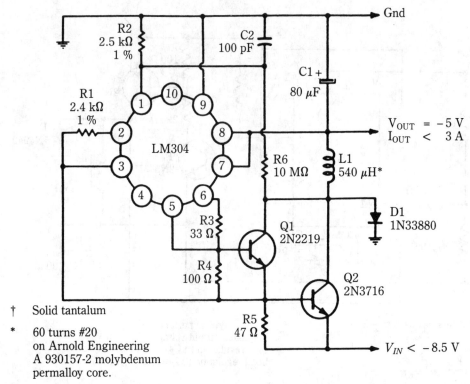

7-11 Negative switching regulator that uses a specially designed IC.

rent source is often dictated in electrochemical applications (such as plating or battery charging). When a linear regulator is used to obtain constant current, the operation can become very inefficient because of the high voltage drop, which must sometimes be accommodated across the series-pass transistor. Current regulation with a switching-type regulator avoids this source of power dissipation.

The switching-type current regulator shown in Fig. 7-12 uses a small sampling resistance, R_1, in series with the load. The constant current in the load can be made adjustable by connecting a 100-Ω potentiometer across R1. Resistor R2 would then be connected to the wiper of this adjustment pot. Notice that the load floats, with respect to the system ground. Alternatively, the regulator (together with the unregulated power supply) could be floated and the load could then be grounded. With the advent of the special ICs for producing pulse-width modulation in switching regulators, it is anticipated that more attention will be directed toward the use of switching regulators in constant-current applications.

Stacked switching regulator

The switching regulator that is illustrated in Fig. 7-13 involves some unique features. First and foremost, the bulk of the load power is supplied directly from the full-wave bridge rectifier, which derives its ac input from a three-phase power line.

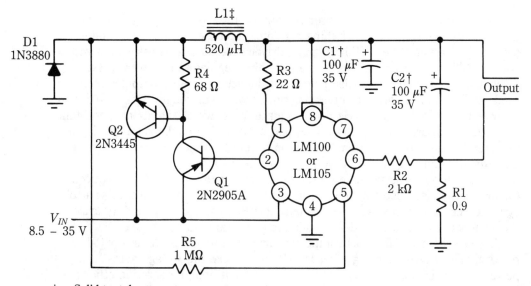

† Solid tantalum
‡ 60 turns = 20 on Arnold Engineering
A 930157-2 molybdenum permalloy core

7-12 A constant-current switching regulator. National Semiconductor Corp.

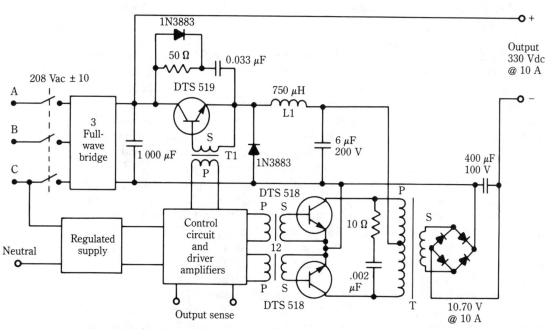

7-13 A 3.3-kw high-efficiency stacked switching regulator. Delco Division of General Motors

The remaining portion of load power is supplied by a switching regulator that is unique in its own right. The switching regulator composes a duty-cycle modulated switching transistor (the DTS519), and a driven inverter that is associated with a pair of DTS518 transistors. Unlike the switching regulators that were previously discussed, the switching transistor does not deliver its filtered output directly to the load. Rather, it provides dc operating power for the driven inverter. The driven inverter is actuated by a 25-kHz nearly square wave, which maintains a constant duty-cycle of approximately 50%. The load contribution of the regulator is then provided by the driven inverter. Regulation occurs because the driven inverter is amplitude modulated by the dc output from the switching transistor. The duty-cycle modulation of the switching transistor is similar to that which is produced in the previously discussed supplies. That is, a feedback loop, which contains a voltage reference and comparator, senses output voltage and varies the duty cycle of the switching transistor in order to maintain a constant output voltage.

These schemes bring about important operating features. Most obviously, a relatively small switching regulator is able to stabilize the voltage across a high-power load. This operation also leads to extraordinarily high operating efficiency. Not only is the full-load efficiency in the vicinity of 95%, but the efficiency remains well above 90% at half load. With this arrangement, the ripple harmonics in the regulated output do not undergo wide changes in frequency or amplitude as regulation occurs. Therefore, it becomes relatively easy to select a driving frequency for the inverter that does not generate offensive interference to sensitive loads or to adjacent instrumentation. Relatively little ripple from the switching transistor finds its way to the rectified-output terminal—the preponderance of output ripple is predictable and constant because the frequency and duty cycle of the inverter remain fixed.

Yet another feature is the minimal physical size of this regulator because no 60-Hz input transformer is required. Such a transformer would be large and heavy at this power level. However, no isolation is between the ac power line and the dc output circuit. Accordingly, be cautious when using this regulating scheme.

Figure 7-14 shows the control circuit for the stacked regulator. Power for the control circuit is derived from a single-stage voltage regulator that uses the DTS1010 power transistor. This regulated supply was shown in block form in Fig. 7-13. The control circuit itself begins with the 50-kHz pulse oscillator, which utilizes the LM555 IC timer. The 2-μs negative-going pulses that are produced by the oscillator serve two purposes. First, these pulses drive the type-D flip-flop, which supplies 25-kHz complementary square-wave drive signals to the bases of the pair of DTS2000 transistors. These transistors compose the driving stage for the DTS518 inverter transistors (Fig. 7-13). In addition, the pulses from the LM555 timer are used to impose a 2-μs *dead time* on the switching transistors of the 25-kHz waveform. This is accomplished via the single DTS2000 stage, the 1N914 diodes, and the associated components. The purpose of the dead time is to prevent simultaneous conduction of the DTS518 inverter transistors.

Duty-cycle modulation of the DTS519 switching transistor is performed in the feedback loop of the system. The feedback loop features the 741 op amp, the LM311 comparator, and the DTS1010 drive transistor. The duty-cycle-modulated

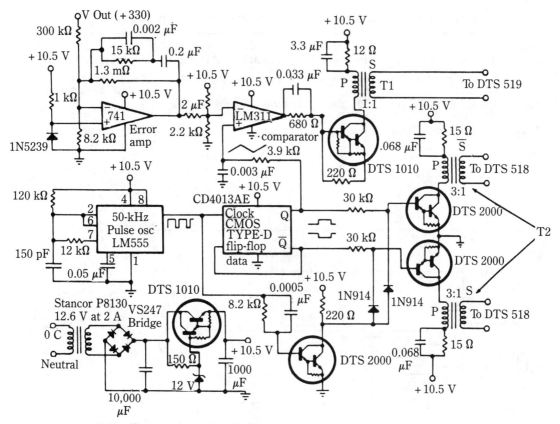

7-14 Control circuit for the 3.3-kw stacked regulator. _{Delco Division of General Motors}

wave is generated in the output circuit of the LM311 comparator. This is the consequence of the two signals that are impressed at its input—a dc level from the 741 error amplifier and a triangular wave that is obtained by integrating one of the outputs from the type-D flip-flop. The reference voltage from a 1N5239 zener diode is applied at the input of the error amplifier.

An important aspect of the control circuitry of Fig. 7-14 is that the DTS1010 and DTS2000 devices are actually monolithic Darlington transistors. This, of course, enhances the performance of every stage that uses these devices. Information that pertains to the electromagnetic components of the stacked regulator circuit is provided in Table 7-1.

A 250-W switching-mode power supply

Figure 7-15 is a block diagram of a 5-V 50-A switching-mode power supply that, at this writing, remains representative of state-of-the-art developments. A constructional and operational feature of this system is that each functional block in the

Table 7-1. Electromagnetic components for the stacked regulator.

Core Device	Description of core	Primary	Secondary	Air gap	Comments
L1	Ferroxcube pot core #45/25-3E1	62 turns no. 12	None	0.0625 in	The air gap prevents magnetic saturation.
T1	Magnetics Inc. EE core #F42510	10 turns no. 19	10 turns no. 19	None	
T2	Magnetics Inc. EE core #F42510	15 turns no. 19	5 turns no. 16	None	Two T2 transformers are required.
T3	Ferroxcube pot core #66/56-3E1	27 turns no. 13	12 turns no. 10	None	Primary is center-tapped. The two half-sections of the primary are bifilarwound.

Courtesy Delco Electronics Division of General Motors Corp.

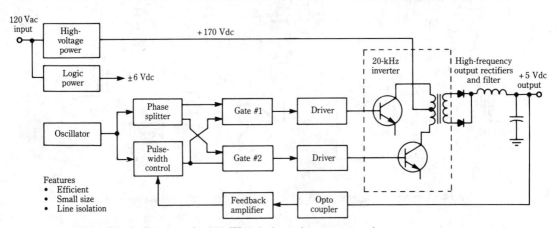

7-15 Block diagram of a 250-W switch-mode power supply. Motorola Semiconductor Products, Inc.

control circuitry is actually implemented by a separate IC. Such architectural organization results in great flexibility and leads to easy debugging and servicing. Admittedly, this arrangement does not represent "the last word" in the use of the IC, because now single ICs that contain all of the control functions needed for a pulse-width (duty-cycle)-modulated switching regulator are available. However, the use of several ICs is a tremendous advance over the old design approach, which used many discrete transistors. It is probable that some designers will favor the use of a separate IC for each function because of the manifold circuit options it affords.

A detailed schematic diagram of the driven inverter and associated power circuits is shown in Fig. 7-16. A salient feature of this switching-mode supply is that no 60-Hz transformer is used. Most of the circuitry is simple and straightforward. The 1N5833 rectifier diodes that are associated with output transformer T1 are Schottky *hot-carrier diodes*. They serve two needs for a supply of this kind. The Schottky diodes have a smaller forward-voltage drop than do conventional silicon-junction diodes. This is very important because the load power is produced at the

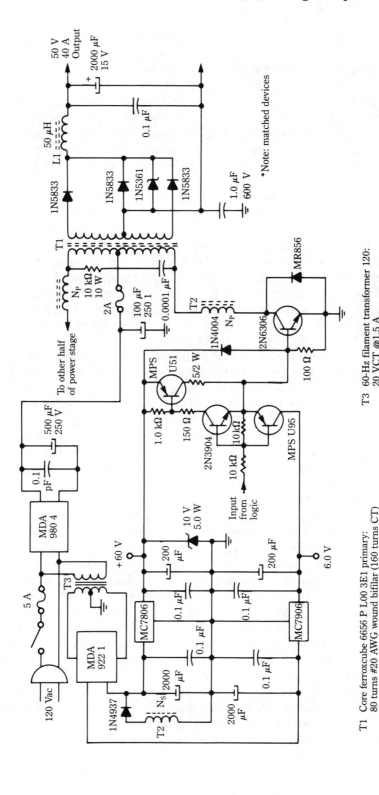

7-16 Schematic of the power circuitry in the 250-W switch-mode supply. *Motorola Semiconductor Products, Inc.*

5-V level. For the same reason, a center-tapped full-wave rectifier is used, rather than bridge rectifiers (in a bridge-rectifier arrangement, two diodes per half cycle always contribute to the overall forward-voltage drop).

Another important feature of the Schottky diodes is the inherently fast response time. No minority charges with their attendant storage effects are involved in the rectification process of these diodes. In a low-voltage high-frequency regulator, much of the efficiency that is gained prior to rectification can be sacrificed in rectifying diodes, which have high voltage drops and which, because of charge storage, continue to conduct for an appreciable portion of the cycle after the applied voltage has reversed.

Two additional diodes are associated with the output circuits. One of these is another 1N5833 type. This is a so-called *freewheeling diode* and it conducts current (from inductor L1) through the load circuit during those intervals when neither rectifying diode is conducting. Such intervals are the 2-μs dead times that were imparted to the original square wave in order to prevent simultaneous conduction of the inverter transistors. As might be anticipated, this freewheeling diode adds to the overall efficiency of the power supply (this technique is not needed when either a true square wave or a sine wave is being rectified). Because the current pulses that are handled by the freewheeling diode are of short duration and have steep wavefronts, the Schottky diode is the logical choice.

The 1N5361 zener diode ordinarily does not participate in circuit action. Its forward bias is greater than that of the Schottky diode, but its reverse-breakdown (zener) voltage is less than the reverse avalanche voltage of the Schottky device. Schottky diodes are vulnerable to destruction from too much energy that is absorbed when they are forced into their reverse-conduction mode. This vulnerability can occur as a result of voltage transients; here, the zener diode provides protection. Although the zener diode is connected directly in parallel with the freewheeling diode, considerable protection is also provided for the rectifying diodes. A possible source of such potentially destructive transients could very likely be momentary faults in the load.

Another unconventional aspect of this switching supply is the presence of the T2 primary windings in the collector leads of the inverter power transistors. Actually, this technique is for *despiking*. The secondary winding of T2 is connected through a diode to the large input capacitor of the logic power supply. This arrangement couples spike energy from the inverter back to the dc supply. These spikes result from leakage reactance in output transformer T1 and from the effects of charge storage in the inverter transistors. The attenuation or elimination of such spikes is desirable to reduce the generation of electrical noise and, perhaps even more importantly, to protect the inverter transistors from catastrophic destruction. This destruction occurs too readily from secondary breakdown when the SOA (safe-operating area) ratings are exceeded.

An interesting feature of the inverter is its driver circuit. One-half of the total circuit is shown in the schematic of Fig. 7-16. The other half is identical. The three transistors that are depicted receive dc operating power from a ±6-V source. This source enables them to deliver either forward or reverse base bias to the 2N6306 output Darlington. Such a provision results in faster and cleaner switching. Because of the output despiking, the SOA is not endangered by this technique.

The control circuit uses CMOS-logic ICs, which are characterized by high immunity to electrical noise. The circuitry for the control functions is shown in Fig. 7-17. Because the IC modules are dual and quadruple types, only two IC modules are needed for the control circuit. In addition are two bipolar transistors, a MOSFET, an optoisolator, and a few diodes, resistors, and capacitors.

The control system starts out with a 40-kHz multivibrator oscillator that utilizes two of the NOR gates of the MC14001 quad 2-input NOR gate IC. This oscillator drives the phase splitter, which is one flip-flop of the MC14013 dual type-D flip-flop IC. The two square-wave outputs, which are produced by the flip-flop, are 180 degrees out of phase with each other. These are basically the proper waveforms to drive the inverter. In this respect, the flip-flop substitutes for a transformer. Unlike a transformer, the flip-flop divides its incoming pulse rate by two. Therefore, the inverter will operate at 20 kHz, rather than at the 40-kHz oscillator frequency.

The output from the phase splitter is then fed to two NOR gates. These gates compose the second half of the MC14001 IC. The purpose of these gates is to provide a means whereby the waveforms that drive the inverter can be duty-cycle modulated. Thus, the logic signal that is applied to inputs 8 and 12 of these gates governs the time in which the outputs are allowed to make their transitions in response to the square waves that are applied to inputs 9 and 13. The timing circuit is the second half of the MC4013 IC. Appropriately, this flip-flop circuit is designated for pulse-width control. More specifically, this flip-flop functions as a resettable monostable (one-shot, multivibrator). The flip-flop is reset when current from the 3N158 FET delivers sufficient charge to the 0.001-μF timing capacitor. The idea here is to produce a delay so that gates #1 and #2 will not be enabled at the same time that the square waves from the phase splitter change their logic states. Therefore, gates #1 and #2 can never deliver a full-duty-cycle square wave to the inverter. Rather, there is a dead time and the waveform is notched. As previously mentioned, the dead time protects the inverter transistors from simultaneous conduction, a destructive phenomenon. Dead time is adjustable by means of the 100-kΩ variable resistance that is associated with the 3N158 FET. The significant aspects of this sequence are shown in the timing diagram of Fig. 7-18. Notice that the dead time imposes a limit to the maximum duty cycle of the inverter drive pulses that are produced by gates #1 and #2.

The MC1741 op amp is used as a feedback (error) amplifier. This stage, together with the zener reference source and transistor input circuit, functions in the same manner as the error amplifiers in linear regulators—a dc voltage is produced when the output voltage of the supply varies from a set value. It is this dc voltage that causes the aforementioned pulse-width control (the MC14013 IC and associated circuitry) to vary the duty cycle of the pulses supplied to the inverter.

The output-voltage sampling technique for this arrangement involves an optocoupler. The optocoupler and the 20-kHz transformer, T1, enable the regulated dc output of the supply to be completely isolated from the 60-Hz power line. The opto-

400 Switching-type regulators that use integrated circuits

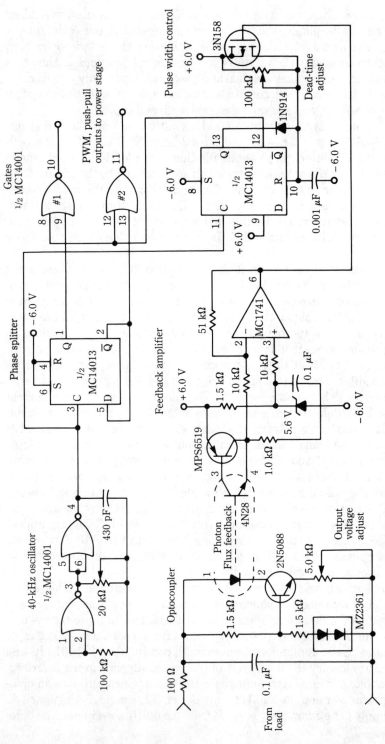

7-17 Control circuit for the 250-W switch-mode power supply. *Motorola Semiconductor Products, Inc.*

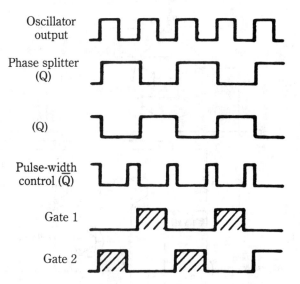

7-18 Timing diagram for control circuit of 250-W switch-mode supply.
Motorola Semiconductor Products, Inc.

coupler contains its own current regulator, which uses the 2N5088 npn transistor, the MZ2361 voltage reference, and the 5-kΩ variable resistor. The current regulator simulates a very-high-value adjustable resistance. At the same time, the constant-current feature stabilizes the operation of the optocoupler.

This switching-mode supply achieves a full-load efficiency of about 82%. Line and load regulation are on the order of $\frac{\pm 1}{2\%}$. The 40-kHz ripple has an amplitude of 20 mV peak-to-peak. Noise spikes of very short duration (with a peak-to-peak level of approximately 200 mV) are also present.

A 100-W switching regulator that uses the flyback principle

The 100-W flyback switching regulator shown in Fig. 7-19 is suggestive of the shunt regulator in linear supplies. The analogy is valid, with respect to the circuit configuration. However, the flyback switcher operates in a more complex manner than does the more conventional series-chopping version. These supplies are ordinarily capable of delivering a higher output voltage than is received from the unregulated dc source. The basic principles are shown in the simple "shunt" switching circuit of Fig. 7-20, together with the waveforms and the mathematics of load-voltage production. Duty-cycle switch control varies the output voltage because a short switch-closing interval does not allow as much energy to be stored in the magnetic field of the inductor as does a longer interval. The average output voltage is a function of the counter EMF that is developed by the inductor when the switch is opened. This counter EMF decreases when the closure time of the switch is shortened.

402 *Switching-type regulators that use integrated circuits*

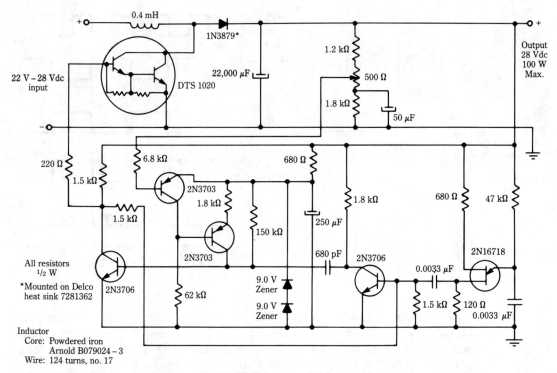

7-19 Schematic diagram of a 100-W flyback switching regulator. Delco Division of General Motors

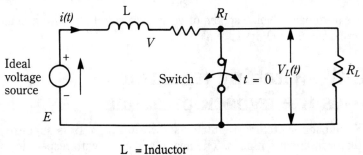

L = Inductor
R_I = Self resistance of inductor
R_L = Load resistance

$$V_L(t) = E\left[\frac{R_L}{R_I+R} + \left(\frac{R_L}{R_I} - \frac{R_L}{R_I+R_L}\right)e^{-\left(\frac{R_I+R_L}{L}\right)t}\right]$$

(A) A simple shunt switching circuit and the equation describing the production of "flyback" voltage.

7-20 Basic principles that are involved in flyback regulators.

7-20 Continued

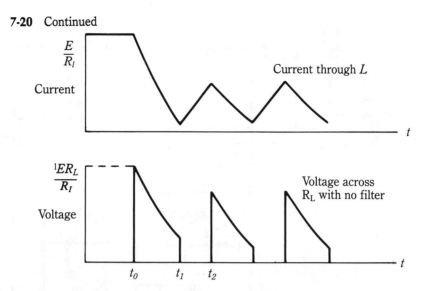

(B) Current and voltage waveforms when the switch is repeatedly opened and closed.

Figure 7-21 is a simplified circuit that shows the way in which the flyback principle is implemented in an electronic regulator. Notice the use of the isolation diode to prevent the switching element from shorting out the load circuit. The multivibrator is a triggered monostable type. Its pulse duration is controlled by the sensor, which can be viewed as a voltage-dependent resistance. Thus, the sensor

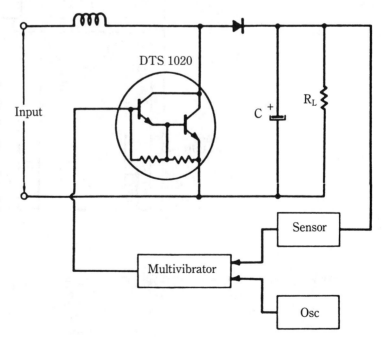

7-21 Simplified circuit of the flyback regulator.

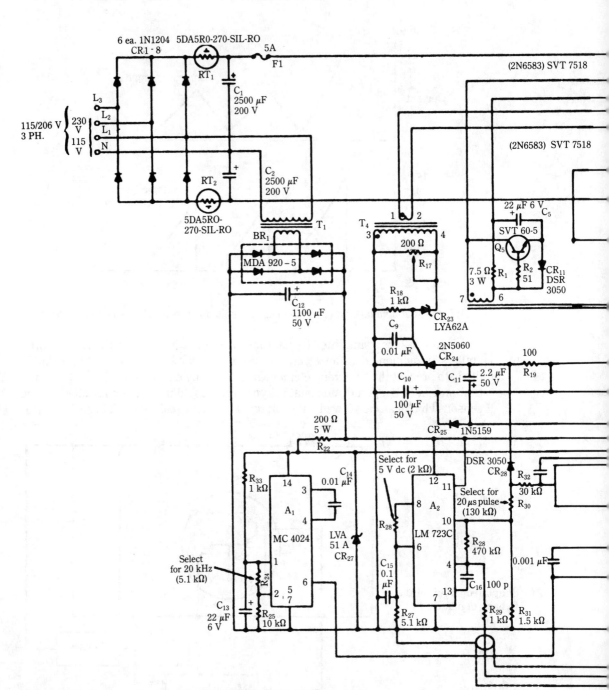

7-22 5-V 200-A switching-type power supply. TRW Power Semiconductors

A 100-W flyback switching regulator that uses the flyback principle

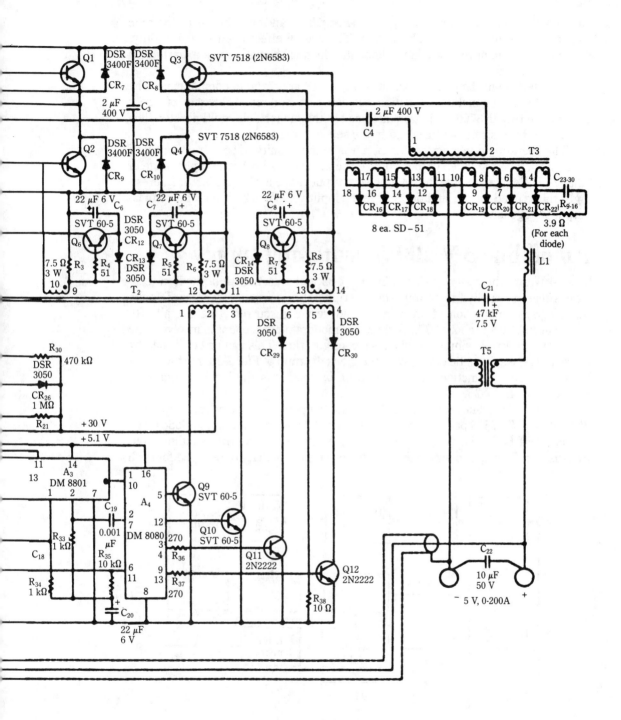

circuitry monitors the output voltage of the regulator and thereby varies the time constant of the monostable multivibrator. The overall effect is pulse-width (duty-cycle) modulation of the switching element, which is shown here as a Darlington power transistor.

The devices in the control circuit of Fig. 7-19 are discrete components. The 2N1671B is a unijunction relaxation oscillator with a repetition rate of about 9 kHz. The two 2N3706 transistors are used in the monostable multivibrator. The two 2N3703 transistors are involved in the sensor circuit.

The operating efficiency of this circuit is in the vicinity of 85% from half load to full load. The regulated output voltage remains at 28 V over the input range of 22 to 28 V. Regulation over the total power-output range is about ±0.3%. Interestingly, this is cited for the situation where the input voltage is 22 V and the nominal output voltage is 28 V!

An off-line 5-V 200-A switching supply

The 1-kW switching supply shown in Fig. 7-22 is representative of modern techniques that are applied to an important peripheral of computer systems. Despite the adverse combination of low-voltage and very-high current, the overall efficiency at full load is about 90%. Although a moderate parts count is involved, the circuitry is relatively simple in this basic configuration. This can be confirmed by correlating the schematic circuit with the block diagram of Fig. 7-23. Further resolution of circuitry functions will follow from inspection of the block diagrams in Figures 7-24 and 7-25.

The high-frequency inverter of this regulated power supply contains the four 2N6583 or SVT 7518 power transistors, the DSR 3400F commutating diodes, and the four SVT 60-5 driver stages with their associated circuitry. Multiwinding input transformer T2 makes possible the isolated drive signals that are required for such

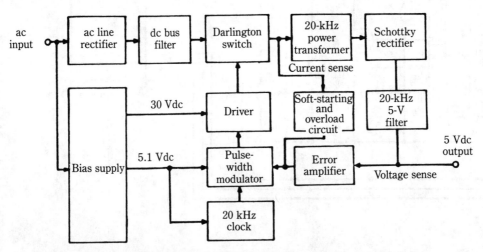

7-23 Block diagram of the 5-V 200-A switching supply. TRW Power Semiconductors

An off-line 5-V, 200-A switching supply

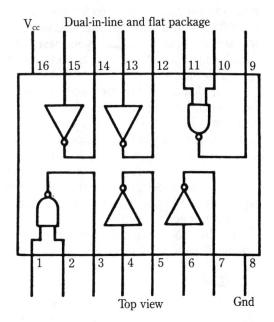

7-24 Block diagram of the DM8601 IC that is used in the circuit of Fig. 7-22.

7-25 Block diagram of the DM8090 IC that is used in the circuit of Fig. 7-22.

a full-bridge inverter. The output transformer is T3. It has 8 secondary windings and connects to 8 Schottky-diode rectifiers, which, in turn, are connected to form a full-wave circuit. By this technique, rectifier voltage drop is made a small percentage of the 5-V output level, which thereby permits high-efficiency operation. Also, awkward hardware problems, which would arise from a single large-conductor winding, are neatly circumvented.

Notice capacitor C4 in the primary circuit of output transformer T3. Because of this series capacitor, the output transformer cannot saturate from inequalities in parameter values of the four inverter output transistors. This measure eliminates destructive current spikes that often plague the ordinary push-pull inverter circuits.

This supply has an interesting power-line rectification circuit. Depending on terminal selection, several ac voltages and formats can be accommodated. A single-phase 115-V line is applied to terminals L1 and N. This connection results in the rectifier-system operating as a voltage doubler. In so doing, it closely simulates *"straight-through" operation* (when a single-phase 230-V line is connected to terminals L1 and L2).

The 208-V three-phase sources can also be used to operate the supply. A three-wire three-phase system would connect to terminals L1, L2, and L3. A four-wire, three-phase system would connect to terminals L1, L2, L3, and N.

Thermistors RT1 and RT2 exhibit high resistance when the power supply is first energized from the line. As current flows through RT1 and RT2, their temperature rises above that of the ambient temperature and their resistance greatly decreases. This action guards against the large current inrush that would otherwise occur because of the initially low impedance of the filter capacitors. Because the conductance of thermistors is a very strong function of the self-generated heat, loss in power-supply operating efficiency is minimal. The thermistors also contribute to the "soft-start" characteristic of the supply in that any tendency for output overshoot is held down. However, most of this function is accomplished electronically by automatically limiting the pulse width of the inverter switching transistors during warmup.

A second internal power supply is used for the low-current requirements of the control and logic circuits. This is the so-called "bias supply" shown in the block diagram of Fig. 7-23. Here, transformer T1 (Fig. 7-22) is a small 115:24-V type and is used in conjunction with bridge-rectifier module MDA 920-5 and filter capacitor C12 to provide a nominal 30 V dc. This voltage is also used to operate zener diode LVA51A in order to develop a second dc operating voltage of 5.1 V.

Allusion has been made to the main rectifying system, the internal or auxiliary dc supply, the bridge rectifier, and the output rectifier and filter. The control and protection functions, which encompass the 20-kHz clock, the error amplifier, the pulsewidth modulator, the driver, and the soft-starting and overload circuit, are described next. All of these functions are shown in the block diagram of Fig. 7-23.

The 20-kHz clock

The basic switching rate of the regulated supply is developed by the MC4024 IC (Fig. 7-22). This IC is essentially a voltage-controlled multivibrator. It is adjusted to generate a fixed 20-kHz square wave by means of selectable resistor R24. This clock, or oscillator, is not embraced within the feedback loop of the supply; it, therefore, produces its 20-kHz 50%-duty-cycle square wave under all conditions. Pin 2 is the frequency-versus-voltage control terminal and pin 6 is the output terminal. Capacitor C14, connected to pins 3 and 4, sets the frequency range so that a convenient dc voltage can be applied to pin 2 in order to develop the 20-kHz switching rate.

An off-line 5-V, 200-A switching supply **409**

With regard to the operation of the 20-kHz clock, as well as to subsequently described circuit functions, refer to the waveform diagrams in Figures 7-26 and 7-27. The 4-V p-p 20-kHz square wave at output pin 6 of the MC4024 IC is depicted at the top of Fig. 7-26.

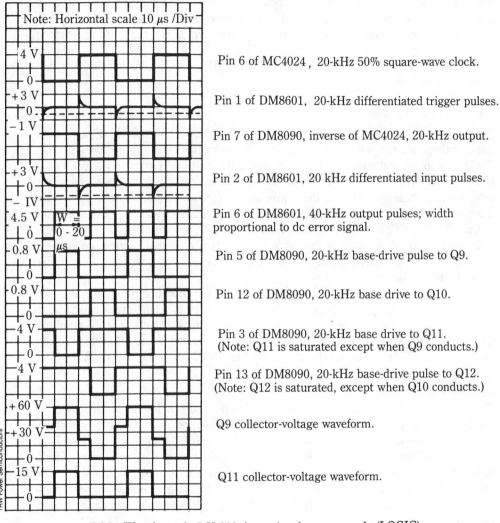

7-26 Waveforms in 5-V 200-A regulated power supply (LOGIC).

The pulse-width modulator

The all-important function of pulse-width modulation is provided by the DM8601 IC. This IC is a *retriggerable monostable multivibrator*. Its essential feature is that it varies the width of rectangular pulses in response to a dc signal from the error amplifier in the feedback path of the supply. More specifically, the leading edges of the rectangular pulses are triggered by the 20-kHz clock wave; the trailing edges of the rectangular pulses are then determined by the error signal.

The *pulse-width modulated wave*, which is available from pin 6 of the DM8601, has a repetition rate of 40 kHz, rather than 20 kHz. This frequency doubling is necessary so that the ultimate switching rate of the regulator will be 20 kHz, because of the bridge configuration of the output inverter transistors. The 40-kHz pulse-width modulated wave is the fifth waveform from the top in the waveform diagram of Fig. 7-26. A simplified block diagram of the DM8601 IC is shown in Fig. 7-24. The error signal is derived from the LM723C IC and is applied to pin 13 of the DM8601 IC.

The error amplifier

The *error amplifier* is an LM723C voltage-regulator IC. This design is a novel technique; otherwise you would expect to encounter an op amp together with a zener diode or another voltage reference. By using the voltage-regulator IC, problems, such as temperature drift, electrical instability from high gain, and malperformance (i.e., latch-up) have all been greatly minimized or eliminated. Schematic diagram Fig. 7-22 shows that the sampled output of the supply is applied to pin 4 of the LM723C. This is the inverting input of its internal voltage comparator. The noninverting input, in turn, is biased with part of the internal reference voltage, which is available at pin 6. Pin 10 is the output terminal, from which the amplified error signal is obtained.

The pin numbers of the LM723C that appear on the schematic diagram pertain to the dual-in-line package, not the metal-can version. Also, direct substitution with an LM723 might be beneficial if an extended range of temperature stabilization is desired. The LM723C is guaranteed for performance over the temperature range of 0° to 70°C, whereas the range for the LM723 is −55° to +125°C.

Driver logic

The *driver logic* is the DM8090 IC and associated circuitry. The best way to visualize the waveform synthesis that is performed by this module and by buffer transistors Q9 through Q12 is to refer to the simplified DM8090 block diagram (Fig. 7-25) and the waveforms of Fig. 7-26. Remember that, besides proper synthesis of individual driving waveforms for the four bridge-connected power switches (Q1, Q2, Q3, and Q4), these drive pulses must be delivered in the right sequence so that ultimately a 20-kHz duty-cycle modulated wave is presented to the output rectifier system.

Schematically, it might not appear that transistors Q5 through Q8 are also part of the driver logic (Fig. 7-22). However, these transistors are not the conventional buffer or drive stages that are ordinarily encountered. Rather, their function is also one of wave shaping. Specifically, these transistors provide a low-impedance source of negative-going base turn-off pulses for the inverter power switches. Thus, the stored charges in transistors Q1 through Q4 are much more rapidly depleted at turn-off time than would be the base if transistors Q5 through Q8 were not in the circuit. This speed-up increases overall efficiency and helps keep the power-switch transistors within their safe operating area (SOA). Also, because the regulation is pulse-width governed, the supply performance can be expected to be better with abruptly terminated current pulses than with pulses that have extended and sloppily defined fall times.

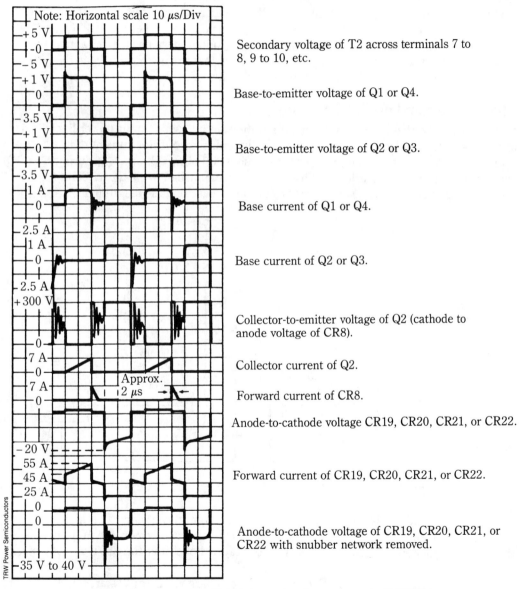

7-27 Waveforms in 5-V 200-A regulated power supply (POWER).

Driver transformer, T2, has two primary windings. Only one of these (the one designated by terminals 1, 2, and 3) provides the requisite pulse logic that is necessary for the proper turn-on sequence of the power switches (Q1, Q2, Q3, and Q4). These pulses are delivered by buffer transistors Q9 and Q10, which are actuated by the DM8090 IC.

Transistors Q11 and Q12 are also buffer amplifiers and their conductive state is also governed by the digital logic from the DM8090 IC. However, the action of Q11 and Q12 on the second primary winding of T2 is not directly related to pulse-width modulation. Rather, this arrangement causes the transformer core flux to short circuit during periods when zero drive is required by the power switches (Q1, Q2, Q3, and Q4). This technique ensures that noise pulses or other spurious signals cannot turn on any of the power switches during periods when they are intended to be in the OFF state. Refer to the waveforms in the power circuitry (Fig. 7-27).

The soft-start and overload circuit

In addition to the thermistors in the main rectifier circuit, which limit power-line inrush current, it is also necessary to specifically limit the collector current of the bridge inverter transistors (i.e., the power switches, Q1, Q2, Q3, and Q4, during start-up and in case of overload). This task is accomplished with the circuitry shown in Fig. 7-28. This is taken directly from the main schematic diagram of Fig.

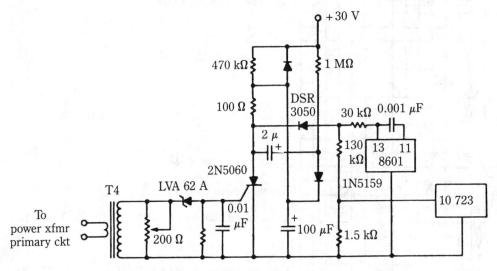

7-28 The soft-starting and overload circuit. TRW Power Semiconductors

7-22, but it is graphically rotated 90 degrees to facilitate discussion. The major components of this circuit are the input current-sensing transformer, the 100-μF time-delay capacitor, the 2N5060 SCR, the four-layer diode, and the 8601 monostable IC.

Prior to start-up, the 100-μF capacitor is discharged. During start-up, it charges through the 470-kΩ resistor from the 30-V dc source. As the capacitor voltage builds, the pulses from the 8601 IC are width controlled so that the supply operates in a current-limited mode for approximately 1 second. This operation is because the pulses are initially narrow and gradually become wider until regulation of the supply is achieved. The start-up performance of the supply is illustrated in Fig. 7-29. This "soft-start" characteristic allows sufficient time for the output filter capacitors to charge and for stability of the drive signals to be obtained before the collector currents of the output transistors can meet the demand. Thus, one of the most common causes of secondary breakdown is eliminated.

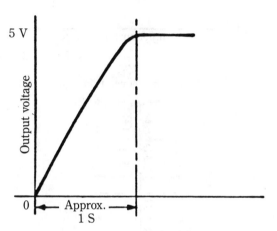

7-29 Start-up performance of the 5-V 200-A regulated supply.
TRW Power Semiconductors

Notice the circuitry position of current-sensing transformer T4 in the main schematic diagram of Fig. 7-22. This transformer monitors the overall current that is supplied to the collectors of the four output transistors (Q1 through Q4). If an overload is detected, the voltage that is induced in the secondary of T4 triggers the SCR and thereby discharges the 100-μF capacitor. This also inhibits the output pulses from the 8601 IC. The power supply is now shutdown, but a timing cycle begins as the 2.2-μF capacitor is charged through the conductive SCR and the 1-MΩ resistor. After approximately 1 second, the voltage that is developed across the 2.2-μF capacitor is sufficient to fire the 4-layer diode, which, in turn, resets the SCR by reverse biasing its anode. The resetting of the SCR frees the circuit to once again function in the soft-start mode (Fig. 7-29). Depending on whether the overload is still present, the power supply will then either attain regulation or the previously described sequence of events will repeat.

The overload characteristics that correspond to this sequence of events are illustrated in Fig. 7-30. This is current foldback and an automatic probing action. When the overload is removed, the probing cycle stops and the power supply operates in its normal regulatory mode.

The specifications for the high-frequency magnetics appear in Fig. 7-31. Transformer T1 is not listed, because it is a "garden-variety" $\frac{115}{24\text{ V}}$ 60-Hz type with a nominal 50-W rating. The electrical parts list for the supply is given in Table 7-2.

414 *Switching-type regulators that use integrated circuits*

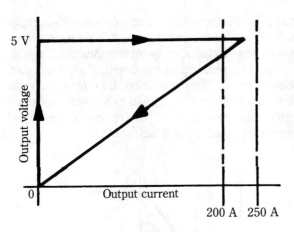

7-30 Overload characteristics of the 5-V 200-A regulated supply. TRW Power Semiconductors

Driver transformer (T2)

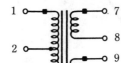

Core:
  TDK, Part number H6A - P22/13z - 52 H, pot core.

Windings:
  (1–2) (2–3)   30T bifilar, #28 single sodereze.
  (4–5) (5–6)   15T bifilar, #30 single sodereze.
  (7–8) (9–10) 5T quad, #24 nylon-covered (kynar).

Terminations:
  Self leads to 14-pin header mounted to core.

Current-sense transformer (T4)

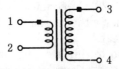

Core:
  Magnetics Inc. Part number 52056 - 1D.

Windings:
  (1–2) Bar primary formed by passing T3 primary lead thru core.
  (3–4) 34T, #24HF

Terminations:
  Self leads, length as required.

Power transformer (T3)

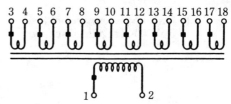

Core:
  Stackpole, part number 50–631, ceramag 24B, Q244.

Windings:
  (1–2) 32T, #12HF
  (3–4) through (17–18) 1T. #8 PVC-covered solid core.

Terminations:
  Self leads, length as required.

Balun transformer (T5)
  Formed by passing output bus bars through Magnetic, Inc. core part number 6 55436 – A2, 1 turn each bus bar.

Filter inductor (L1)
  Formed by winding 2 turns of positive bus bar through 2 Magnetics Inc. cores part number 55436 – A2.

7-31 High-frequency magnetics for the 5-V 200-A regulated supply. TRW Power Semiconductors

Table 7-2. Parts list for 5-volt 200-ampere regulated supply.

Component	Part No.	Vendor
Semiconductors		
Q1,2,3,4	SVT7518 or 2N6583	TRW Power Semiconductors
Q5,6,7,8,9,10	SVT60-5	TRW Power Semiconductors
Q11,12	2N2222	Motorola
CR1,2,3,4,5,6	1N1204	G.E.
CR7,8,9,10	DSR3400F	TRW Power Semiconductors
CR11,12,13,14 26,28,29,30	DSR3050	TRW Power Semiconductors
CR15,16,17,18	SD-51, 1N6098	TRW Power Semiconductors
CR19,20,21,22	SD-51, 1N6098	TRW Power Semiconductors
CR23	LVA62A	TRW Power Semiconductors
CR24	2N5060	Motorola
CR25	1N5159	Motorola
CR27	LVA51A	TRW Power Semiconductors
A1	MC4024	Motorola Voltage-Controlled Multivibrator
A2	LM723C	National Voltage Regulator
A3	DM8601	National Monostable Multivibrator
A4	DM8090	National Dual NAND, Quad Inverter
BR1	MDA920-5	Motorola
Resistors		
RT1,2	5 Ω (25 °C)	Thermistor, 5DA5RO-270-SIL-RO Rodan Industries
R1,3,6,8	7.5 Ω, 5 W	TRW/IRC type PW5
R2,4,5,7	51 Ω, 1/2 W	TRW/IRC type GBT 1/2
R18,29,33,34	1 kΩ, 1/2 W	TRW/IRC type GBT 1/2
R9,10,11,12, 13,14,15,16	3.9 Ω, 1/2 W	TRW/IRC type GBT 1/2
R17	200 Ω Trim pot	TRW/IRC type 20-207-200
R19,23	100 Ω, 1/2 W	TRW/IRC type GBT 1/2
R20,28	470 kΩ, 1/2 W	TRW/IRC type GBT 1/2
R21	1 MΩ, 1/2 W	TRW/IRC type GBT 1/2
R22	200 Ω, 5 W	TRW/IRC type PW-5
R24	5.1 kΩ, 1/2 W*	TRW/IRC type GBT 1/2
R25,35	10 kΩ, 1/2 W	TRW/IRC type GBT 1/2
R26	2 kΩ, 1/2 W*	TRW/IRC type GBT 1/2
R27	5.1 kΩ, 1/2 W	TRW/IRC type GBT 1/2
R30	130 kΩ, 1/2 W*	TRW/IRC type GBT 1/2
R31	1.5 kΩ, 1/2 W	TRW/IRC type GBT 1/2
R32	30 kΩ, 1/2 W	TRW/IRC type GBT 1/2
R36,37	270 Ω, 1/2 W	TRW/IRC type GBT 1/2
R38	10 Ω, 1/2 W	TRW/IRC type GBT 1/2

Table 7-2. Continued.

Component	Part No.	Vendor
Capacitors		
C1,2	2500 μF, 200 V	Sprague type 36D
C3,4	2 μF, 400 V 10%	TRW Capacitors type TRW-28
C5,6,7,8,13,20	22 μF, 6 V	TRW Capacitors type 935
C9,14	0.01 μF, 100 V 10%	TRW Capacitors type X1263UW
C10	100 μF, 50 V	Sprague type 30D
C11	2.2 μF, 50 V	TRW Capacitors type 935
C12	1.1 KμF, 50 V	Sprague type 39D
C14,17,18,19	0.001 μF, 100 V 10%	TRW capacitors type X1263UW
C15	0.1 μF, 100 V 10%	TRW capacitors type 663UW
C16	100 pF, 300 V	Elmenco type DM
C21	47 kμF, 7.5 V	Sprague type 432-D
C22	10 μF, 50 V 10%	TRW Capacitors type X463UW
C23,24,25,26,27,28,29,30	0.1 μF, 50 V	ARCO type TCD-104Z
Transformers		
T1	UP8609/F40	TRW/UTC transformers
T2	Special driver	TRW/UTC transformers
T3	Special power	TRW/UTC transformers
T4	Special current	TRW/UTC transformers
L1	Special filter	TRW/UTC transformers
T5	Special balun	TRW/UTC transformers
Fuses		
F1	5-A fast blow	Bussman type 3AG

*Selected value.

A 200-kHz 50-W switching regulator that uses a power MOSFET

Besides schematic simplicity, the 200-kHz regulator shown in Fig. 7-32 features appreciably more compact packaging and lighter weight than 20-kHz designs of similar electrical rating. Because power MOSFETS are rapidly sliding down the "learning curve," the cost of such a supply is (or is destined to shortly be) competitive with circuits that use bipolar transistor switching elements. Dynamic response and recovery time can be expected to be much better than that which is obtained from 20-kHz regulators, because the 20-kHz switching rate excels over earlier 2-kHz circuits. Also, ripple with a fundamental frequency of 200 kHz is generally a less troublesome form of EMI than that which is derived from 20 kHz. Table 7-3 compares this 200-kHz regulator with like-rated 20-kHz regulators that use bipolar transistors.

A 200-kHz 50-W switching regulator that uses a power MOSFET

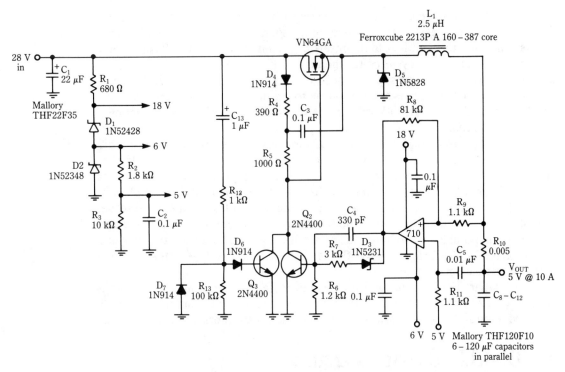

7-32 A 200-kHz 50-W switch-mode regulated supply that uses a power MOSFET. Siliconix, Inc.

**Table 7-3.
50-W regulator comparisons between 20- and 200-kHz designs.**

	20 kHz	200 kHz
Approximate recovery time for a 40% change in load	100 μsec	10 μsec
Inductor core	3019 pot core $\sim$0.85 in^3, 1.2 oz	2213 pot core $\sim$0.31 in^3, 0.43 oz
Capacitors	8 $\times$ 220 μF 1.0 in^3	6 $\times$ 120 μF 0.45 in^3

The circuit of Fig. 7-32 is self-oscillating with necessary positive feedback being provided by resistor R9. The LM710 is a high-speed comparator that is caused to operate in its hysteresis mode by resistor R8. A novel feature of this circuit is the bootstrap network (diode D4, resistors R4 and R5, and capacitor C3). This network enables sufficient gate-drive voltage to be obtained without the need for a separate dc source. Bipolar switching regulators generally do not have a similar problem with their base-drive requirements. This follows from the fact that the bipolar transistor only needs about 0.7 V to drive it into saturation, whereas the power MOSFET often requires 10 or 12 V of gate drive.

The LM710 voltage comparator is powered by positive 18- and 6-V sources. This technique dispenses with the need for a negative dc source. The 5-V source that is connected through R11 to the inverting terminal of the LM710 is the reference voltage. The duty cycle of the nominal 200-kHz oscillation varies as the sensed output voltage of the supply tends to be greater or less than 5 V; the nature of this duty-cycle variation is self-correcting to maintain the 5-V output. Capacitor C5 allows modulation of the 5-V reference so as to use the self-correcting action to materially decrease the ripple voltage.

Although the topography of the Q2, Q3 circuitry might initially suggest a differential amplifier, only Q2 is involved as a driver stage for the power MOSFET. Transistor Q3 is part of a soft-start circuit. When the supply is first turned on, the initial dc-current transient through capacitor C13 turns Q3 on, which thereby disables Q2. As C13 charges, the forward bias at the base of Q3 gradually decreases and Q3 finally becomes inactive. Simultaneously, Q2 becomes an effective driver stage and the regulatory action of the supply assumes control of the supply's output voltage. This technique spares the load from initial voltage transients that would otherwise occur at start-up.

The six paralleled output capacitors, C6 through C12, are connected primarily to achieve a low ESR so that the overall impedance will be on the order of 10 mΩ (it might be rewarding to experiment with polypropylene capacitors in this application).

A 500-kHz flyback-circuit switching power supply

While classic bipolar designs were struggling to exceed the popular 20-kHz switching rate, the availability of power MOSFETS immediately made leap-frog jumps into very-high-frequency switching feasible. An example is the simple and cost-effective 35-W regulated supply shown in Fig. 7-33. The 0.5-MHz switching rate of this flyback circuit is accomplished in a straightforward manner—no trick circuitry or critical adjustments are involved. Indeed, several salient features are worthy of mention:

- Because of the flyback principle, the output voltages can be, and are, higher than the dc input voltage. Thus, the flyback regulator is, among other things, a dc step-up transformer. Specifically, a nominal 12 V from an unregulated dc source is sufficient for proper operation of the regulated 20-V outputs. Actually, the allowable input-voltage range is 12 to 16 V.
- Dual-polarity outputs are provided without the need for separate transformer windings.
- The 500-kHz switching rate is likely to circumvent noise problems that are often troublesome at 20 kHz. Any 500-kHz energy or harmonics thereof that appear in the load are likely to yield easily to simple bypass and filter techniques (this is true; both in the load and in the supply). Although 500 kHz is more radiation-prone from a pc board than 20 kHz, shielding is more effective at the higher switching rate.

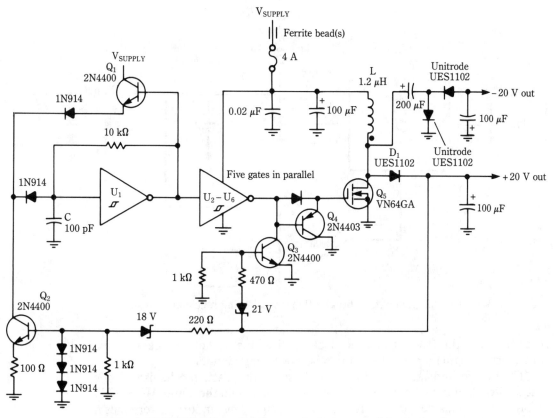

7-33 A 35-W 500-kHz switching supply that uses a flyback circuit.

- The expected size reduction in magnetics and in filter capacitors results in exceptionally compact packaging.
- The electrical ruggedness of the power MOSFET is superior to that of an RF-type bipolar transistor, which would be required to simulate the operation of this regulator. Also, the cost is lower—especially considering the drive circuitry for a bipolar transistor.

Figure 7-34 is the basic block diagram of the 500-kHz flyback regulator. A fixed-frequency clock is used to operate the power MOSFET as a pulse-width modulated switch. Thus, the duty cycle of the clock oscillator is modulated by the error signal. Energy is stored in inductor L during the ON time of the power MOSFET. When the MOSFET is switched to its OFF state, the energy stored in inductor L is transferred to output capacitor C, where it develops a reasonably steady dc voltage. This same flyback principle is used in automobile ignition systems and in the high-voltage supply of TV sets. As might be suspected, such a voltage tends to be somewhat "wild." It is very much a function of turn-off time, supply voltage V_s, and the load. It also depends on the value of the inductor, the distributed capacitance that is associated with it, and the eddy current and hysteresis losses in its

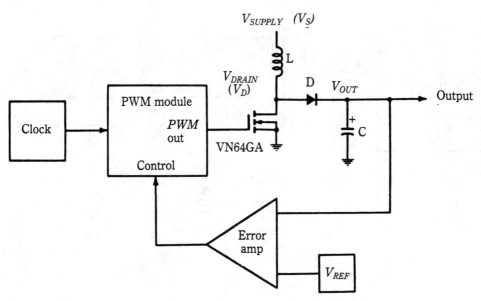

7-34 Block diagram of the 500-kHz flyback switching regulator. Siliconix, Inc.

core. However, the PWM feedback loop stabilizes the output voltage at a fixed value, just as regulation is achieved in other switching circuits.

The waveform diagrams of Fig. 7-35 illustrate this basic mechanism. Notice that the length of time that the clock is allowed to maintain the power MOSFET in its ON state governs the peak current in the inductor. This, in turn, determines its energy storage and, therefore, the voltage level that can be developed across the output capacitor. Accordingly, pulse-width modulation, under control of the feedback-loop error signal, is able to automatically control the output voltage (i.e. to regulate it).

Refer to Fig. 7-33. Gate U1 is connected as a relaxation oscillator to provide the nominal 500-kHz clock signal. The "natural" duty cycle of the clock oscillation is 50% (i.e., a square wave is generated). However, because of the way that U1 is situated in the feedback circuit, the 50% duty cycle is maximum; lower duty cycles are produced (narrower pulses) to regulate the output voltage. The duty-cycle variation is caused by diverting charge from the timing capacitor, C. This task is done by Q1, its associated diodes, and by Q2.

Actually, Q2 is the comparator. Its conductivity to ground varies with the deviation of the positive output from 20 V. Thus, Q2 is able to affect the duration of charge in capacitor C. However, such control should only be effective on the trailing portion of the clock's pulse-train. This requirement is satisfied by the action of transistor Q1, which only allows collector voltage for Q2 after the leading edge of the pulse has been generated. The timing action of Q1 is precise in this respect because its base is controlled from the output of U1 itself. Such control of the trailing edge of the pulse train constitutes the pulse-width modulation that is needed to achieve regulation.

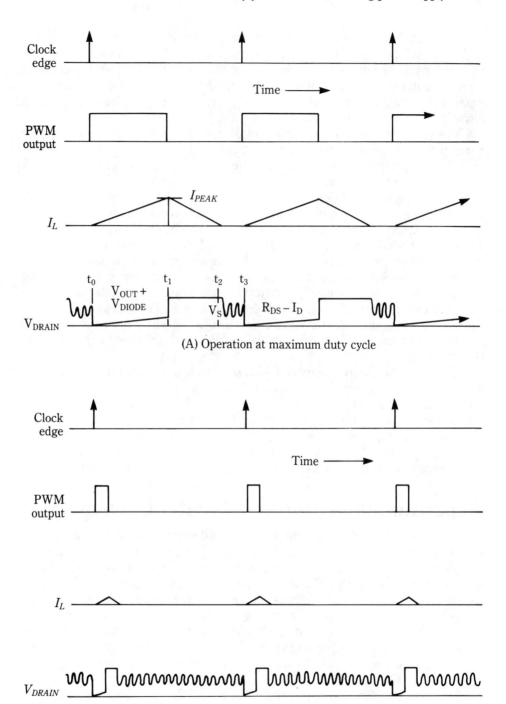

7-35 Relevant waveforms in the 500-kHz flyback switching regulator. Siliconix, Inc.

Gates U2 through U6 are five gates in parallel and they form the driver for the MOSFET switch. The parallel combination provides the low-impedance source that was made necessary by the input capacitance of the MOSFET gate. This arrangement, together with emitter-follower Q4, helps maintain the integrity of the pulses that are applied to the gate. (The six gates, U1 through U6, are contained in a single National Semiconductor 74C14 CMOS IC module.)

The circuitry that involves transistor Q3 and the 21-V zener diode does not participate in pulse processing during ordinary operation. If, however, a fault should develop in the feedback path or in the regulatory function of the supply, Q3 would prevent any appreciable rise in the output voltage (Q3 would then become active and deprive the MOSFET of gate drive). As previously pointed out, the output voltage of flyback regulators tends to be wild when not properly stabilized.

Some experimentation is in order for inductor L. The designated inductance of 1.2 microhenries is predicated upon 35 W of total load power and an assumed efficiency of 80%. However, it is not easy to take into account the nonideal characteristics of an actual inductor (such as stray capacitance, nonlinearity, core loss, and ohmic resistance). It is especially important to avoid pronounced effect from magnetic saturation. A toroidal core of high-frequency ferrite material should be used. Toroidal cores with an outer diameter of $1^1/_2$ inches and with a cross-sectional area of about $^1/_{10}$ square inch should be satisfactory. Wire size should not be smaller than 22 gauge.

The 100-μF output filter capacitors are solid-state tantalums. Trouble is almost sure to be encountered with the use of ordinary electrolytic types, although some that are intended specifically for high-frequency applications could conceivably be satisfactory—especially if paralleled by small ceramic capacitors.

This regulator, together with the previously described 200-kHz circuit, represents incipient power-supply technology of the 1990s. During this decade, considerable effort will be expended on switchers in the 100- to 500-kHz range.

Experimenter's prototype switching supply

The circuit shown in Fig. 7-36 embodies many desirable features of the various regulated supplies that were previously discussed. Although it is "bare bones" in simplicity, control and operational sophistications are readily added via the IC pulse-width modulation IC. The flexibility of this supply is such that it is a candidate as a basic system for a wide range of power-supply applications. For example, it can be made to operate over a switching rate range of 20 to 200 kHz with minimal changes in components. At the same time, its power-handling capability ranges from several tens of watts to more than several hundred watts—again with minimal circuit modification. A parts list for the experimenter's prototype power supply is shown in Table 7-4.

The basic circuitry was devised by International Rectifier Corporation to illustrate the ease and versatility of their high-voltage MOSFETs. These devices are not subject to second-breakdown destruction and their drive requirements are not critical, relative to those of bipolar transistors. Clean switching performance

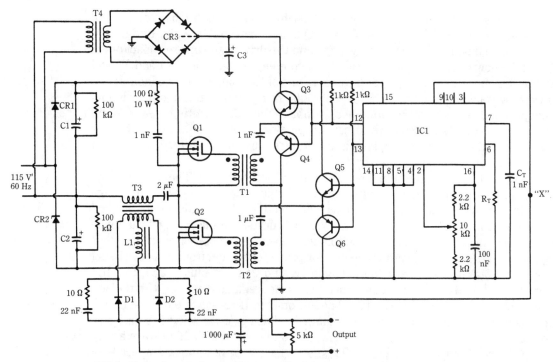

7-36 Experimenter's prototype regulating power supply. International Rectifier Corp.

Table 7-4. Parts list for experimenter's prototype power supply.

Item	Description
C1, C2	350 µF 250 Vdc electrolytic
C3	500 µF 50 V electrolytic
CR1, CR2	1N3211 silicon pn-rectifier diodes
CR3	MDA90-2 or 4 1N4001 diodes
D1, D2	75HQ045 diodes
IC1	SG3524 Silicon General
Q1, Q2	IRF300 transistors
Q3, Q5	MPS3694 transistors
Q4, Q6	2N9249 transistors
R_T	15 kΩ for 36 kHz; 4.75K for 100 kHz; and 1.6 kΩ for 200 kHz
T4	Small 12-V filament-type transformer (50 or 100 va)

extends well into the several-hundred-kHz region because MOSFETs do not exhibit the charge-storage phenomena of bipolar transistors. Monolithically self-contained diodes render the bypass or "free-wheeling" diodes (ordinarily associated with the switching devices) unnecessary.

A bridge amplifier, which involves transistors Q3 through Q6, provides the drive and the MOSFET switches, Q1 and Q2, are connected in a half-bridge configuration. Thus, both the switching stage and the drive stage utilize capacitors to block the flow of dc through the windings of coupling transformers T1 and T2 and output transformer T3. This action removes one of the main sources of spikes and switching transients that is found in push-pull arrangements and in other circuits that permit dc in transformer windings. Although the safe-operating area of MOSFETs is not nearly so sensitive to catastrophic destruction as is that of bipolar transistors, the presence of spikes is nonetheless undesirable—they cause EMI and are not always easy to filter from the output.

Diodes D1 and D2, the output rectifiers, are Schottky diodes. They provide high rectification efficiency from high-frequency switching waveforms. The line-voltage rectification system is a voltage doubler with rectifier diodes CR1 and CR2 and capacitors C1 and C2. The nominal 300 V that is developed makes practical implementation much easier than with low-voltage high-current bipolar transistors.

The most straightforward procedure for proper operation from this power supply is to break the feedback loop at point X and connect pins 1 and 9 of IC1 together. Then, the system should function as a nonregulated supply with the output voltage under control of the 10-kΩ potentiometer. If reasonable results are not forthcoming, recheck the connections and the phasing of the windings on transformers T1 and T2. Under proper operating conditions, a switching wave across the primary of output transformer T2 should have a peak-to-peak amplitude of approximately 300 V.

Because of oscilloscope grounding conflicts, it might not be convenient to directly make this measurement. However, if the transformation ratio of T3 is taken into account, the scope can be connected to the temporarily freed secondary terminals of T3.

After operation has been restored to the regulating mode, you will find that the dc output voltage can be adjusted by both the 5-kΩ and the 10-kΩ potentiometers. The retention of both of these adjustments actually provides a wider adjustment range than would be otherwise attainable. In general, a given output-voltage level can be obtained from various positions of the two potentiometer wipers. It might be possible to optimize regulation, ripple, response, and stability by experimenting with the two adjustments.

Finally, added performance features can be incorporated by appropriately using either the unused or the grounded pins of IC1. Examples of such features are soft-start, current limit, electronic shutdown, and various programming formats. The switching rate is governed by the time constant of R_T and C_T. The output voltage is considerably influenced by the turns ratio of the windings on output transformer T3. Available output power (up to several hundred watts) is primarily governed by the copper and core parameters of T3 and by the rectifying diodes and capacitors of the ac line-voltage doubler. For even greater power, Q1 and Q2 can be paralleled with similar devices—no ballast resistors are needed and no provisions will probably be needed for increased drive. Interestingly, the output Schottky rectifiers are likewise amenable to direct paralleling to extend the output current capability of the supply.

Summarizing, it is clear that the use of dedicated components frees the designer and experimenter from awkward situations and formidable problems that, at one time, appeared to be intrinsically associated with the switching regulator.

The current-mode IC controller

A perusal of IC controllers for switch-mode power supplies reveals essentially three evolutionary processes since the debut of these handy subsystems. First, their switching-rate capability has progressively increased; once limited to the 20- to 50-kHz region, 1-MHz controllers are now readily available. Secondly, these controllers continually accumulate more "bells and whistles" as time goes by. Thus, such a dedicated IC module not only provides the basic PWM function, but might incorporate such features as soft-start circuitry, undervoltage and overvoltage protection, automatic shutdown from overload, thermal shutdown, synchronization, optional single-ended or push-pull drive, etc.

A third-path of technological progress has also been evident in the deviation from the "plain vanilla" mode of operation. Thus, new modes of basic operation have been introduced that significantly alter the way in which switch-mode regulators work. The most prominent of these newer ideas have been implemented in the current-mode controller and in the resonant-mode controller. These represent important departures from the long-enduring methods of triggering switch-mode performance in regulated power supplies.

The *current-mode control technique* is probably the least radical change. To begin with, these controllers simply incorporate an additional operating mechanism in the traditional PWM IC controller. That is, output voltage stabilization against load variations occurs in very nearly the same manner as we have been accustomed in erstwhile ICs. The term *current-mode regulator* might be misleading; a sample of output voltage is compared to a reference-voltage and thereby produces an error signal, which, in turn, varies the duty cycle to produce regulation. The manner by which line regulation is achieved in the current-mode controller does, however, differ from the operation of the ordinary PWM controller.

Before further exploring the changes in current-mode controllers, divert your attention briefly to an important concept of feedback amplifiers, of which regulation circuitry can be considered a special case. A useful analogy can be established between a feedback audio amplifier and the basic circuitry of a regulated power supply.

Figure 7-37A depicts a simplified audio amplifier with a single-feedback path that returns a fraction of the output voltage to the amplifier's input. With sufficient feedback, an interesting thing happens—the overall voltage gain of the amplifier then becomes virtually independent of the no-feedback gain. To put it another way, the overall voltage gain of such an amplifier chain (Fig. 7-37A) becomes very nearly independent of the voltage gain of any of the three amplifying stages. Illustrating this important behavior is the modified amplifier of Fig. 7-37B. Here, additional feedback paths have been added. These paths must necessarily reduce the voltage gain that could be developed if the original feedback path was not present.

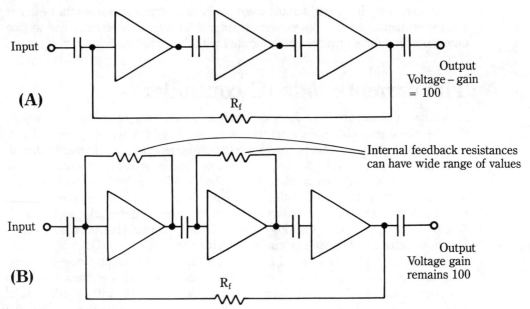

7-37 An analogy for demonstrating an important feedback characteristic. Overall amplifier gain is determined by all-embracing feedback path.
(A) Amplifier with all-embracing feedback path.
(B) Same amplifier with additional "internal" feedbacks; the voltage-gain of the amplifier is not changed.

Yet, because of the presence of the original feedback path, the overall voltage gain of the modified amplifier remains essentially unchanged! Notice that the original feedback path embraces the entire amplifier.

This behavior was one of the major breakthroughs in electronic technology, because it enabled amplifiers to be built so that the voltage gain was a function only of the fraction of feedback voltage, not of the gain of individual stages, dc voltages, temperature, or device characteristics. This capability of the most-embracing feedback path to maintain overall voltage gain, despite the addition of other "internal" feedback paths, has its counterpart in the so-called *current-mode controller* for switching power supplies.

It is natural to ponder why someone might add an additional feedback path to the one that senses output voltage in the more-ordinary PWM controller for switch-mode supplies. The answer has to do with *line regulation*. In controllers that do not utilize the current-mode principle, a line-voltage change (transient) should be prevented from appearing as a change of dc output voltage because the servo action of the regulator doesn't care why the output voltage tries to change. The servo action is "dumb" and is merely ready to generate an error signal to counteract the tendency for such change. This, indeed, had been the prevailing concept for many years. However, one thing had either been overlooked or been passively accepted as inevitable.

For slow changes of line voltage, the "ordinary" PWM switch-mode controller (a dedicated control IC that is not designated as a current-mode controller) effectively regulates the dc output voltage against both load variation and line-voltage change. It follows from the preceding allusion of "stupidity" in the generation of the corrective error signal; any tendency for the output voltage to change is counteracted, no matter the cause. However, for fast line-voltage changes or for line transients, the disturbance must traverse the power switch, possibly a transformer, and certainly the filter section before hitting the output terminal of the supply. This trip requires time and it is likely that the line disturbance will already be history by the time that corrective action is initiated! Thus, appreciable phase difference can exist between the error signal and the disturbance itself. The practical consequence is that line regulation will be poor for fast changes in line voltage. Even worse, the line transients will find the supply "transparent" and be passed to the load.

Clearly, it would be a significant improvement if this fault in regulator controllers could be circumvented. It has already been alluded that it can be done with the current-mode technique. In current-mode operation, the conventional (ordinary) PWM controller is referred to as a *voltage-mode regulator*. Remember, however, that current-mode operation still uses the same error-generating technique as voltage-mode operation does for regulating load voltage. This implies that the current-mode IC regulator has an added feature—an independent means of line regulation.

Figure 4-6A shows the generalized scheme of producing pulse-width modulation in voltage-mode controllers. Essentially, an op amp or a comparator is driven simultaneously by a triangular or saw-tooth wave and the dc error signal. The PWM wave emerges with its duty cycle as a function of the error signal.

In a current-mode controller, an internally-derived sawtooth would not be used. Rather, a sampled current ramp from the power switch would be impressed (along with the dc error signal) at the input of the op-amp PWM modulator. The ramp is the result of current buildup in the inductor or transformer winding of the power-switch circuit. Pulse-width modulation occurs in much the same manner as when the ramp is supplied by the internal oscillator.

Now, contemplate what the current-mode technique does differently. If line voltage would increase, the sampled current ramp through the power switch would have a steeper slope (i.e. it would rise to a given level in less time). This slope would narrow the duty cycle of the PWM wave (which drives the power switch) and thereby counteract the tendency for increased output voltage at the load. This corrective action occurs instantly so a belated correction by the output-sensing feedback circuit is unnecessary. Indeed, the current-ramp sampling scheme is actually a *feed-forward technique*, although it has been referred to as an *additional feedback path* in the generic sense. It follows that in the event of a decrease in line voltage, the ON portion of the PWM wave widens in order to counteract ahead of time any tendency for the load voltage to fall. Figure 7-38 compares the much-used voltage-mode technique with the newer current-mode scheme.

The operating features of the current-mode IC controller are:

- Current-mode control instantaneously corrects line-voltage variations. By the same token, line-transients are prevented from reaching the load.

- The technique amounts to pulse by pulse current limiting. The power switch is protected from peak currents, whether the cause is a short circuit, transformer saturation, or a strong line transient. Because of this inherent feature, current-limiting amplifiers usually are not necessary in the circuitry.
- Load regulation occurs in essentially the same manner as in the voltage-mode controller. This example is similar to the gain of the audio amplifier in Fig. 7-37B, which was unaffected by additional "internal" feedback paths.
- A surprising characteristic of current-mode operation is that the overall stability of the regulator is much improved—the filter no longer exerts a very adverse effect on the phase margin of the feedback circuit.
- Current-mode operation provides flux balancing in the transformer of push-pull circuits, which eliminates the tendency for saturation in this transformer.
- Current-mode operation can enhance current balancing between paralleled supplies.

A disadvantage of current-mode control is the tendency for a runaway condition when it is associated with a half-bridge switching circuit. Other switch-circuit topographies seem to work well, however.

In the current-mode control scheme, the sawtooth or ramp that was derived from an internal oscillator was no longer used as one of the input signals to the pulse-width modulator. Instead, a ramp that was sampled from the current in the power-switch circuit was substituted. However, the current-mode controller still uses an internal oscillator or clock to time the basic pulse rate of the switching process. It accomplishes this via logic circuits that are interposed between the pulse-width modulator and the power-switch driver(s). This information was not mentioned earlier to avoid confusion while the focus was on more basic matters.

In the simplest implementation of the current-control mode, a resistor is inserted in the circuit of the power switch to provide a voltage that represents the sought current ramp. This resistor dissipates some power and it can be the source of reduced overall efficiency. Of course, the idea is to make this sampling resistance as small as possible. An amplifier can be helpful in this regard. Via other means, the current ramp can be sampled. A transformer can be used; appropriately, such a transformer would be a "current transformer," with perhaps a single turn on the primary side and a number of turns on the secondary. This transformer would circumvent the power dissipation of a sampling resistor, but it could cause unexpected troubles from leakage inductance or from stray resonance.

A more elegant way to dispense with the simple sampling resistor is with a special power MOSFET as the power switch. This dedicated power MOSFET is the Motorola SENSEFET. In this device, a tiny percentage of drain-source cells are brought out to a fourth lead ("mirror" terminal). The current that is available at this terminal is minuscule, but it accurately represents the actual drain-source current. A high resistance to ground is necessary to cause this tiny current to

The current-mode IC controller 429

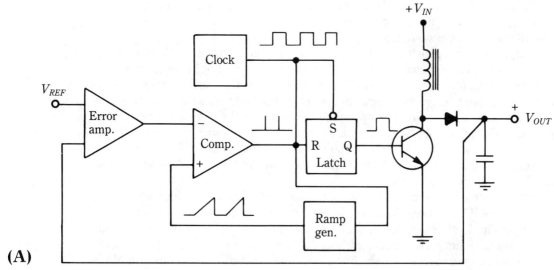

(A)

Voltage-mode feedback

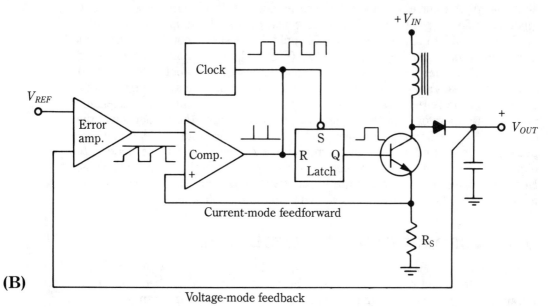

(B)

Voltage-mode feedback

7-38 Voltage-mode vs. current-mode control of a regulator.
(A) Voltage-mode: Comparator sees error signal and ramp derived from internal clock or oscillator.
(B) Current-mode: Comparator sees error signal and ramp derived from inductor charging current in power-switch circuit.

develop a usable voltage drop, but the power dissipation involved is negligible compared to that when a sampling resistance is inserted directly in the source lead of the MOSFET power switch.

Figure 7-39 is a nearly complete switching regulator that uses a Motorola control IC. In this circuit, some of the "bells and whistles" are not connected so that the basic regulatory function is not detracted from. The traditional voltage mode is used in order to achieve load regulation. (Notice the resistance network in the dc-output section and the circuit connection to the input of the error amplifier). Additionally, the transformer ramp current is sampled across series resistance R_s; the voltage that represents this ramp is applied to one input of the PWM comparator, instead of an oscillator-derived sawtooth. This statement at-once defines this control IC as a current-mode type.

Now contemplate the regulating supply shown in Fig. 7-40, which utilizes the same control IC. Load regulation is obtained in essentially the same way as in the supply of Fig. 7-39—even though sampling is done via an optoisolator, rather than with a resistive network. More significant is the unique method of incorporating current-mode operation. The sampling resistance is not directly in the path of the power-switch current. Rather, it is associated with a "mirror" terminal of the special power MOSFET, known as a SENSEFET. As previously explained, such an arrangement circumvents power dissipation in the sampling resistance because an almost-negligible current flows in the mirror-electrode circuit. Otherwise, current-mode operation works essentially the same as it does in the supply of Fig. 7-39.

Next, consider the partial regulated-supply that is shown configured around a Cherry Semiconductor CS-1524 control IC (Fig. 7-41). Even though the amplified voltage, which is derived from a current-sensing resistance, is applied to the PWM comparator, this is not a current-mode arrangement. First, dc output current, not a current ramp is sensed. Second, its intended effect is to limit peak current by inhibiting comparator operation, which it does by restricting the ON portion of the duty cycle. In performing this protective measure, it does not provide pulse-to-pulse current limiting, as does the current-mode system. The current-mode technique is superior in that its response to overload is practically instantaneous. However, current-mode designs are not always needed and the current-limiting technique (Fig. 7-41) is satisfactory in many practical applications.

The resonant-mode IC controller

In some form, resonant phenomena has had a long association with power supplies, inverters, and converters. Often, the association has not been beneficial because various undesired circuit resonances and stray reactances can be the root of dangerous spikes, RFI, EMI, energy dissipation, "mysteriously" destroyed power semiconductors, and various operational malperformance. On the other hand, resonance can be deliberately deployed to produce desirable circuit and performance features. In some implementations, sustained high-Q resonance is involved. Examples of such operation are the high-voltage RF power supplies (Figs. 4-41 and 4-42). In other cases, only a single- or a partial-cycle is allowed to

The resonant-mode IC controller 431

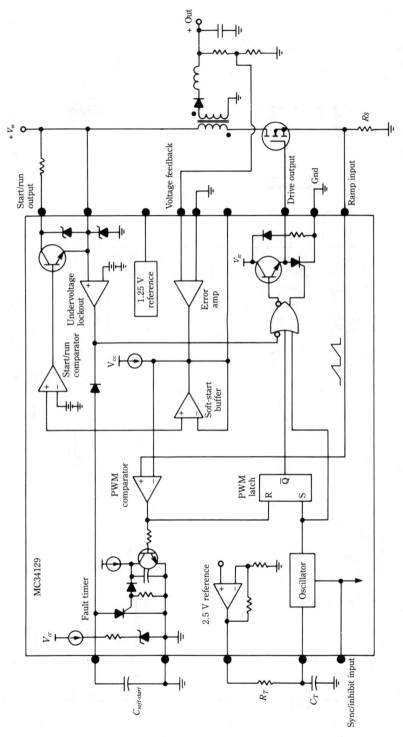

7-39 Typical switch-mode regulated supply with current-mode control. Notice that the PWM comparator is ramped from a sample of power-switch current rather than from the oscillator voltage.

432 *Switching-type regulators that use integrated circuits*

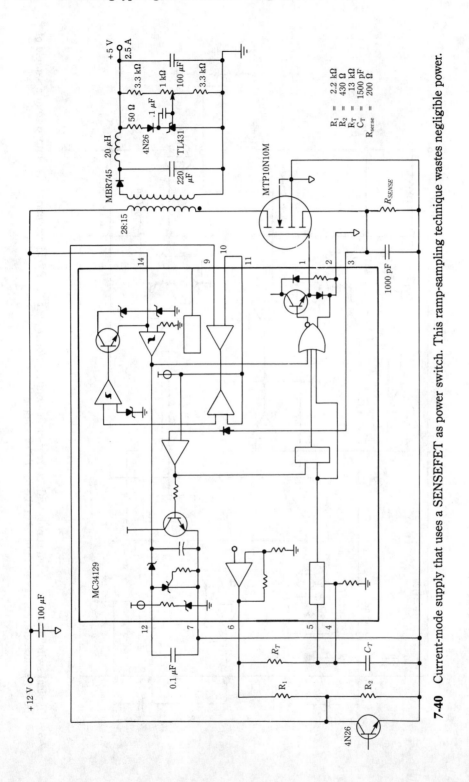

7-40 Current-mode supply that uses a SENSEFET as power switch. This ramp-sampling technique wastes negligible power.

The resonant-mode IC controller 433

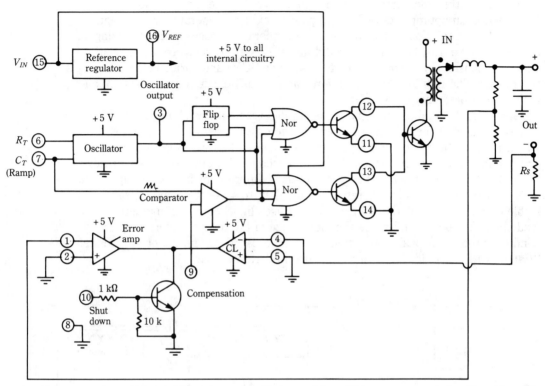

7-41 Voltage-mode switcher with peak current-limit provision. This arrangement should not be mistaken for a current-mode supply.

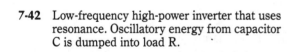

7-42 Low-frequency high-power inverter that uses resonance. Oscillatory energy from capacitor C is dumped into load R.

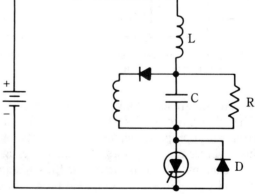

manifest itself. Both series and parallel "tank circuits" are encountered. Often, the energy storage in the LC elements is initiated by shock excitation.

The SCR circuit of Fig. 7-42 provided hundreds or thousands of watts of fairly good sine-wave power to a load, R. In this arrangement, L and C are the resonating elements and the SCR delivers short-duration pulses that are terminated by self-commutation (the other circuit elements are involved in the commutation process). By selecting an appropriate SCR, high power could be generated up to about 35 kHz. Voltage regulation was adequate for a number of purposes, such as driving an ultrasonic apparatus, but the circuit did not work well with electronic regulation. Other versions of this resonant inverter used two SCRs and various commutating techniques. The root idea in all of them was to shock-excite a resonant tank and deliver the stored energy to a load.

An example of an inverter that uses parallel resonance is shown in Fig. 7-43. The series-feed choke, L_S gives this circuit the characteristic of a current source, which is much better for driving fluorescent lamps than a voltage source. Both driven and self-oscillatory versions have been used for this purpose. At frequencies in the 20- to 60-kHz region, the light output of fluorescent lamps can approach 15% greater than for 60-Hz operation. Altogether, this use of resonance produces stable operation from fluorescent lamps and uses light, compact magnetics. The feed choke can be about 7 times the inductance of the total transformer primary-winding (a more-formal estimation procedure is given in chapter 4 under "Ballparking the series inductor in the current-driven inverter").

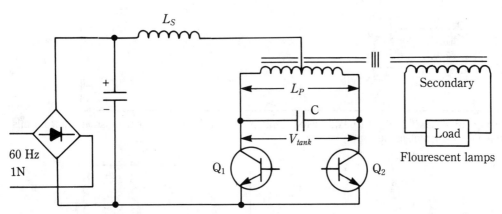

7-43 Resonant-mode inverter that uses parallel-resonant tank circuit. As a result of L_S, this inverter is "seen" as a current source by the load.

Another resonant inverter is the electronic fluorescent-lamp ballast (Fig. 7-44). This circuit uses series LC resonance. If the transistors are alternately switched from an appropriate square-wave source, an increasing sinusoidal voltage will build across the capacitors until the lamps ignite. However, the resistances of the active lamps, which appear across the capacitors, thereby lowers the Q of the resonant circuits. This, in turn, lowers the allowable current through the lamps. Thus, ballast action occurs without the need for bimetallic switches.

Although the described uses of resonance have been eminently valuable, much time passed before concerted attempts were made to utilize the resonance principle in dedicated control ICs for regulated power supplies. Once developed and marketed, however, resonant-mode supplies provided some compelling advantages

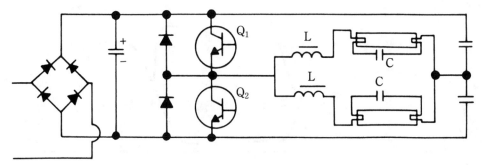

7-44 Resonant-mode inverter that uses series-resonant tank circuits. Ballast action is provided for fluorescent-lamp loads.

over the long-used PWM technique. The salient features of these unique switch-mode power supplies are:

- Simultaneous presence of current through and voltage across the power switch can be virtually eliminated. This dramatically reduces one of the main sources of power dissipation in high-frequency switch-mode supplies.
- Much less RFI and EMI are developed in such supplies.
- High operating frequencies can be readily attained (1 MHz and higher) because stray capacitance and stray inductance can be used as part of the resonant "tank".
- The use of a voltage-controlled oscillator in place of the PWM comparator provides flexible design and contributes to feedback stability at high frequencies.
- Although resonant-mode control tends to cause higher peak currents in the power switch than does PWM control, overall electrical stress on this device is probably less in resonant-mode control. This difference is because of relatively low spike energy in the switching wave, as borne out by the low RFI and EMI.

The primitive resonant-mode circuits shown have been known and exploited for many years. Yet, the resonant-mode regulated power supply did not become commercially available until relatively recently. This was partly as a result of the long-enduring 20-kHz PWM switcher, which truly gave yeoman service for a wide variety of applications. To the imaginative designer, it had been obvious that the resonant-mode could be expected to allow one hundred times this "standard" switching rate. However improved power switches, better passive components, and (most importantly) dedicated control ICs were needed for resonant-mode operation. Even though the desirable features of higher switching rates were fully realized, initial impetus went to extending PWM frequencies. Admittedly, unsuspected progress was made here and some tendency was to again neglect development of the resonant-mode technique. Finally, however, one attribute of resonant-mode operation could no longer be swept under the rug: the near-elimination of simultaneous voltage across, and current through, the power switch (Fig. 7-45).

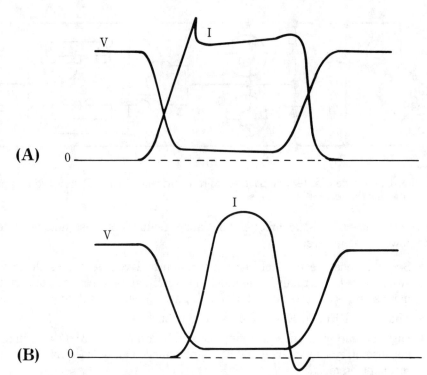

7-45 Comparison of PWM and resonant-mode switching waveforms.
(A) PWM waveform displays regions of simultaneous voltage and current.
(B) Resonant-mode waveform shows more ideal zero-current switching.

After a little lost motion, it was realized that pulse-width modulation was not suitable for the new approach. Rather, an FM technique, in which frequency was varied and pulse-duration remained constant, was very satisfactory. The basic idea was to work on one side or the other of true resonance. Then, the sign and amplitude of the error signal would vary according to departure from resonance. Thus, regulation could be made to occur in response to how far the switching rate departed from resonance of the tank circuit. A nice thing about this technique is that stray, rather than physical, inductance or capacitance could be used. Thus, what might have been bothersome in trying to extend the frequency of a PWM circuit, could be beneficially exploited in the resonant mode!

The block diagram of Fig. 7-46 depicts the basic arrangement of the functional blocks of resonant-mode regulated power supplies. This technique is quite straightforward; moreover, it has much in common with the PWM regulators. The heart of the scheme is the voltage-controlled oscillator (VCO) and the one-shot (monostable multivibrator). Because of these circuits, the resonant tank is shock-excited by a constant width variable repetition-rate pulsetrain. The error amplifier simply controls the frequency of the pulsetrain so that the dc output voltage is regulated. At the nominal dc output voltage, the pulse frequency can be either a little lower or a little higher than the ringing frequency of the resonant tank.

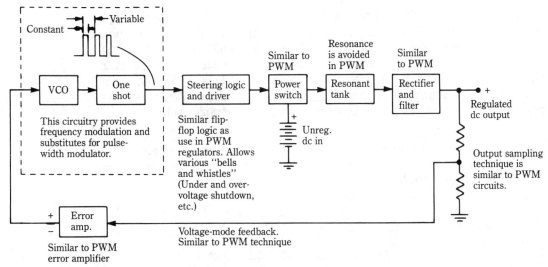

7-46 Basic resonant-mode regulated supply. Notice similarities and differences from PWM techniques.

An example of a resonant-mode regulated power supply

As might be expected, the heart of such a regulated supply is a dedicated IC (Gennum GP605) that was made specifically for resonant-mode regulation in high-frequency power supplies. An example of one way that this IC can be used is shown in the simplified circuit of Fig. 7-47. The salient features of this 125-W regulated supply are:

- It is an off-line supply. Strapped options enable either 115- or 230-V ac use.
- It uses the resonant mode, which is determined by the series resonance of inductor L1 and capacitor C7. This resonance is about 750-kHz. The nominal operating frequency is approximately 600 kHz, but it varies above and below this frequency in order to produce regulation.
- The power switches, Q1 and Q2, are arranged in a half-bridge configuration.
- An optoisolator (U3) is used in the voltage-mode feedback path to electrically isolate the load from the line.
- A driver circuit is inserted between the control IC and the power switches to obtain faster charge of the input capacitance of Q1 and Q2. This has nothing to do with the resonant mode per se, but this technique is useful at higher frequencies. If the supply was designed for lower-frequency operation (100 kHz, for example) this driver might not be needed.
- Interestingly, this circuit also has current-mode feedback via T2. Thus, the overall scheme is not unlike that in the so-called current-mode PWM supplies.

The actual schematic diagram of the 125-W resonant-mode power supply is shown in Fig. 7-48. At first glance, the circuit might appear complex. This is primarily because of many household functions and "bells and whistles," such as soft start, overvoltage and undervoltage shutdown, remote shutdown, etc. Here, the concern is only with the functions that are pertinent to resonant-mode operation. Accordingly, it will best serve your purpose to refer to the simplified circuit of Fig. 7-47, which covers the basic operating principle of this resonant-mode power supply.

Table 7-5. Parts list for 125-watt resonant-mode power supply.

Resonant power supply	Parts list
C1,C2	Capacitor al. el. 330 µF 200 V
C3,C4	Capacitor met. polypropolene 1 µF 200 V
C5,C6	Capacitor cer. NPO 47 pF 1 kV
C7	Capacitor cer. NPO 8200 pF 1 kV
C8,C10,C11,C18,C30	Capacitor cer. X7R 0.22 µF 25 V
C9,C31	Capacitor tant. conformally coated 4.7 µF 35 V
C16,C17,C21,C27	Capacitor tant. conformally coated 4.7 µF 16 V
C12,C13,C14,C15	Capacitor tant. conformally coated 220 µF 10 V
C19,C24	Capacitor cer. X7R 0.022 µF 25 V
C20,C26	Capacitor cer. NPO 1000 pF 25 V
C22	Capacitor cer. NPO 150 pF ±5% 25 V
C23	Capacitor cer. NPO 100 pF ±5% 25 V
C25	Capacitor cer. X7R .047 µF 25 V
C28,C29	Capacitor film 1000 pF 250 Vac 'Y' type
D1	Diode bridge 4 A 600 V Gen. Instr. KBU4J
D2,D3	Diode Schottky IN5823
D4,D5	Diode UFRD Amperex BYV29-500
D7	Diode IN4001
D8,D9	Diode UFRD Motorola MUR105
D10,D11,D14	Diode IN4148
D12,D13	Diode Schottky Motorola MBR3045PT (each)
D15	DiodeZener IN967A
F1	Fuse 5A 250 Vac slo. blo.
L1	Inductor 5 µH ±5%, Core micrometals T51-8/90 12 turns of #28 AWG
L2	Inductor 7.5 µH ±10%. Core: Arnold Eng. A-445146-2 or Magnetics 55344-A2. 7 turns of 5 strands in parallel #20 AWG.
L3	Inductor 330 µH 85 ma. TDK EL0606 SKI-331K
Q1,Q2	MOSFET IRF840
Q3,Q5	MOSFET Supertex VP0104N3 p-channel 8 Ω
Q4,Q6	MOSFET Supertex VN1306N3 n-channel 8 Ω

Table 7-5. Continued.

Resonant power supply — **Parts list**

Ref	Description
R1	Thermistor NTC 3 A Ametek SG220
R2,R3	Resistor 33 kΩ 5% 0.25 W
R4,R5	500 Ω 10% 1 W Corning FP1
R6	Thermistor PTC 3 kΩ Midwest 220Q32214
R8,R9	Resistor 100 kΩ 5% 0.25 W
R10,R11	300 kΩ 5% 0.25 W
R12	360 kΩ 1/2 W 400 V
R13,R18	3 kΩ 5% 0.25 W
R14	17.4 kΩ 1% 0.25 W
R15	240 Ω 5% 0.25 W
R16	56 Ω 5% 0.25 W
R17	10 kΩ 5% 0.25 W
R19	9.31 kΩ 1% 0.25 W
R20	1 MΩ 5% 0.25 W
R21,R22	470 Ω 5% 0.25 W
R23	910 Ω 5% 0.25 W
R24	1 kΩ 5% 0.25 W
R25	Trim pot. 200 Ω
R26	Resistor 20 kΩ 5% 0.25 W
T1	PWR. transformer. Core: magnetics DF43622-UG. (1-2) = 12 turns of #30 AWG, six strands in parallel, split 6 turns + 6 turns. (3-4) = (4-5) = 1 turn of 0.005" copper foil, sandwiched between two halves of the primary (1-2). (6-7) = (7-8) = 3 turns of #30 AWG
T2	Transformer, current sense Core: Magnetics 40705-TC-F (1-2) = 1 turn of #28 AWG (3-4) = 40 turns of #32 AWG
T3	Transformer, gate drive. Core: magnetics 40705-TC-F. (1-2) = (3-4) = 20 turns of #32 double-insulated wire, wound bifilar.
U1	Linear regulator LM78M12
U2	FM controller Gennum GP605
U3	Optocoupler Motorola MOC604A
U4	Shunt regulator TI 431CLP

Referring to Fig. 7-47, the first thing you must know is the identity of the resonant circuit or "tank." This identity is not always obvious in resonant-mode supplies. Stray parameters sometimes are used instead of physical inductors or capacitors. Also, elements that serve purposes other than as resonators are often mixed in with the tank-circuit proper. In Fig. 7-47, the series-resonant tank is composed of L1 and C7. The other elements that are associated with this series-resonant tank are so-sized that they perform their intended functions with negligible effect on resonance. For example, the primary winding of output transformer, T1, has a high enough inductance that it exerts minimal disturbance on the capacitor, C7, of the resonant tank. Nonetheless, T1 is able to transfer stored energy in C7 to

440 *Switching-type regulators that use integrated circuits*

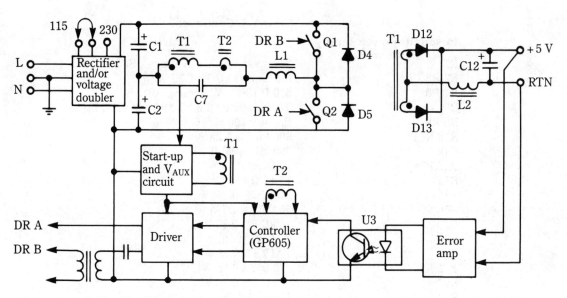

7-47 Simplified circuit of the 125-W resonant-mode power supply. Gennum Corp.

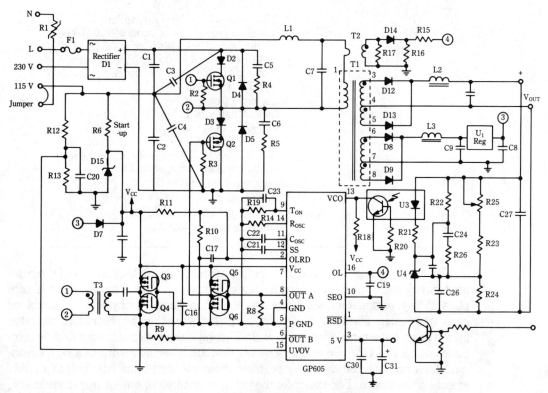

7-48 Schematic diagram of the 125-W resonant-mode power supply. Gennum Corp.

the output rectifier and filter circuit(s). At the same time, the primary of the current transformer has such a low inductance that it does not materially disturb resonance or adversely affect the function of T1. This current transformer samples peak current in the primary winding of T1, which provides a current-mode function for the power supply.

At this point, it is only natural to ponder the roll of Q in the resonant circuit. In a series-resonant tank, high Q is derived from high inductance and small capacitance. It might seem desirable to aim for high Q resonance, but it becomes increasingly more difficult to extract energy as Q is made higher (this situation also prevails in the RF power stage of a transmitter where too high of a tank-circuit Q increases difficulties in transferring power to the antenna). On the other hand, excessively-low Q is undesirable in the power supply because the peak currents in the power switches could become dangerously high. All in all, a compromise is sought so that Q is made low enough to enable effective power transfer, but high enough to reasonably approximate a sine wave in the secondary of the power transformer.

Because power switches Q1 and Q2 are alternately turned on and off, power transformer T1 is worked in a bipolar mode—its primary winding is first impressed with a positive-, then a negative-going half sine wave. The secondary winding of T1 then develops a full sine wave of induced voltage. Rectification is imparted by Schottky diodes D12 and D13. L2 together with C12 compose a conventional filter for the dc output voltage. This switching and resonating process gives rise to waveshapes that are shown in Fig. 7-45. Thus, neither Q1 nor Q2 suffer high switching losses from simultaneous voltage and current.

The main operating characteristics of resonant-mode operation have been covered without detailing the various housekeeping functions that IC controllers provide or make available. This style is in harmony with the basic objective of this book—to guide capable electronics workers. One of the many auxiliary functions that is built into the GP605 control IC is relevant to resonant-mode operation, and it can hardly be ignored as just another housekeeping function. This pertains to all of the circuitry that is associated with pin 16 (OL), including the current transformer, T2.

As previously mentioned, T2, with its single-turn primary winding, has a negligible effect on the resonant tank of L1 and C7. However, T2 does monitor the peak current in the primary winding of T1, the output transformer. The voltage that is induced in the 40-turn secondary of T2 accurately represents the monitored peak current in the primary of the output transformer, and therefore, in power switches Q1 and Q2. As the result of rectification and partial filtering, the voltage that is delivered by the current transformer is made to resemble a sawtooth before it is applied to shutdown circuitry within the control IC. This overall situation clearly resembles the current-mode feature in some PWM control ICs. Indeed, it is referred to as *current-mode operation in the resonant-mode regulator*.

However, this current-mode feature does not provide exactly similar performance to current-mode circuitry in PWM controllers. In the PWM situation, current-mode operation manifests itself as a feed-forward technique and thereby it improves line regulation; it also provides pulse-to-pulse limiting of peak-current.

442 *Switching-type regulators that use integrated circuits*

The current-mode operation just described for resonant-mode regulators does not directly enhance line regulation; it does, however, limit peak current.

The last statement is of great importance to the successful implementation of resonant-mode operation because of the inherent tendency of this regulation mode to produce high peak currents—on the order of 3 or 4 times greater than in equally rated PWM supplies. This current can be safely handled with appropriately selected power switches. However, a reduced margin of safety remains from the effects of transients, shorts, overloads, or saturation in cores of magnetic components. Usually, the protection that is provided by the current-mode feature had best be viewed as mandatory, rather than merely as an optional convenience.

An interesting switching supply is shown in Fig. 7-49. This arrangement uses both high and low frequencies in the regulating process. MOSFET switches are turned on and off at a 200-kHz rate, whereas the output transformer and the rectifying system operate at 20 kHz and at a constant duty cycle. At first, a transition appears to be needless in the overall scheme. Why switch at 200 kHz if the output energy is to be transformed to 20 kHz; the entire regulator could just as well perform at 20 kHz. However, this design possesses some worthwhile advantages over a "straight" switcher that operates all the way through at 20 kHz.

Notice that the pulse-width modulator is used to produce a pulse-width modulated "carrier," at 200 kHz. The pulse-width modulation, on the other hand, repre-

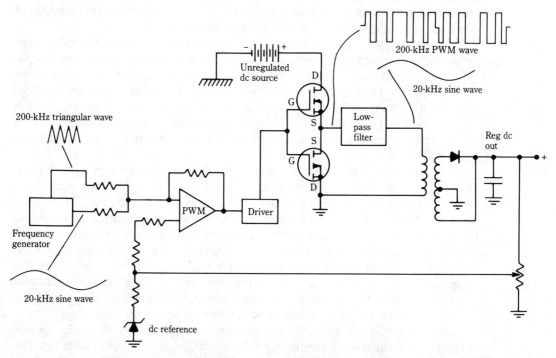

7-49 An unconventional switch-mode power supply. In this hybrid arrangement, the power switches, output transformer, and the rectifier-filter systems are used efficiently.

sents a 20-kHz sine wave. Demodulation of this type of modulated carrier is easily accomplished via a low-pass filter. Ideally, the low-pass filter would have a cut-off frequency of slightly higher than 20 kHz. In practice, a cut-off frequency of 50 or 75 kHz (for example) will still work satisfactorily and reduce some physical size and cost. Also, by designing the cut-off frequency comfortably higher than 20 kHz, the stability of the feedback loop of such a regulator is improved.

The switching elements are complementary-symmetry MOSFETs. They switch more efficiently at 200 kHz than all but the most expensive bipolar transistors at 20 kHz. This is especially true with recent MOSFETs that have very low ON-resistances. These elements make an appreciable contribution to the overall efficiency of the supply. Another gain in overall efficiency stems from the improvement in transformer utilization factor because of the constant duty-cycle sine wave that is involved in energy transfer. Finally, it is easier to obtain good operation from the rectifier at the lower frequency. Probably the salient feature of this regulation scheme is the reduction in RFI and EMI. In conventional 20-kHz switchers, the varying duty cycle of the output energy generates abundant high-frequency harmonics that are not readily attenuated by the filtering system. In this "hybrid" regulator, the output duty cycle remains constant, which makes it much easier to optimize the effectiveness of the output filter. Also, the reverse-recovery time of the rectifier is not as critical a factor when the duty cycle of the rectified wave is constant.

Of course, the trade-off that is incurred in this design approach is that the physical size of the output transformer and the output filter is necessarily larger than would be the case for straight-through 200-kHz operation. At the same time, the constant duty-cycle operation of the output transformer enables it to be made smaller than required in conventional 20-kHz switching supplies.

In the example cited, the carrier frequency was 200 kHz and the modulation frequency was 20 kHz. This example enabled a comparison to be made with conventional 20-kHz switching supplies, which have become very common. If, however, you assume a modulation frequency of 40 or 50 kHz, the gain in operating efficiency over 40- or 50-kHz conventional switchers becomes even more meaningful. The bipolar transistor power switches in conventional switchers generally exhibit relatively high switching losses.

Another point that is worthy of consideration is the waveshape of the modulating signal. Although a sine wave was used for illustrative purposes, Square-waves can lead to further enhancement of overall efficiency, primarily through better utilization of the rectifier. In actual practice, a waveshape that resembles a trapezoid more than a square wave works well—it avoids transient problems that are bound to occur with waves that have high rise and fall rates. This waveshape helps hold down electrical noise. Although the square wave has the desirable feature that peak, average, and RMS values are one and the same, this is substantially true for a "slowed-down" square wave (i.e. one that just begins to appear trapezoidal when a wavetrain of 5-cycles are viewed on the oscilloscope screen).

Notice that this basic scheme is not altogether "blue sky" in nature. It is actually derived from similar circuit concepts that were tried out and marketed in stereo amplifiers, where a high power-to-volume ratio was necessary. However, it was difficult to keep manufacturing costs down as a result of the need to practice inordinate shielding, bypassing, and isolation techniques to prevent spurious frequencies from contaminating the reproduced audio. It was amply demonstrated, however, that this class-D high-frequency carrier amplification and subsequent "demodulation" by a low-pass filter merited further development. Problems encountered in stereo work would not necessarily be recognized as such in a switch-mode regulated supply. One technique that appears to relax the burden that is imposed on the low-pass filter is to derive the modulating low-frequency signal from a divided-down sampling of the high-frequency carrier.

Resist the temptation to use IGBTs as switching devices. Although the actual ac power that is delivered to the output transformer is invested in a 20-kHz wave, the switching rate is, nonetheless, at the higher carrier (200 kHz). This rate lies in the domain of the power MOSFET, not the IGBT or the ordinary bipolar transistor.

Sample list of dedicated ICs for switching supplies

It is easy enough to infer from our previous discussion of switching-type regulating supplies that the dedicated control IC has become the heart of these circuits. Many advantages are gained from the use of these controllers; much time, effort, and expense will generally be saved through their use. Relatively little experimentation can be anticipated for desired performance. It is as if the several dollars that were paid for the purchase of such ICs also provide the services of competent engineers and technicians who have previously done the design and troubleshooting tasks. On top of it all, you usually receive a number of "free" bells and whistles—so-called housekeeping functions that would tax the abilities of the best of us to implement from scratch. The self-protective features that are incorporated in these dedicated control ICs provide reliable insurance against destruction to the supply or load from transients, overvoltage, undervoltage, overloads (including short-circuits, excessive temperature-rises, or start-up surges).

The need to protect from undervoltage, often a blessing to the linear-type supply, arises in the switching-type supply from the increased duty cycle that is brought about by undervoltage operation. Inordinately high duty cycle or ON time, will be accompanied by higher-than-normal input line current. Unless fuses or circuit breakers react in time, damage can result from such operation. Added insurance is provided by the IC's internal electronics. Usually, the duty cycle is automatically decreased in the event of undervoltage.

The sample list of control ICs that are depicted in Table 7-6 are intended for PWM switch-mode power supplies. Both voltage-mode and current-mode types are included. Two of these ICs have optional voltage- or current-control modes. Both, single-ended and push-pull drive provisions will be noted. For those ICs with complementary drive capability, a push-pull power switch will deliver half the output frequency of a single-ended frequency capability in touting the product. One of these ICs contains the power switch on the chip—certainly a portender of forthcoming technology for low and moderate-powered regulated supplies.

Table 7-6. A sample list of PWM switch-mode control ICs.

HS7107
National Semiconductor
Hybrid 7-A multimode high-efficiency switching regulator.
Mode: voltage
Output type: uncommitted transistor (1)
Max output current 8000 mA
Max output voltage 100 V
Max operating frequency 300 kHz
Input voltage limits 100, 10 V
Undervoltage lockout: no
Soft-start facility: no

The device is housed in a T03 package, which allows 25 W maximum power dissipation. The HS7067 is similar except for a supply voltage range of 10 to 60 V.

SG3530
Silicon General
High-speed current mode PWM controller.
Mode: current
Output type: totem pole (1)
Max output current 2000 mA
Max output voltage 17 V
Max operating frequency 2000 kHz
Input voltage limits 17, 13 V
Undervoltage lockout: yes
Soft-start facility: yes

This device is similar to the SG3528 except that the duty cycle is variable from 0 to 100%.

SG3528
Silicon General
High-speed current mode PWM controller.
Mode: current
Output type: totem pole (1)
Max output current 2000 mA
Max output voltage 17 V
Max operating frequency 2000 kHz
Input voltage limits 17, 13 V
Undervoltage lockout: yes
Soft-start facility: yes

This device allows a maximum duty cycle of 50%. Many protection features are built in allowing use as a versatile control chip.

MAX641/2/3
Maxim
These devices are fixed-output 5 V 10 W step-up switching regulators
Mode: voltage
Output type: open drain (1)
Max output current 2000 mA
Max output voltage 0 V
Max operating frequency 45 kHz
Input voltage limits 17, 2 V
Undervoltage lockout: no
Soft-start facility: no

Low-battery comparator is provided on chip. Output voltage can be varied using a voltage divider. The power switch is provided on chip.

µA78S40
Fairchild
Universal switching regulator subsystem.
Mode: voltage
Output type: totem pole (1)
Max output current 1500 mA
Max output voltage 40 V
Max operating frequency 100 kHz
Input voltage limits 40.3 V
Undervoltage lockout: no
Soft-start facility: no

A freewheel diode is provided on-chip. The device is capable of switching 1.5 A of output current.

L4970
SGS Thomson
10 A switching regulator.
Mode: voltage
Output type: mosfet (1)
Max output current 10000 mA
Max output voltage 50 V
Max operating frequency 500 kHz
Input voltage limits 50, 7 V
Undervoltage lockout: yes
Soft-start facility: yes

The output is configured for buck (step-down) operation. It incorporates a built-in current sense resistor for over-current protection. L4972 is a 2-A version in a 20-pin DIL package. L4974 is a 3.5 A version in a 20-pin DIL package.

Table 7-6. Continued.

LM3578
National Semiconductor
8-pin switching regulator with duty-cycle variable up to 90%.
Mode: voltage
Output type: uncommitted transistor (1)
Max output current 750 mA
Max output voltage 34 V
Max operating frequency 100 kHz
Input voltage limits 40, 2 V
Undervoltage lockout: no
Soft-start facility: no

MAX638
Maxim
Fixed 5-V step-down switching regulator.
Mode: voltage
Output type: open drain (1)
Max output current 525 mA
Max operating frequency 65 kHz
Input voltage limits 17, 2 V
Undervoltage lockout: no
Soft-start facility: no

This device includes the switching power transistor on chip. The output voltage can be varied from 5 V with only an additional voltage divider.

MAX634
Maxim
Micropower switching regulator.
Mode: voltage
Output type: open drain (1)
Max output current 525 mA
Max output voltage −20 V
Max operating frequency 75 kHz
Input voltage limits 17, 2 V
Undervoltage lockout: no
Soft-start facility: no

Inverting regulator for operation from a battery supply. It includes a low-battery indicator circuit. The power switch is on chip.

UC3823
Unitrode
High-frequency PWM controller.
Mode: voltage/current
Output type: totem pole (1)
Max output current 500 mA
Max output voltage 30 V
Max operating frequency 1000 kHz
Input voltage limits 30, 9 V
Undervoltage lockout: yes
Soft-start facility: yes

This is a widely used, versatile device which may be operated in either voltage or current mode.

UC3841
Unitrode
Programmable off-line PWM controller.
Mode: voltage
Output type: open collector (1)
Max output current 400 mA
Max output voltage 40 V
Max operating frequency 500 kHz
Input voltage limits 32, 8 V
Undervoltage lockout: yes
Soft-start facility: yes

Similar to UC3840 but includes a number of refinements. Very flexible device with protection and monitoring circuitry for most applications.

UC3840
Unitrode
Programmable off-line PWM controller.
Mode: voltage
Output type: uncommitted transistor (1)
Max output current 400 mA
Max output voltage 40 V
Max operating frequency 500 kHz
Input voltage limits 32, 8 V
Undervoltage lockout: yes
Soft-start facility: yes

All control functions, monitoring and protection functions that are usually required in a power supply are included. The UC3841 supersedes this device for new designs.

TL594
Texas Instruments
PWM control circuit.
Mode: voltage
Output type: uncommitted transistor (2)
Max output current 250 mA
Max output voltage 41 V
Max operating frequency 100 kHz
Input voltage limits 41, 7 V
Undervoltage lockout: yes
Soft-start facility: yes

This device contains two op amps with their outputs or-ed so that the one with the greatest output voltage controls the output pulse width. This is a relatively simple, but useful device.

MC34060
Motorola
PWM control circuit.
Mode: voltage
Output type: uncommitted transistor (1)
Max output current 250 mA
Max output voltage 42 V
Max operating frequency 200 kHz
Input voltage limits 42, 0 V
Undervoltage lockout: no
Soft-start facility: no

Two error amplifiers are provided along with a minimum dead-time control input.

CS593
Cherry Semiconductor
PWM control circuit.
Mode: voltage
Output type: uncommitted transistor (2)
Max output current 250 mA
Max output voltage 41 V
Max operating frequency 300 kHz
Input voltage limits 41, 7 V
Undervoltage lockout: yes
Soft start facility: no

Similar to TL494/CS494 except that current-limit comparator with 80-mV offset is included instead of a standard op amp. This device is a relatively simple general-purpose controller.

UC3847
Unitrode
Current-mode PWM controller.
Mode: current
Output type: totem pole (2)
Max output current 200 mA
Max output voltage 40 V
Max operating frequency 500 kHz
Input voltage limits 40, 8 V
Undervoltage lockout: Yes
Soft-start facility: no

This device is similar to the UC3846 but offers complementary outputs.

TL495
Texas Instruments
PWM control circuit.
Mode: voltage
Output type: uncommitted transistor (2)
Max output current 250 mA
Max output voltage 41 V
Max operating frequency 100 kHz
Input voltage limits 41, 7 V
Undervoltage lockout: no
Soft-start facility: no

Predecessor to the TL595. This device does not feature an undervoltage lockout, but in other ways is very similar.

TL494
Texas Instruments
PWM control circuit.
Mode: voltage
Output type: uncommitted transistor (1)
Max output current 250 mA
Max output voltage 41 V
Max operating frequency 100 kHz
Input voltage limits 41, 7 V
Undervoltage lockout: no
Soft-start facility: no

Predecessor to TL594. This device does not feature an undervoltage lockout, but in other ways is very similar.

Table 7-6. Continued.

TL493
Texas Instruments
PWM control circuit.
Mode: voltage
Output type: uncommitted transistor (2)
Max output current 250 mA
Max output voltage 41 V
Max operating frequency 100 kHz
Input voltage limits 41, 7 V
Undervoltage lockout: no
Soft-start facility: no

This device is similar to TL494, except that one error amplifier has a 0.08-V offset for use in current-limit sensing.

UC3846
Unitrode
Current-mode PWM controller.
Mode: current
Output type: totem pole (2)
Max output current 200 mA
Max output voltage 40 V
Max operating frequency 500 kHz
Input voltage limits 40, 8 V
Undervoltage lockout: yes
Soft start facility: no

This device includes a shutdown pin and most of the circuitry required to control a power-supply using a half- or full-bridge configuration.

TL595
Texas Instruments
PWM control circuit.
Mode: voltage
Output type: uncommitted transistor (2)
Max output current 200 mA
Max output voltage 40 V
Max operating frequency 100 kHz
Input voltage limits 41, 7 V
Undervoltage lockout: yes
Soft-start facility: no

This device contains a 39-V zener diode so that the device can provide a housekeeping supply by using the diode as a shunt regulator. A pulse-steering input which can be used to inhibit operation is provided.

UC3850
Unitrode
Switching power-supply control system
Mode: voltage/current
Output type: uncommitted transistor (2)
Max output current 100 mA
Max output voltage 40 V
Max operating frequency 200 kHz
Input voltage limits 40, 5 V
Undervoltage lockout: yes
Soft-start facility: yes

This device is a high-functionality controller that can also be used in current-mode control systems. All the usual protection circuitry is included.

SG3529
Silicon General
Regulating PWM controller.
Mode: voltage
Output type: uncommitted transistor (2)
Max output current 100 mA
Max output voltage 60 V
Max operating frequency 400 kHz
Input voltage limits 40, 7 V
Undervoltage lockout: yes
Soft-start facility: no

This is a modified SG3524B with uncommitted input to the PWM comparator to simplify feed-forward operation. Notice the transconductance-type error amplifier, which has an output impedance of approximately 4 MΩ.

SG3526B
Silicon General
Regulating PWM circuit.
Mode: voltage
Output type: totem pole (2)
Max output current 100 mA
Max operating frequency 500 kHz
Input voltage limits 35, 8 V
Undervoltage lockout: yes
Soft-start facility: yes

This device is an improved version of SG3526 featuring much faster shutdown and a more accurate-voltage reference, along with other enhancements.

SG3526
Silicon General
Regulating PWM circuit.
Mode: voltage
Output type: uncommitted transistor (1)
Max output current 100 mA
Max operating frequency 350 kHz
Input voltage limits 35, 8 V
Undervoltage lockout: yes
Soft-start facility: yes

This device contains all the control and protection circuitry that is required for most power-supply designs. Note the transconductance-type error amplifier, which has an output impedance of approximately 4 MΩ.

SG3527A
Silicon General
Regulating PWM circuit.
Mode: voltage
Output type: totem pole (2)
Max output current 200 mA
Max operating frequency 500 kHz
Input voltage limits 35, 8 V
Undervoltage lockout: yes
Soft-start facility: yes

This device is similar to the SG3525A, except that the outputs are complementary.

SG3525A
Silicon General
Regulating PWM circuit.
Mode: voltage
Output type: totem pole (2)
Max output current 200 mA
Max operating frequency 500 kHz
Input voltage limits 35, 8 V
Undervoltage lockout: yes
Soft-start facility: yes

This device is a widely used control chip. Notice the transconductance-type error amplifier, which has an output impedance of approximately 4 MΩ.

RC4191
Raytheon
Micropower switching regulator.
Mode: voltage
Output type: open collector (1)
Max output current 150 mA
Max output voltage 30 V
Max operating frequency 75 kHz
Input voltage limits 30, 4 V
Undervoltage lockout: no
Soft-start facility: no

This device is designed for battery operation. A low battery comparator and output are provided.

SG3524B
Silicon General
Regulating PWM circuit.
Mode: voltage
Output type: uncommitted transistor (2)
Max output current 100 mA
Max output voltage 60 V
Max operating frequency 400 kHz
Input voltage limits 40 V
Undervoltage lockout: yes
Soft-start facility: no

This is an improved version of SG3524 with greater drive capability and an undervoltage lockout circuit. This device is very widely used. Note the transconductance-type error amplifier, which has an output impedance of approximately 4 MΩ.

NE5562
Philips
PWM control circuit
Mode: voltage
Output type: totem pole (1)
Max output current 100 mA
Max output voltage 0 V
Max operating frequency 600 kHz
Input voltage limits 16.9 V
Undervoltage lockout: yes
Soft-start facility: yes

Device contains overvoltage protection, two current limiters, feed-forward control, demagnetization sense input, loop-fault protection and other features.

Table 7-6. Continued.

Unitrode
Current-mode PWM controller.
Mode: current
Output type: totem pole (1)
Max output current 1 mA
Max output voltage 30 V
Max operating frequency 500 kHz
Input voltage limits 30, 16 V
Undervoltage lockout: yes
Soft-start facility: no

This device along with UC3844 and UC3845 are simple, easy to use controllers. UC3844 and UC3845 are limited to a maximum duty cycle of 50% whereas the UC3842 and UC3843 go up to 100%. The undervoltage-lockout thresholds are 16 V (on) and 10 V (off) for the UC3842 and 3844, 8.5 V (on), 7.9 V (off) for the UC1843 and UC1845.

SG3524
Silicon General
Regulating PWM circuit.
Mode: voltage
Output type: uncommitted transistor (1)
Max output current 50 mA
Max output voltage 40 V
Max operating frequency 300 kHz
Input voltage limits 40, 8 V
Undervoltage lockout: no
Soft-start facility: no

This device is a basic control chip. The transconductance-type error amplifiers have a 4 MΩ output impedance.

NE5560
Philips
PWM control circuit.
Mode: voltage
Output type: uncommitted transistor (1)
Max output current 40 mA
Max operating frequency 100 kHz
Input voltage limits 18, 9 V
Undervoltage lockout: yes
Soft start facility: yes

Clamped to the supply voltage through two diode drops (1.4 V). This device includes a minimum dead-time control, a current-limit circuit and remote on/off along with an overvoltage protection or demagnetization input (to prevent core-walking).

NE5561
Philips
PWM control circuit.
Mode: voltage
Output type: open collector (1)
Max output current 20 mA
Max operating frequency 100 kHz
Input voltage limits 18, 9 V
Undervoltage lockout: yes
Soft-start facility: no

Limited to the positive supply voltage plus two diode drops (1.4 V). Soft start can be implemented with a few external components. A current-sense comparator is provided for cycle-by-cycle current limiting.

The sample control ICs that are listed in Table 7-7 are intended for the newer breed of resonant-mode regulated supplies, which enables attainment of both high switching rates and high efficiency. Before the availability of these designated-control ICs, a number of practical obstacles slowed the practical implementation of resonant-mode regulating supplies. Notice that these controllers utilize FM, rather than PWM, in order to produce regulation; housekeeping functions, similar to those of PWM control ICs, are provided.

Sample list of dedicated ICs for switching supplies

Table 7-7. Resonant-mode control ICs.

Cherry Semiconductor Corp. types CS-360 and CS-3805A

Description

The CS-360 IC controller is designed for use in resonant and quasiresonant-mode topologies. The architecture is configured to allow operation in the following variable-frequency control methods: fixed ON-time, fixed OFF-time, and a combination of fixed ON/OFF times.

It contains the standard complement of "house-keeping" functions, including a programmable UVLO, soft-start, and current limiting.

The precision voltage-controlled oscillator is specifically designed to offer a high degree of linearity (typ. 2%) over a frequency range of 100 kHz to 1 MHz, while accurately clamping the minimum and maximum frequency to user selected values. Control of the output driver "deadtime" is externally programmable with a single resistor.

The temperature-compensated one shot delivers a well-controlled pulse width to the dual 1-A totem-pole output drivers and is retriggerable, allowing it to function in the three different control methods.

Features:

- 1-MHz VCO with user-programmable min, max frequencies
- Temperature-compensated one shot
- Programmable output "deadtime"
- UVLO with low start-up current and alternative start/stop thresholds
- Latched overcurrent protection
- Soft-start

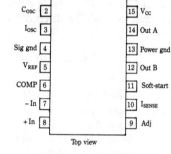

Pin connections (Top view)

Description

The CS-3805A, which uses fixed on-time, variable-frequency control, is specifically intended for resonant-mode power supply control applications. Two complementary outputs are capable of directly driving power MOSFETs.

Opening a normally grounded-control pin puts the CS-3805A into single-ended operation. In this mode, the frequency is doubled and the two outputs are identical so that they can be paralleled for increased drive capability.

Included on the chip are peripheral functions (such as soft-start, undervoltage/overvoltage lockout, remote shutdown, and overload shutdown with delayed restart). All shutdown modes are synchronous (the last output pulse is completed), and default to soft-start once the fault condition is removed.

Absolute maximum ratings

Parameter	Values and units
Supply voltage	20 V
Undervoltage/overvoltage input	−0.4 V to 6 V
Overload input	−0.4 V to 6 V
Remote shutdown	−0.4 V to V_{CC}

Table 7-7. Continued.

Absolute maximum ratings

Parameter	Values and units
VCO input	-0.4 V to V_{CC}
Operating temperature range: CS-3805A	$0\,°C \leq T_A \leq 70\,°C$
Storage temperature range	$-65\,°C \leq T_S \leq +150\,°C$
Lead temperature (soldering, 10 sec.)	260 °C
Junction temperature	150 °C
Power dissipation at $T_A \leq 70\,°C$ (derate 9 mW/°C for $T_A > 70\,°C$)	720 mW

Features:

- 1-MHz maximum operating frequency
- Synchronous overload shutdown with delayed restart
- Synchronous overvoltage, undervoltage, and remote shutdown
- Soft start
- Single-ended or complementary outputs
- Drives power MOSFETs directly (0.8 A peak)

Pin connections

$\overline{RSD}$ 1	16 $\overline{OL}$
$\overline{OLRD}$ 2	15 $\overline{UVOV}$
V_{REF} 3	14 R_{OSC}
A gnd 4	13 VCO
P gnd 5	12 SS
$\overline{Out\ B}$ 6	11 C_{OSC}
V_{CC} 7	10 $\overline{SEO}$
$\overline{Out\ A}$ 8	9 T_{ON}

Gennum Corporation type GP605

Features

- frequency range of 1 kHz to 2 MHz
- operating frequency range (min. and max.) set by a resistor and a capacitor
- pulse width set by a resistor and capacitor
- synchronous overload shutdown with delayed restart
- synchronous overvoltage, undervoltage and remote shutdown
- soft-start
- single-ended or complementary outputs
- drives power MOSFETs directly (0.8-A peak)
- low-cost 16-pin DIP or SOIC

Circuit description

The GP605 utilizes frequency modulation instead of pulse-width modulation to achieve regulation. The pulse width is held constant while the frequency is varied over an operating range set by a resistor and capacitor. A feedback voltage controls the switching frequency of the two complementary outputs, which are capable of driving power MOSFETs directly.

Opening a normally grounded control pin puts the GP605 into single-ended operation. In this mode, the frequency is doubled and the two outputs are identical so they can be paralleled for increased drive capability.

The high operating frequency of up to 2 MHz results in significant reductions in the size of the required magnetic and capacitive components in the power supply. This leads to dramatic savings in volume, weight, and manufacturing cost of switching power supplies.

Included on the chip are peripheral functions such as soft-start, undervoltage/overvoltage lockout, remote shutdown and overload shutdown with delayed restart. All shutdown modes are synchronous (the last output pulse is completed), and default to soft-start once the fault condition is removed.

Unitrode integrated circuits type UC1860, UC2860, UC3860

Features

- 3 MHz VFO linear over 100:1 range
- 5 MHz error amplifier with controlled output swing
- Programmable one-shot timer—down to 100 ns
- Precision 5-V reference
- Dual 2A-peak totem-pole outputs
- Programmable-output sequence
- Programmable undervoltage lockout
- Very low start-up current
- Programmable fault management and restart delay
- Uncommitted comparator

Description

The UC1860 family of control ICs is a versatile system for resonant-mode power-supply control. This device easily implements frequency-modulated fixed-on-time control schemes as well as a number of other power-supply control schemes with its various dedicated and programmable features.

The UC1860 includes a precision voltage reference, a wide-bandwidth error amplifier, a variable frequency oscillator operable to beyond 3 MHz, an oscillator-triggered one-shot, dual high-current totem-pole output drivers, and a programmable toggle flip-flop. The output mode is easily programmed for various sequences such as A, off, B, off; A and B, off; or A, B, off. The error amplifier contains precision output clamps that allow programming of minimum and maximum frequency.

The device also contains an uncommitted comparator, a fast comparator for fault sensing, programmable soft-start circuitry, and a programmable restart delay. Hiccup-style response to faults is easily achieved. In addition, the UC1860 contains programmable under-voltage lockout circuitry that forces the output stages low and minimizes supply current during startup conditions.

Index

A
ac voltage regulation, 231-234
alternator, regulating permanent-magnet charging current, 236-237
amplifiers
  ac feedback, 110-113
  differential, 14
  error signal, 327
  hi-fidelity improvements, 4-8
  IC operational, 327-330
  input, 327
  operational, 119, 330-331
  sensing, 327
avalanche breakdown, 248
avalanching, 248

B
battery, characteristics of giant storage, 2-3
beryllium oxide, 102-103
bipolar voltage-regulated power supplies, 174-176
  dual output tracking, 174-176
Boschert regulated power supplies, 129
breakdown voltage, characteristics, 248-251
bridge-circuit inverter, 196-197

C
capacitive coupling, 204
capacitors, 386
  electrolytic, 305-315
  electrolytic vs. polypropylene, 313
  filter, 303-315
  output, 114, 141-143
  polypropylene, 313
  solid-state tantalum, 309
  tantalum electrolytic, 308-309
  tantalum foil, 309
  tantalum wet-slug, 309
chokes
  magnetically biased, 298-301
  selecting, 292-296
clock, 20-kHz, 408-409
commutation, 28, 288
comparator, 327
components, 247-326
connector-pin filters, 148-149
constant-current operation, current limiting, 77-81
constant-current voltage-regulated power supplies, 15
constant-voltage transformer, 172, 301
convection cooling, 96
converters, 190-194
  digital-to-analog, 223
core saturation, 196
corona, 151
corona balls, 151
crowbar technique, 75-77
Curie temperature, 295
current-driven inverter, 238, 241-244
  ballparking series inductor, 244-245
  waveforms, 241
current-foldback operation, current limiting, 77-81
current limiting, 64, 77-81
current-mode control technique, 425
current-mode IC controller, 425-430
current-regulated power supplies, 10-11
  5-A, 339-340
  combined with voltage-regulated power supplies, 21
  dynamics, 125-127
  line-voltage regulation, 57-58
  load-current regulation, 53-55
  low-current circuits, 355-357
  testing/measuring techniques, 15-17
  typical applications, 12-13
  voltage limiting, 25

D
Darlington transistors, 28
  giant, 276-278
Datel regulated power supplies, 130
dc power, backup source, 377-378
despiking, 398
devices, 247-326

devices (cont.)
  germanium, 154-155
  IGBT, 318-326
  transcalent power, 291-292
diamond, 103
differential amplifier, 14
digital-to-analog converters, 223
digitally-controlled regulated power supplies, 222-228
  systems-oriented dc power, 228-229
diodes, 134-141, 153
  freewheeling, 382, 398
  hot-carrier, 136, 396
  LM113 energy-gap reference, 266-267
  photon-emitting, 19
  rectifier, 134-136, 139, 152-161
  reverse-biased base-emitter, 361
  Schottky, 136, 153, 257-262, 396, 398
  silicon, 136
  zener, 15, 231, 248-257, 359, 398
dissipation control, 165-172
  double regulation, 170
  limiting techniques, 166-170
  preregulating transistor, 171-172
distributed power systems, 372-377
drift, 61, 62, 133
driven inverter, 194-195, 289
driver logic, 410-412
duty cycle, 421
dynamic load, 20-21

E

Eccles-Jordan multivibrator, 7
effective series inductance (see ESI)
effective series resistance (see ESR)
electric field, 144
electrical conductivity, 102
electrical insulators, 102
electrical noise, 149-152
electrochemical reactions, 11-14
electrolytic capacitors, 306-315
  impedance vs. frequency curves, 305
  pitfalls, 310-312
electromagnetic interference (see EMI)

electromagnetic radiation, 96, 204
electronic feedback, 6
electronic filter, 8
EMI, 134-141, 144, 304
  output capacitor as contributor, 141-143
energy-gap principle, 214-216
energy-gap reference diode, LM113, 266-267
error amplifier, 410
error signal, 120-121
ESI, 141, 145, 304, 307, 313
ESR, 141-143, 145, 304, 307, 313

F

Faraday's law, 13
farads, 377
feed-forward technique, 427
feedthrough filters, 148-149
ferrite, 296-298
ferrite beads, 145-147
ferroresonant constant-voltage transformer, 172-174, 301-303
ferroresonant transformer, 172
field-emission breakdown, 248
filter capacitors, 303-315
  technological breakthrough, 312-315
filters
  connector-pin, 148-149
  electronic, 8
  feedthrough, 148-149
floating-field alternator, 369
Fluke high-voltage regulated power supplies, 130-131
foldback characteristics, 21
forced-air cooling, 96
  effects, 99
freewheeling diode, 382, 398

G

gate turn-off (see GTO)
germanium devices, 154-155
GTO silicon-controlled rectifiers, 287-291
  electrical characteristics, 290
GTO thyristors, 289
  symbols/definitions, 320

H

harmonic distortion vs. power output, 5
heat pump, thermoelectric, 315-317

heat radiation, effects, 99-101
heat removal, 58-59 (see also thermal conductivity)
  basic aspects, 92-99
heatsinks, 104
  power transistors, 99
hi-fidelity amplifiers, improvements, 4-8
high resistance, simulating, 14
high-voltage regulated power supplies, 30-31
  applications/processes involving, 31
  electrical noise, 149-152
  Fluke, 130-131
  solid-state RF, 212-213
hot-carrier diode, 136, 396
hysteresis loop, 296-298

I

IC controller
  current-mode, 425-430
  resonant-mode, 430-436
IC operational amplifiers, 327-330
IC-regulated power supplies
  linear, 361-364
  paralleling, 370-372
  three-terminal, 267-276
  three-terminal, adjustable voltage applications, 349
  three-terminal, fixed-voltage, 347-349
  three-terminal, protection techniques, 81-82
  three-terminal, versatility, 353-354
ICs, 444-453
  Cherry Semiconductor CS-360/CS-3805A, 451-452
  Cherry Semiconductor CS593, 447
  Fairchild µA78S40, 445
  Gennum Corporation GP605, 452-453
  inverter-type switching-regulated power supplies, 270-272
  Maxim MAX634, 446
  Maxim MAX638, 446
  Maxim MAX641/2/3, 445
  Motorola MC34060, 447
  National Semiconductor HS7107, 445
  National Semiconductor LM3578, 446

overvoltage protection, 73-77
Philips NE5560, 450
Philips NE5561, 450
Philips NE5562, 449
Raytheon RC4191, 449
SGS Thomson L4970, 445
Silicon General SG3524, 450
Silicon General SG3524B, 449
Silicon General SG3525A, 449
Silicon General SG3526, 449
Silicon General SG3526B, 448
Silicon General SG3527A, 449
Silicon General SG3528, 445
Silicon General SG3529, 448
Silicon General SG3530, 445
switching regulators, 379-453
Texas Instruments TL493, 448
Texas Instruments TL494, 447
Texas Instruments TL495, 447
Texas Instruments TL594, 447
Texas Instruments TL595, 448
Unitrode, 447, 450
Unitrode UC1860/UC2860/ UC3860, 453
Unitrode UC3823, 446
Unitrode UC3840, 446
Unitrode UC3841, 446
Unitrode UC3846, 448
Unitrode UC3850, 448
IGBT, 318-326
  specifications, 324-325
impedance, dynamic output, 107
inductive coupling, 138
inductors, 299, 386
  selecting, 292-296
input amplifier, 327
insulated gate bipolar transistor (see IGBT)
interference (see also noise)
  electrical noise, 149-152
  EMI/RFI, 134-141
  noise benefits, 143
  noise spikes, 145-148
  shielding, 144-145
  transference of electrical, 138
inverters, 190-194, 398
  bridge-circuit, 196-197
  current-driven, 238, 241-244
  driven, 289, 194-195
  improving design, 237-244
  parallel resonance, 434
  push-pull, 220
  saturable-core oscillation frequency, 193
  series resonance, 435

voltage-driven, 238, 240

J

JFET, 179-185
  important characteristics, 179-180
  using, 180-185
junction field-effect transistor (see JFET)

K

KEPCO regulated power supplies, 132

L

Lambda 100-volt regulated power supplies, 132
light-emitting diodes (LEDs)
  low-voltage, 359-360
  stabilizing light output, 18
line-operated regulated power supplies, combining linear/ switching techniques, 22, 221
line-voltage regulation, 55-58
  combined with load-current regulation, 59-61
  Paraformer transformer, 207
linear-regulated power supplies, 262
  using ICs, 327-378
  using LM105 IC, 340
linear voltage ramps, 14
load-current regulation, 53-55
  combined with line-voltage regulation, 59-61
  Paraformer transformer, 207
load power, 24

M

magnetic field, 144
magnetic shunt, 172, 302
master supply, 85
Meissner oscillator, 210-211
metal-oxide semiconductor field-effect transistors (see MOSFET)
metal-oxide varistor, characteristics, 69
microprocessors
  power supplies, 29-30
  power/current formats, 30
minimum input-output voltage differential, 349
modulation, pulse-width, 218-221, 409-410

MOSFET, 154, 157-161, 185 (see also power MOSFET)
  high-rating, 282-287
MOSPOWER FET, 278-281
multivibrator, 6
  Eccles-Jordan, 7

N

noise spikes, attenuation with ferrite beads, 145-148

O

oersteds, 297
Ohm's law, 95, 119, 204, 245
operational amplifiers, 330-331
optoisolators, 217
oscillator, self-excited, 382
output capacitor, 114, 141-143
output impedance, dynamic, 107
overload circuit, 412-416
overvoltage protection, 73-77
ozone, 151

P

Paraformer transformer, 174, 203, 303
  characteristics, 206-209
  construction, 205
  line-voltage regulation, 207
  load-current regulation, 207
peltier effect, 315
periodic and random deviation (PARD), 122
photon-emitting diodes, constant-current sources, 19-20
plus or minus (±) designation, 50-51
polypropylene capacitor, 313
Power/Mate regulated power supplies, 132-133
power MOSFET, 278-281, 318-326
  battery charger, 354
  electrical characteristics, 280, 284, 286
  performance curves, 281, 283, 285, 287
  vs. IGBT structures, 321
power-regulated power supplies, 23-25
power supplies (see regulated power supplies)
  microprocessor, 29-30
  regulation techniques, 26-29
  stability, 61-62

458   Index

power supplies (cont.)
  UPS (see uninterruptible power supplies)
power switches, 190-194
  basic, 200-202
power transistors
  applications guide, 83-84
  evaluating/selecting, 82-84
pulse-width modulated wave, 410
pulse-width modulation, 218-221
  methods for implementing, 219
pulse-width modulator, 409-140
  regulating Silicon General SG1524/SG2524, 272-276
push-pull inverter, 220

R

radiation cooling, 96
radio-frequency interference (RFI), 134-141
rectifier diodes, 134-136, 139
  GTO silicon-controlled, 287-291
  synchronous, 152-161
regulated power supplies, 1-46
  ±15-V dual, 345-347
  advantages/disadvantages, 1-2
  applications, 41-42
  based on parametric power conversion, 203-206
  Boschert, 129
  combined use of switching and dissipative, 186
  constant torque from dc motor, 34
  constant-speed motor control, 34
  current (see current-regulated power supplies)
  Datel, 130
  digitally controlled, 222-228
  dual-polarity outputs from single-ended, 364-366
  dual-polarity with independently adjustable outputs, 36, 366
  dynamic characteristics, 107-163
  extending transmitting-tube life, 40-41
  features, 45-46
  future problems, 43-45
  high-voltage (see high-voltage regulated power supplies)
  interconnecting, 84-92

  interconnecting, automatic parallel operation of two voltage-regulated power supplies, 87
  interconnecting, automatic series operation of two voltage-regulated power supplies, 89
  interconnecting, automatic tracking operation of two voltage-regulated power supplies, 88
  interconnecting, dual-output tracking operation of two voltage-regulated power supplies, 91
  interconnecting, parallel operation by parallel programming, 85
  interconnecting, parallel operation by separate pass-element circuits, 85
  interconnecting, parallel operation of current-regulated power supplies, 91
  interconnecting, series-connected voltage-regulated power supplies, 88
  JFET in dissipative, 179-185
  KEPCO, 132
  Lambda 100-volt, 132
  line-operated, 221-222
  linear (see linear regulated power supplies)
  microprocessor, 29-30
  miscellaneous uses, 33-36
  multiple output, 229-230
  overvoltage protection using special IC, 73-77
  power (see power-regulated power supplies)
  Power/Mate, 132-133
  protection techniques, 63-67
  regulation criteria, 62-63
  regulation techniques, 26-29
  resonant-mode, 437-444
  RF (see RF-regulated power supplies)
  shunt-current (see shunt-current regulated power supplies)
  specifications, 129-134
  stabilizing light intensity, 36-37
  stabilizing RF output level of TWT, 37-38
  stabilizing temperature, 35

  static characteristics, 47-105
  switching (see switching-regulated power supplies)
  systems-oriented digital control of dc power, 228-229
  temperature-effect on output, 59
  three-terminal IC (see IC-regulated power supplies)
  transient protection, 68-73
  UPS, 230-231
  using power op amps, 337-339
  voltage (see voltage-regulated power supplies)
regulation
  ac voltage, 231-234
  additional criteria, 62-63
  basic concepts, 50-53
  current, 355-357
  line-voltage, 55-58, 207
  load-current, 53-55, 207
  specifications pitfall, 161-163
  relative, 61
  thermal, 58
  voltage, 355-357
regulation techniques
  dissipation control, 165-172
  implementing, 165-245
relative regulation, 61
remote-sensing terminals, preventing problems, 39-40
resistance, simulating high, 14
reverse-biased base-emitter diode, 361
RF-regulated power supplies, 210-212
  advantages/disadvantages, 210
  solid-state high-voltage, 212-213
RF spray radiation, 150
ripple, 121-123
  PARD, 122
  practical considerations, 122-123
RMS sensing, 235

S

safe-operating area (SOA), 269
saturable-core inverter, oscillation frequency, 193
Schottky diode, 136, 153, 257-262, 396, 398
self-excited oscillator, 382
sensing amplifier, 327
shielding, 144-145

shunt voltage-regulated power supplies, 177-179
shunt current-regulated power supplies, 17-18
silicon diodes, 136
silicon-controlled rectifier GTO, 287-291
silicon-controlled switches, 288
slave unit, 85
slew rate, 127-129
snubber circuit, 288
stabilization factor, 62-63
static characteristics, 47-105
   dynamic behavior, 110
static-magnetic regulating transformer, 172, 301
switching transistor, 381
switching-regulated power supplies, 185-186
   100-W using flyback principle, 401-406
   200-kHz 50-W using power MOSFET, 416-418
   250-W, 395-401
   500-kHz flyback-circuit, 418-422
   characteristics, 186
   constant-current, 391-392
   dc, 189-190
   dedicated ICs, 444-453
   experimenter's prototype, 422-425
   IC details/operation, 381-383
   IC determination of output-filter inductor/capacitor, 383
   ICs for inverter-type, 270-272
   negative, 390-391
   off-line, 216-218
   off-line 5-V 200-A, 406-416
   phase-controlled, 186-188
   stacked, 392-395
   using ICs, 379-453
symmetry correction circuit, 198-200

T

tantalum capacitors
   electrolytic, 308-309
   foil, 309
   solid-state, 309
   wet-slug, 309
temperature
   capacitance, 315
   coefficient, 58
   convection cooling, 96

Curie, 295
electrochemical reactions, 11-14
electrolytic capacitors, 312
ferrite-core material, 296
forced-air cooling, 96, 99
heat radiation, 99-101
heat removal, 58-59, 92-99 (see also thermal conductivity)
radiation cooling, 96
rising, 58
stabilizing using regulated power supply, 35
temperature-compensated zener diodes, 252-255
thermal bonding, improving, 105
thermal circuits, usage, 96-99
thermal conductivity, 101-105
thermal impedance, 93-96
thermal regulation, 58
thermal resistance, 93-96
thermocouples, 315
thermoelectric heat pump, 315-317
thermometer, electronic, 16-17
thyristors
   GTO, 289
   GTO symbols/definitions, 320
transcalent power devices, 291-292
transconductance
   regulator circuits, 115
   transistors, 117-118
   tube/transistor circuits 115-118
   voltage gain for pentodes, 117
   voltage gain for triodes, 116-117
transformers, 239, 300
   constant-voltage, 172, 301
   ferroresonant, 172
   ferroresonant constant-voltage, 172-174, 301-303
   magnetically biased, 298-301
   Paraformer, 174, 203, 205, 206-209, 303
   static-magnetic, 172, 301
transients, 68-73
   responses, 123-124
   zener diode as suppressor, 255-257
transistors
   giant Darlington, 276-278
   heatsinks for power, 99
   low voltages, 361
   power, 82-84

preregulating, 171-172
switching, 381
thermal information, 97
transconductance, 115-118
transmitting tubes, extending life, 40-41
traveling wave tube (TWT), stabilizing RF output level, 37-38
tubes, 211
   transconductance, 115-118
   transmitting, 40-41
   traveling wave (TWT), 37-38
   vacuum, 15, 379

U

uninterruptible power supplies (UPS), 32-33, 230-231

V

vacuum tubes, 379
   stabilized emission, 15
volt-ampere, characteristics of arc welder, 26
voltage
   ac regulation, 231-234
   breakdown characteristics, 248-251
   LED, 359-360
   linear ramps, 14
   nominal line, 55
   regulation, 55-56
   regulation for low-voltage circuits, 355-357
   reverse-biased base-emitter diode, 361
   zener diode, 360
voltage comparator, 264-266
voltage reference
   developing, 214-216
   zener diode, 15
voltage-driven inverter, 238
   inductive load, 240
voltage-mode regulator, 427
voltage-regulated power supplies, 328-330
   1.2-V, 200-μA, 358-359
   5-A, 339-340
   723 IC, 340-344
   ac feedback amplifier, 110-113
   ac with electronic sensing, 234-236
   advantages, 3
   automobile with dedicated IC, 367-370

voltage-regulated power supplies (*cont.*)
  bipolar, 174-176
  combined with current-regulated power supplies, 21
  constant current, 15
  current-mirror op amp, 331-333
  feedback arrangements, 114-115
  high-performance using ICs, 336-337
  ideal, 9-10
  line-voltage regulation, 55-56
  LM105 positive, 262-266
  low, 360-364
  low dynamic impedance, 2
  op amp, 330-333
  op-amp added features, 335
  op-amp with current-booster stages, 333-335
  operation, 108-109
  retrospect, 119-120
  series-type, 109
  shunt, 177-179
  simulated capacitor, 8-9
  solid-state for automobile alternators, 27
  solid-state for automobile dc generators, 28
  summary, 10
  using MC1560/1561 IC, 344-345
  using phase-controlled switching, 187

W

waveforms, 75, 436

current-driven inverter, 241
diode current, 140
phase-controlled switching regulator, 188
regulated power supply, 411
wet-slug tantalum capacitor, 309

Z

zener breakdown voltage, 248
zener diode, 231, 248-257, 359, 398
  dynamic impedance vs. breakdown voltage, 252
  low voltages, 360
  LVA, 251-252
  temperature-compensated voltage reference, 252-255
  transient suppressor, 255-257
  voltage reference, 15